P9-DEZ-858

Radar Reflectivity of Land and Sea

ARTECH HOUSE, INC.
610 WASHINGTON ST.
DEDHAM, MA 02026

Originally published by Lexington Books, 1975.

International Standard Book Number: 0-89006-130-0
Library of Congress Catalog Card Number: 75-13435

Radar Reflectivity of Land and Sea

Maurice W. Long
Consulting Engineer
and Physicist
Atlanta, Georgia

Artech House, Inc.

To Beverly

Maurice W. Long is a consulting engineer and physicist with address at 1036 Somerset Drive, N.W., Atlanta, Georgia 30327. He received the B.E.E. degree from the Georgia Institute of Technology, the M.S.E.E. from the University of Kentucky, and the M.S. and Ph.D. degrees in Physics from Georgia Tech. For most of the 1946-1975 period, Dr. Long was employed at the Engineering Experiment Station, Georgia Institute of Technology, where he held various positions including Radar Branch Head, Electronics Division Chief, and Director. He has taught electrical engineering courses at Georgia Tech and the University of Kentucky, and during 1966 he was a Liaison Scientist with the U.S. Office of Naval Research in London. Dr. Long is a Fellow of the Institute of Electrical and Electronics Engineers and is a member of the U.S. Commission F of the International Union of Radio Science. His other publications include technical and scientific papers in the fields of radar, antennas, propagation, microwave components, and data analysis.

Contents

Preface

This book was prepared to provide reflectivity data useful for extending the utility of radar sensing through the development of improved techniques and methodologies. Information is presented on radar echo characteristics of value for navigation and terrain classification, and for determining sea surface characteristics. Known characteristics of radar reflectivity of land and sea are presented through summaries of a large body of experimental data and, where possible, the fundamental causes for the various observed characteristics are outlined and interrelated.

The emphasis in this book is on microwave backscatter. Forward scatter is discussed because under many conditions forward scatter from land or sea strongly influences radar performance. Discussion of longer wavelengths (frequencies of a few megahertz or more) is also included because in the past several years doppler radars operating at these wavelengths have shown great promise for sensing conditions over huge areas of the sea from distances of a thousand miles or more. Rapid advances are foreseen for such ground-based over-the-horizon radars that should provide for a continuous surveillance capability with relative immunity to effects of weather.

The book is well suited for first-year graduate students or for college seniors having strong backgrounds in applied electromagnetics. It is the author's belief that the contents of the book present the current state of experimental knowledge in radar backscatter and therefore the book should have significant value for the design engineer and for the user of remote sensing radar data.

1975

Acknowledgments

This book contains material that has not been available previously, information from research reports not widely distributed, and material borrowed from technical journals and books.

The author is indebted to the editors of the Institute of Electrical and Electronics Engineers for permission to reproduce material from records and digests of conventions and symposia, the *Proceedings*, the *Transactions on Antennas and Propagation*, the *Transactions on Aerospace and Electronic Systems*, and the *Transactions of Geoscience Electronics*.

Other material from journals includes a table from *Undersea Technology* by permission of Compass Publications, Inc. and figures from *Profile* by permission of Goodyear Aerospace Corporation.

Grateful acknowledgment is made to McGraw-Hill Book Company for permission to use material from the following: RADAR SYSTEM ENGINEERING, L.N. Ridenour; RADAR AIDS TO NAVIGATION, J.S. Hall; PROPAGATION OF SHORT RADIO WAVES, D.E. Kerr; RADAR ASTRONOMY, J. V. Evans and T. Hagfors; and RADAR DESIGN PRINCIPLES, F.E. Nathanson.

Permission is also appreciated for use of material from books by other publishers. Figures 4-6 through 4-9 are from *Airborne Radar* by D. J. Povejsil, R. S. Raven, and Peter Waterman, copyright 1961 by Litton Educational Publishing, Inc. and reprinted by permission of Van Nostrand Reinhold Company; figure 4-15 is from *The Scattering of Electromagnetic Waves from Rough Surfaces* by Petr Beckmann and André Spizzichino, copyright 1963 by Pergamon Press Limited and reprinted by permission of Pergamon Press Limited; and figures 2-3 and 2-4 are from *The Log Normal Distribution* by J. Aitchison and J. A. C. Brown, copyright 1969, and reprinted by permission of Cambridge University Press.

Appreciation is also expressed for use of a figure borrowed by permission of the Controller of Her Britannic Majesty's Stationery Office. The figure is included as part of figure 6-42, is acknowledged as British Crown Copyright, and is from a Royal Radar Establishment Memorandum written by Geoffrey Bishop.

As mentioned earlier, considerable information was borrowed from research reports including those of the Engineering Experiment Station at Georgia Tech, the United States Naval Research Laboratory, the Ohio State University, the United States Army Electronics Laboratory, the Environmental Research Institute of Michigan, the University of Kansas,

Lincoln Laboratories of the Massachusetts Institute of Technology, and the Research Institute of Swedish National Defence (Försvarets Forsjningsanstalt).

A number of persons reviewed the manuscript and their comments and suggestions are appreciated. The author wishes to specifically acknowledge the comments and encouragement of E. R. Flynt, H. A. Corriher, F. B. Dyer, R. C. Johnson, G. W. Ewell, R. D. Hayes and J. L. Eaves of the Engineering Experiment Station at Georgia Tech, and of Isadore Katz of the Johns Hopkins Applied Physics Laboratory.

A special word of appreciation is due Betty Yarborough who contributed much personal time to assisting with the manuscript in almost every phase of its creation. I also wish to thank Martha Shoemaker for her work on the illustrations, and to acknowledge my appreciation to Claudine Taylor, Deborah Ariail, and Bonnee Wettlaufer for their assistance with typing, proofing, and editing.

Foreword

Radar Reflectivity of Land and Sea has been revised and updated by means of the foreword and its bibliography, plus appendices. Added material includes discussions on spatial and temporal statistics, comparisons between lognormal and Weibull distributions, dependencies of tree echo spectra on wavelength, and certain averages for low angle land data. References to land and sea clutter models, in addition to sources of new data for microwaves and millimeter waves, are included in the extensive bibliography.

The original manuscript was completed in early 1975, and since then there have been a number of books[a] that give significant attention to radar reflectivity of land and sea including Barton (1975b), Reeves (1975), Schleher (1978), Blake (1980), Skolnik (1980), and Ulaby, Moore and Fung (1982).

In 1975 the most frequently used probability density functions for describing clutter were the Rayleigh, Ricean and lognormal. The Weibull has now gained widespread acceptance and, as a consequence, appendix A on the Weibull distribution and its relationship to the lognormal has been added. Today, the Weibull and the lognormal distributions are widely used to describe clutter and target statistics when large RCS values occur with higher probability than with the Rayleigh function. The Weibull may, in fact, be generally more useful than the lognormal for describing spatial clutter statistics. Even so, the lognormal continues to be used extensively. Among its attractive features is simplicity resulting from the fact that a straight line on ordinary normal probability paper yields a lognormal cumulative distribution, providing the data are graphed in logarithmic units (e.g., decibels).

Some years ago Fishbein, Graveline, and Rittenbach (1967) observed that the shape of the noncoherent doppler spectra of tree echo at X band dropped off slower with doppler frequency than was generally expected (see sec. 5.3). Later, as discussed in appendix B, Currie, Dyer and Hayes (1975) observed that the spectral shapes at 9.5, 16, 35, and 95 GHz all decreased slowly (consistent with Fishbein *et al*) with increases in doppler frequency. Further, they observed that the half-power bandwidths of their measured spectra do not increase as rapidly as directly with radar frequency as one would attain for a collection of independent, randomly moving scatterers.

---

[a]See complete reference data at end of Foreword

Since this book was first published, radars having high spatial resolutions have become almost commonplace through use of millimeter waves, synthetic aperture techniques and high speed digital processing. Available radar spatial resolutions have thereby become considerably smaller. As a consequence of interest in smaller available range-azimuth cells, the clutter models now often include effects of both point and distributed scatterers, and greater general awareness exists regarding distinctions between spatial statistics (fixed time) and temporal statistics (fixed position). Further, the Weibull distribution has emerged as a principal statistic for describing clutter.

Clutter sources are actually comprised of a continuum of scatterer types, from point sources, i.e., discrete clutter, at one extreme to diffuse scatterers, i.e., distributed clutter, at the other extreme. Examples of discrete clutter sources include water towers, buildings, tree trunks, floating debris, and sea-wave prominences. Depending on the radar application, a discrete clutter source can of course be either undesirable clutter or a target of interest. Distributed clutter is caused by a large number of separate scatterers comprised of, e.g., tree canopy, grass, small sea waves, wind ripples, rain.

There have been a number of excellent papers on the general subject of echo amplitude statistics including Chen and Morchin (1977); Ekstrom (1973); Lind (1976); and Schleher (1976; 1978, pp. 37-52). Most of the statistical data reported are for short-term echo fluctuations in amplitude from a given range-azimuth cell of resolution. Recent publications giving land statistics (see appendix B) include Currie, Dyer, and Hayes (1975); Hayes (1979); Schleher (1976; 1978, pp. 37-52); and Sekine *et al* (1981; 1982). Papers on sea echo variations include Chen, Havig, and Morchin (1977); Croney, Woroncow and Gladman (1975); Fay, Clarke, and Peters (1977); Jakeman and Pusey (1976; 1977); Maalϕe (1982); Olin (1982); Rivers and Tuley (1977); Schleher (1975; 1976; 1977; 1978, pp. 37-52); Trunk (1977), and Ward (1982).

Clutter, as seen from a particular fixed point, varies both in time at a given range and spatially. Spatial and temporal distributions are generally different. For land, temporal distributions (observed spatial grid or footprint fixed) are generally considered to be Ricean (sec. 5.2) which includes the Rayleigh as one limit and a non-fluctuating echo as the other limit. However, for X band and higher frequencies Currie, Dyer, and Hayes (1975) reported lognormal shapes (appendix B), and Hayes (1979) reported that at 95 GHz the temporal distributions, although usually lognormal, are sometimes Weibull in shape. Spatial distributions for land usually range from the Rayleigh to large standard deviation distributions that have been described both by the lognormal and the Weibull (appendix A).

Since sea waves propagate through a given spatial grid, temporal and spatial sea statistics are closely related. As in the case of land, the data on temporal sea statistics are extensive. The temporal statistics are usually reported as Rayleigh for depression angles of five degrees or more and for large cell areas. For smaller angles and cell sizes, the temporal statistics have distributions with shapes that fall between the Rayleigh and the log-normal. Obviously, conditions that cause temporal sea echo variations with large standard deviations will also cause spatial sea echo variations with large standard deviations. Therefore, depending on sea surface details, spatial sea echo distributions are expected to lie between Rayleigh and log-normal.

An interesting account of measurements on the persistence of sea spikes (strong, discrete sources of echo) versus polarization and sea state for high resolution, noncoherent radar at small grazing angles is given by Kalmykov and Pustovoyesko (1976). Readers desiring more information on discrete sea echo should also see "spiky echo" in subject index, and Bishop (1973), Lewis and Olin (1977), Lind and Bergkvist (1975), and Olin (1982).

Significant papers on discrete land clutter which were not previously referenced in the subject book include Edgar, Dodsworth, and Warden (1973); McEvoy (1972a; 1972b); Rigden (1973); Riley (1971); and Ward (1971). Generally speaking, discrete clutter sources are "sprinkled" more or less randomly throughout the distributed clutter. Even for built-up industrial areas, discrete clutter seems to be distributed randomly both in space and amplitude. Thus, there seems to be little apparent spatial correlation for discrete clutter. On the other hand, as will now be discussed, for distributed clutter there is, depending on cell size, spatial correlation between the temporal averages in nearby range-azimuth cells.

Rigden (1973) reported the results of a C-band, high resolution (3 meters in range) study of rural terrain. The azimuthal beamwidth was 1.5 degrees and the ranges used were between 7 and 11 km. Rigden measured average (temporal) RCS versus range and displayed statistics of these RCS values that help to illustrate the continuum in clutter types between point "targets" (discrete clutter) and distributed scatterers. From Rigden's graphs one may see that (1) median patch lengths for the larger RCS sources are short and these patch lengths are very small compared to the median patch separations and, on the other hand, (2) median patch lengths for the smaller RCS sources are relatively long and the median patch separations are short with respect to these patch lengths. Thus (1) sources of point clutter can have large RCS values with sparse spatial density and little spatial correlation, and (2) sources of distributed clutter usually have relatively small RCS values with substantial spatial correlation in the temporal averages of range-azimuth cells.

As with distributed clutter, discrete clutter can be expected at all radar wavelengths. Generally, in urbanized areas, discretes with RCS exceeding $10^3 m^2$ are commonplace and numerous. There are, of course, progressively fewer discretes that exceed larger RCS values. Values of RCS as large as $10^6 m^2$ have been observed. Buildings and water towers are sources of very large RCS values, especially at small depression angles.

The variation of echo strength with radar range, e.g., standard deviation of variations with range, usually increases for decreases in grazing angle. Possible reasons for this increased variation or spikyness include:

On land, echo from strong reflecting objects such as buildings, water tanks, fences, tends to be largest for line-of-sight near horizontal.

At sea, echo from wave crests is enhanced (relative to other sea echo) because of the classical intererence effect, i.e., multipath, at the smaller grazing angles.

On land, lowlands which are usually flat or gently undulating surfaces are shadowed (masked) by the highlands, e.g., hills or mountains.

At sea, the troughs are shadowed by the crests.

The examples above relate to two basically different types of changes that can occur to a statistical distribution: (1) large RCS values that occur with small probability are added (due to the introduction of strong discrete scatterers), and (2) a substantial fraction of the small RCS values are deleted (due to shadowing). A mixture of the two types of changes is also possible. For example, consider a heavy sea viewed with a high resolution radar. Then for a small grazing angle large individual waves can produce large RCS values (with small probability in a given range-azimuth cell) and considerable shadowing. The examples above help to illustrate why this author feels it unlikely a single simple statistic will ever be found that describes all echo fluctuations.

The simpler clutter models describe only an average or median clutter level, but the more comprehensive models express various parameters, such as temporal and spatial variations, in terms of detailed statistics. An early and yet comprehensive model is the IITRI (Illinois Institute of Technology Research Institute) or ORT (Overland Radar Technology) model. This model seems to be the first one that included virtually all known scattering parameters, including terrain type, incidence angle, average RCS per unit cell area, temporal statistics, spatial statistics, spatial correlation, plus discrete clutter. A digest of the IITRI, or ORT, model is given by Schleher (1978, pp. 52-58). Key features include:

Lognormal statistics for range dependencies of distributed clutter, with medians and standard deviations being functions of terrain type and incidence angle.

Rayleigh temporal statistics for distributed clutter.

Cell-to-cell correlation of the temporal averages, with the correlation distances depending on terrain type.

Discrete clutter that is lognormally distributed, with medians specified by a cumulative distribution and with values up to $10^6 m^2$.

Various equations have been developed from a combination of measurements data, scattering theories and statistical techniques that describe expected values of σ°, radar cross section normalized to illuminated surface area, for land and sea. Schleher (1978, pp. 58-61) has published an outline of a sea clutter model developed in the late 1960s by Wayne Rivers and his Georgia Tech colleagues. Equations are given for horizontal (HH) and vertical (VV) polarizations, and the equations express the expected average for σ° in terms of radar wavelength, wind speed and/or wave height, wind direction and incidence angles with respect to the sea. The model has been continually updated through the use of additional clutter data. Presently there are two Georgia Tech sea clutter models, one for the 1 GHz to 10 GHz region (Ewell, Horst, and Tuley, 1979) and the other for the 10 GHz to 100 GHz region (Horst, Dyer, and Tuley, 1978). The above models are applicable to small grazing angles of 15 degrees and smaller. Chan and Fung (1977) have developed a comprehensive sea echo theory and data-based model for determining σ° at the larger depression angles applicable to spacecraft radar.

Henn, Pictor, and Webb (1982) give a review of prior efforts at land clutter modeling for small grazing angles. One approach for developing prediction equations for σ°, began by Hayes and Dyer (1973) and followed by Trebits, Hayes, and Bomar (1978), is to express σ° by an equation which is a function of grazing angle and wavelength, and to choose constants for the equation on the basis of clutter type (trees, crops, concrete, urban, etc). Another approach for obtaining σ° equations, developed principally from microwave data, was originated by Zehner and Tuley (1979) and has since been modified by Currie and Zehner (1982) through use of millimeter-wave data up to 95 GHz. For this modelling, equations for σ° versus grazing angle are developed based on empirically determined constants that depend on clutter type and radar frequency. Moore, Soofi, and Purduski (1980) have also developed a model that relates σ° to incidence angle and radar frequency for describing general North American terrain during the summer months, for snow-covered terrain, and for sea ice. This model is applicable to a broad range of incidence angles, exclusive of near grazing incidence, and to the 1 - 18 GHz frequency regime. There are few σ° data for land which are valid for grazing angles of less than 5 degrees (see Appendix C). However, the model of Zehner and Tuley (1979) is based on a limited data base down to 0.8 degree grazing and it is therefore considered valid for small angles. Deloor, Jurriens, and Gravesteijn (1974) is a source which includes reflectivity data for crops at grazing angles between one and 10 degrees.

The reflectivity of dry, freshly fallen snow can be quite small. However, usually snowpack has been melted and refrozen a number of times (old metamorphic snow) and this type of snow can have strong reflective properties. Consequently the reflectivity of snow can vary widely. At 1 GHz or less (Stiles and Ulaby, 1980a; Ulaby and Stiles, 1980) the free water content of snow has little effect on σ°, but for microwaves and millimeter waves the wetness of snow will usually cause a substantial reduction in σ°. For example, the equations by Currie and Zehner (1982), described above, give the following predictions: σ° for dry snow exceeds that for wet snow by 9 dB at X band, by 11 dB at 35 GHz and by 5 dB at 95 GHz. The X band data used were from Stiles and Ulaby (1980b); the 35 GHz data used were from Currie, Dyer, and Ewell (1976), Hayes *et al* (1979), Hayes, Currie, and Scheer (1980), and Stiles and Ulaby (1980b); and the 95 GHz data used were from Hayes *et al* (1979).

The University of Kansas Remote Sensing Laboratory's Sea Ice Group has been engaged in backscatter studies of sea ice for several years (Delker, Onstott, and Moore, 1980; Onstott, *et al,* 1982). Additionally, Ulaby, Moore, and Fung (1982, pp. 908-910) include a brief review of the sea ice results. According to the authors, σ° for sea ice can vary by more than 20 dB depending on ice type and condition. Ulaby, Moore, and Fung also state therein that the magnitude of σ° for sea ice is usually much larger than it is for lake ice.

There have been a number of recent publications on bistatic reflectivity and forward scattering including Barton (1975a, 1976, 1977a, 1979); Cornwell and Lancaster (1979); Ewell and Zehner (1982); Lammers and Hayes (1980); Larson *et al* (1978); Lennon and Papa (1980); Mrstik and Smith (1978); and Papa, Lennon, and Taylor (1980a, 1980b). Noteworthy earlier papers on bistatic scattering, and not previously referenced, include Kell (1965) and Pidgeon (1966). A report by Shotland and Rollin (1976) contains calculated specular reflection coefficients for water versus frequency for horizontal, vertical and circular polarizations. Lammers and Hayes (1980) have measured effects of multipath at 35, 98, and 140 GHz for snow in addition to their snow backscatter measurements already mentioned. Deep interference nulls due to multipath were usually observed with the 0.5 to 2 degree grazing angles used. The reflecting surfaces included freshly fallen snow, old metamorphic snow, and sleet over gently rolling grass-covered terrain. Although backscatter and forward scatter for land and sea are relatively well understood, much is yet to be learned about the general bistatic scatter problem.

Additional data are now available on the dielectric properties of land and sea surfaces for radar frequencies. For example, the already mentioned 1976 report by Shotland and Rollin includes useful data on the dielectric properties of water versus frequency. Further, Hoekstra and Delaney (1974) have reported on the dielectric properties of soils versus water content and frequency, and Stiles and Ulaby (1981) give measured and calculated data on the dielectric properties of snow and ice.

Papers on the average reflectivity at various microwave frequencies for vegetation canopy and the physical scattering mechanisms of canopy include Rosenbaum and Bowles (1974); Ulaby, Bush, and Batlivala (1975); Bush and Ulaby (1976); Ulaby and Batlivala (1976); and Ulaby (1980).

New and important subject areas are attenuation and backscattering caused by dust storms over desert regions. According to Goldhirsh (1982), a heavy dust storm over a desert can be expected to cause signal attenuations that are negligible for L band radar, borderline for S band radar and severe for X band radar. Goldhirsh also gives calculation results for expected backscatter cross sections under various assumptions regarding degree of dust storm/antenna beam filling.

Research continues to be active in the subject area of over-the-horizon (OTH) radar at HF frequencies. Maresca and Barnum (1982a) have performed modelling on sea echo in the 7 to 25 MHz region. Sea echo spectra have been calculated as functions of transmit frequency, sea state, radar/wind angle, and the results have been compared with expected signals from ships. Maresca and Barnum (1982b) have also made analyses for estimating ocean surface wind speed based on the doppler spectra of radar signals when operating at frequencies within the 7 to 25 MHz region.

Jones *et al* (1975) and Jones, Schroeder, and Mitchell (1977) have by means of aircraft measurements investigated the effects of sea surface conditions on average reflectivity. These investigations were at the relatively large depression angles of interest to NASA. Taylor (1975) reported on a ground-based radar investigation of the echo from sea-wave patterns, and Sittrop(1977) reported on the sea clutter dependency on windspeed.

In the subject area of polarization, Long (1977) presented a model for land and sea comprised of depolarizing and non-depolarizing scatterers, Lee (1978) published laboratory data on the polarization properties of scattering from windwave water surfaces, and Bush and Ulaby (1978) reported that improved agricultural crop identification is possible if a dual polarization rather than a unipolarization radar is used.

Use of circular polarization has long provided a useful method for reducing the average echo strength of rain relative to that of aircraft. Hendry and

McCormick (1974), through use of a tunable polarizer, have vividly shown how rain suppression when using circular polarization is affected by rain rate. Hogg and Chu (1975) show both experimental and theoretical results on the effects of rain rate on relative phase shift and relative attenuation versus frequency for two orthogonal polarizations when propagating through rain. The paper by Antar *et al* (1982) represents an entry into the extensive recent literature on the depolarization caused by ice and rain.

In recent years there has been considerable emphasis on the development of technology for short-range, narrow-beam radars. For this reason there has been much interest in learning more about the dependencies of the reflectivities of land and sea surfaces on frequency at the millimeter wavelength frequencies of 35 GHz and higher. Papers containing data on reflectivity for millimeter waves include Sackinger and Byrd (1973); Currie, Dyer, and Hayes (1975); Currie, Dyer, and Ewell (1976); Hayes, Dyer, and Currie (1976); Dyer, Currie, and Applegate (1977); Dyer and Currie (1978); Horst, Dyer, and Tuley (1978); Trebits, Hayes, and Bomar (1978); Currie (1979); Hayes (1979); Hayes *et al* (1979); Trebits, Currie, and Dyer (1979); Hayes, Currie, and Scheer (1980); Lammers and Hayes (1980); Stiles and Ulaby (1980a, 1980b); Ulaby and Stiles (1980); and Currie and Zehner (1982). There are few calibrated millimeter data for either land or sea at grazing angles less than a few degrees.

Bush, Ulaby, and Peake (1976) addressed the subject of variability and reliability of ground echo data taken with ground-based radars. Their comments help to underscore the difficulty and patience required for obtaining experimental data that have accurate absolute calibrations, and which are usually obtained under less-than-ideal field conditions.

R. K. Moore (1978) gives a useful review of land and sea data at incidence angles applicable to airborne and spacecraft radars. Although airborne and spacecraft radar results have been used extensively for some years in geology, land use studies, and in ice and oil-spill mapping, research investigators continue to uncover potential, new operational applications for radar. Some of the research results for remote sensing include soil moisture measurements (Ulaby, Batlivala, and Dobson, 1978; Ulaby, Bradley, and Dobson, 1979), estimates of water content in snow cover (Stiles and Ulaby, 1980a; Ulaby and Stiles, 1980), agricultural crop identification (Bush and Ulaby, 1978), plant stress (Ulaby and Bush, 1976), plant moisture content (Attema and Ulaby, 1978).

A review of events relating to land and sea scatter would be incomplete without underscoring the dramatic results in remote sensing attained with space radar. The first radars for remote sensing were altimeters that conducted global monitoring from the satellites Skylab and GEOS-3. Skylab was launched in 1973 and its radar altimeter operated for several months; GEOS-3 was launched in 1975 and its radar altimeter operated for more than three years. The Skylab and GEOS-3 altimeters, and the Seasat altimeter discussed in the following paragraph, each operated at 13.9 GHz (McGoogan, 1975). Moore and Young (1977) reported on estimating ocean surface wind speeds by using amplitude measurements from Skylab, and Gower (1979) reported on computing ocean wave heights from GEOS-3 data.

The Seasat radar altimeter has provided extensive, world-wide data on sea surface topography, wave heights, and wind speeds (Townsend, 1980; Apel, 1982). For example, Apel (1982) gives elevation contours of the earth's seas at 2 meter intervals and wave height and wind speed versus location with each averaged over a three-month period during 1978. Three types of measurements were made:

1. Time delay between the transmission and reception of compressed pulses (3 ns) which, along with precision satellite orbit data, was used to determine large-scale topography of the sea surface;
2. Broadening of the leading edge of the received pulses which was used to determine wave height; and
3. Amplitude of the received pulses which was used to determine radar cross section per unit area of the sea for calculating wind speed.

Precision of measurement has been given as 0.7 meters for elevation contours, as under 0.3 meter or 10 per cent for wave heights up to 8 meters, and plus or minus 2 meters per second for wind speeds up to 25 meters per second.

Spaceborne synthetic aperture radars (SARs) with fine range and cross-range resolution provide very useful images of the earth. The first space-based imaging radar was aboard satellite Seasat and operated for three months during 1978. The second imaging radar, the-shuttle-imaging-radar A (SIR-A), was aboard U.S. Space Shuttle Columbia during its second flight in November 1981. Radar images have proven to be particularly useful for surface mapping, e.g., geological exploration and terrain classification, of features dependent on surface roughness and/or slope variation (Elachi, 1982; Elachi and Granger, 1982). Additionally, useful imagery has been obtained with both spaceborne radars for ocean surface waves and their patterns. Other significant developments include Seasat imagery having been used to show the feasibility of precision, large-scale polar ice mapping and tracking, and effects of elevated soil moisture due to rainfall can be easily seen on Seasat images over large regions of the midwestern U.S. farm belt, where the terrain is flat (NASA, 1982, fig. 2-5(a)). At the opposite extreme of water content, SIR-A has penetrated sands in hyperarid regions of the eastern Sahara Desert and the images have revealed radar echoes from previously unknown subsurface features such as buried valleys, geologic structures and possibly Stone Age occupation sites (McCauley, *et al,* 1982).

The Seasat and SIR-A radar operating parameters are identical in most respects except that the grazing angles were 70° (incidence angle 20°, i.e., radar beam 20° from local horizontal) and 40° (incidence angle 50°), respectively. Both radars operated at L band (23 cm wavelength), with horizontal (HH) polarization, and the range and cross-range resolutions were about 40 meters. Differences in incidence angle cause differences in echo strength sensitivity to terrain surface slope and small-scale surface roughness. For example, use of a near normal incidence angle (Seasat) stresses echo strength sensitivity to surface slope and use of an intermediate incidence angle (SIR-A) stresses echo strength sensitivity to small-scale surface roughness. Therefore, in general the Seasat and SIR-A images are more sensitive, relatively, to effects of topography and local surface roughness, respectively. In fact, a Seasat/SIR-A image (see e.g., Elachi, *et al,* 1982, fig. 15) formed from a combination of the two (Seasat and SIR-A) equal resolution images provides information of importance to geologists which is not otherwise available.

An important feature of shuttle experiments is that the radars can be returned from space, recalibrated and refurbished for continuing observations. For example, the SIR-A was returned from space, and is presently being upgraded with improved instrumentation. The upgraded radar, SIR-B, will include digital signal processing capabilities and the ability to vary incidence angle from 15° (75° from grazing) to 60° (30° from grazing). SIR-B is scheduled for launch in August 1984, and is expected to provide data useful in the fields of geography, geology, hydrology, oceanography, agronomy, and botany (NASA, 1982). Although data from SIR-A are continuing to be processed (see, e.g., Cimino and Elachi, 1982; Ford, Cimino, and Elachi, 1983), it is now quite apparent that space shuttle imaging radars provide excellent sensors for investigating specific land and sea surface areas distributed over the earth, and investigating the effects of the environment on these surface areas at particular points in time.

Beginning in January 1976, this book was used a number of times as the text for a continuing education course at the Georgia Institute of Technology, and in 1979 it was also used as the text for a somewhat similar course offered by the Technology Service Corporation (TSC) in Washington, DC.

The author expresses here his appreciation to those persons who have reported errors and suggested improvements. Special thanks are given to the lecturers who have used this book with the Georgia Tech and TSC courses including H. A. (Archie) Corriher, Jr., N. C. (Nick) Currie, F. B. (Fred) Dyer,

Jerry L. Eaves, H. Allen Ecker, George W. Ewell, R. D. (Bob) Hayes, Margaret M. Horst, R. C. (Dick) Johnson, D. J. (Don) Lewinski, J. I. (Jim) Metcalf, R. K. (Dick) Moore, Fred E. Nathanson, E. K. (Ed) Reedy, C. E. (Chuck) Ryan, Jr., M. T. (Mike) Tuley, and S. P. (Steve) Zehner. Most of the persons named here were and still are members of the research staff with the Engineering Experiment Station at Georgia Tech. Allen Ecker and Jim Metcalf are now employed, respectively, at Scientific-Atlanta Incorporated and the Air Force Geophysics Laboratory, Hanscom Field, Massachusetts. Dick Moore and Fred Nathanson are, respectively, with the University of Kansas, Lawrence, and the Washington, DC, Division of Technology Service Corporation.

In this revised edition of the book, appendices have been added with figures borrowed from research reports of the Engineering Experiment Station at Georgia Tech. Permission is hereby acknowledged for use of those figures which are from reports by N.C. Currie, F.B. Dyer, R.D. Hayes, W.K. Rivers, M.T. Tuley, and S.P. Zehner.

References

Antar, Y.M.M., A. Hendry, J.J. Schleska, and R.L. Olsen, "Measurements of Ice Depolarization at 28.56 GHz Using the COMSTAR Beacon Simultaneously with a 16.5 GHz Polarization Diversity Radar," *IEEE Transactions on Antennas and Propagation*, vol. AP-30, pp. 858-866, September 1982.

Apel, J.R., "Some Recent Scientific Results From the Seasat Altimeter," *Sea Technology*, pp. 21, 23-25, 27, October 1982.

Attema, E.P.W., and F.T. Ulaby, "Vegetation Modeled as a Water Cloud," *Radio Science*, vol. 13, pp. 357-364, 1978.

Barton, D.K., *Radar Resolution and Multipath Effects*, Artech House, Inc., Dedham, Massachusetts, 1975a.

Barton, D.K., *Radar Clutter*, Artech House, Inc., Dedham, Massachusetts, 1975b.

Barton, D.K., "Radar Multipath Theory and Experimental Data," *Radar 77 Conference Proceedings*, IEEE Conference Publication No. 155, London, October 1977a.

Barton, D.K., *Frequency Agility and Diversity*, Artech House, Inc., Dedham, Massachusetts, 1977b.

Barton, D.K., "Low-Altitude Tracking Over Rough Surfaces I: Theoretical Predictions," *EASCON-79 Record*, Institute of Electrical and Electronic Engineers, October 1979.

Bishop, G., "Radar Sea Clutter," *Determination and Use of Radar Scattering Characteristics*, AGARD Lecture Series No. 59, 1973.

Blake, L.V., *Radar Range-Performance Analysis*, D.C. Heath, Lexington, Massachusetts, 1980.

Bush, T.F., and F.T. Ulaby, "Radar Return from a Continuous Vegetation Canopy," *IEEE Transactions on Antennas and Propagation*, vol. AP-24, pp. 269-276, May 1976.

Bush, T.F., and F.T. Ulaby, "An Evaluation of Radar as a Crop Classifier," *Remote Sensing of the Environment*, vol. 7, pp. 15-36, 1978.

Bush, T.F., F.T. Ulaby, and W.H. Peake, "Variability in the Measurements of Radar Backscatter, *IEEE Transactions on Antennas and Propagation*, vol. AP-24, pp. 896-899, November 1976.

Chan, H.L. and A.K. Fung, "A Theory of Sea Scatter at Large Incident Angles," *Journal of Geophysical Research*, vol. 82, pp. 3439-3444, August 20, 1977.

Chen, P., T.F. Havig, and W.C. Morchin, "Characteristics of Sea Clutter Measured from E-3A High Radar Platform," *Proceedings of the IEEE 1977 National Aerospace and Electronics Conference*, Dayton, Ohio, May 17-19, 1977.

Chen, P., and W.C. Morchin, "Detection of Targets in Noise and Weibull Clutter Backgound," *Proceedings of the IEEE 1977 National Aerospace and Electronics Conference*, Dayton, Ohio, May 17-19, 1977.

Cimino, J.B., and C. Elachi, "Shuttle Imaging Radar-A (SIR—A) Experiment," Jet Propulsion Laboratory Publication 82-77, December 15, 1982.

Cornwell, P.E., and J. Lancaster, "Low Altitude Tracking Over Rough Surfaces II: Experimental and Model Comparisons," *EASCON-79 Record*, Institute of Electrical and Electronic Engineers, October 1979.

Croney, J., A. Woroncow, and B.R. Gladman, "Further Observations on the Detection of Small Targets in Sea Clutter," *The Radio and Electronics Engineer*, vol. 45, pp. 105-115, March 1975.

Currie, N.C., "Characteristics of Millimeter Radar Backscatter From Wet/Dry Foliage," *1979 IEEE-APS International Symposium*, Seattle, Washington, June 1979.

Currie, N.C., and S.P. Zehner, "Millimeter Wave Land Clutter Model," *Proceedings of Radar 82*, IEE Conference Publication 216, London, England, 18-20 October 1982.

Currie, N.C., F.B. Dyer, and G.W. Ewell, "Characteristics of Snow at Millimeter Wavelengths." *Digest of the 1976 IEEE/APS International Symposium*, October 1976.

Currie, N.C., F.B. Dyer, and R.D. Hayes, "Radar Land Clutter Measurements at Frequencies of 9.5, 16, 35, and 95 GHz," Engineering Experiment Station, Georgia Institute of Technology, Technical Report No. 3, Contract DAAA25-73-C0256, March 1975.

Delker, C.V., R.G. Onstott, and R.K. Moore, "Radar Scatterometer Measurements of Sea Ice: The Sursat Experiment," The University of Kansas Remote Sensing Laboratory Report RSL TR331-17, August 1980.

DeLoor, G.P., A.A. Jurriens, and H. Gravesteijn, "The Radar Backscatter from Selected Agricultural Crops," *IEEE Transactions on Geoscience Electronics*, vol. GE 12, pp. 70-77, 1974

Dyer, F.B., and N.C. Currie, "Environmental Effects on Millimeter Radar Performance," *AGARD/NATO Symposium on MM and Sub-MM Wave Propagation and Circuits*, September 1978.

Dyer, F.B., N.C. Currie, and M.S. Applegate, "Radar Backscatter from Land, Sea, Rain, and Snow at Millimeter Wavelengths," *RADAR-77 Digest*, London, October 1977.

Edgar, A.K., D.J. Dodsworth, and M.P. Warden, "The Design of a Modern Surveillance Radar," *Proceedings of Radar — Present and Future*, IEE Conference Publication No. 105, London, pp. 8-13, October 1973. Reprinted in Barton (1975b).

Ekstrom, J.L., "The Detection of Steady Targets in Weibull Clutter," *Proceedings of Radar — Present and Future*, IEE Conference Publication No. 105, London, pp. 221-226, October 23-25, 1973. Reprinted in Barton (1975b).

Elachi, C., "Radar Images of the Earth from Space," *Scientific American*, pp. 54-61, December 1982.

Elachi, C., and J. Granger, "Spaceborne Imaging Radars Probe 'In Depth'," *IEEE Spectrum*, pp. 24-29, November 1982.

Elachi, C., *et al*, "Shuttle Imaging Radar Experiment," *Science*, vol. 218, pp. 996-1003, 3 December 1982.

Ewell, G.W., M.M. Horst, and M.T. Tuley, "Predicting the Performance of Low-Angle Microwave Search Radars — Targets, Sea Clutter, and the Detection Process," *Proceedings of IEEE Oceans 79*, September 1979.

Ewell, G.W., and S.P. Zehner, "Bistatic Sea Clutter Near Grazing Incidence," *Proceedings of Radar 82*, IEE Conference Publication 216, London, 18-20 October 1982.

Fay, F.A., J. Clarke, and R.S. Peters, "Weibull Distribution Applied to Sea Clutter," *Proceedings of Radar 77*, IEE Conference Publication 155, London, pp. 101-104, October 1977.

Fishbein, W., S.W. Graveline, and O.E. Rittenbach, "Clutter Attenuation Analysis," U.S. Army Electronics Command, Technical Report ECOM-2808, March 1967. Reprinted in Schleher (1978).

Ford, J.P., J.B. Cimino, and C. Elachi, "Space Shuttle Columbia Views the World With Imaging Radar: the SIR-A Experiment," Jet Propulsion Laboratory Publication 82-95, January 1, 1983.

Goldhirsh, J., "A Parameter Review and Assessment of Attenuation and Backscatter Properties Associated with Dust Storms over Desert Regions in the Frequency Range of 1 to 10 GHz," *IEEE Transactions on Antennas and Propagation*, vol. AP-30, pp. 1121-1127, November 1982.

Gower, J.F.R., "The Computation of Ocean Wave Heights from GEOS-3 Satellite Radar Altimeter Data," *Remote Sensing of the Environment*, vol. 8, pp. 97-114, 1979.

Hayes, R.D., F.B. Dyer, and N.C. Currie, "Backscatter from Ground Vegetation Between 10 and 100 GHz," *Digest of the 1976 IEEE/APS International Symposium*, October 1976.

Hayes, R.D., "95 GHz Pulsed Radar Return from Trees," *EASCON 79 Conference Record*, pp. 353-356, October 1979.

Hayes, R.D., and F.B. Dyer, "Land Clutter Characteristics for Computer Modelling of Fire Control Radar Systems," Engineering Experiment Station, Georgia Institute of Technology, Technical Report No. 1, Contract DAAA 25-73-C-0256, May 1973.

Hayes, R.D., N.C. Currie, and J.A. Scheer, "Reflectivity and Emissivity of Snow and Ground at MM Waves," *1980 IEEE International Radar Conference*, Washington DC, April 1980.

Hayes, D.T., U.H.W. Lammers, R. Marr, and J.J. McNally, "Millimeter Wave Backscatter from Snow," *1979 IEEE-APS International Symposium Record,*, p. 499, October 1979.

Hendry, A., and G.C. McCormick, "Deterioration of Circular Polarization Clutter Cancellation in Anisotropic Precipitation Media," *Electronics Letters*, vol. 10, pp. 165-166, 16 May 1974.

Henn, J.W., D.H. Pictor, and A. Webb, "Land Clutter Study: Low Grazing Angles (Backscattering)," *Proceedings of Radar 82*, IEE Conference Publication 216, London, 18-20 October 1982.

Heokstra, P., and A. Delaney, "Dielectric Properties of Soils at UHF and Microwave Frequencies," *Journal of Geophysical Research*, vol. 79. pp. 1699-1708, April 10, 1974.

Hogg, D.C., and T.S. Chu, "The Role of Rain in Satellite Communications," *Proceedings of the IEEE*, vol. 63, pp. 1308-1331, September 1975.

Horst, M.M., F.B. Dyer, and M.T. Tuley, "Radar Sea Clutter Model," *Proceedings of Institution of Electrical Engineering International Conference on*

Antennas and Propagation, London, November 1978.

Jakeman, E., and P.N. Pusey, "A Model for Non-Rayleigh Sea Echo," *IEEE Transactions on Antennas and Propagation*, vol. AP-24, pp. 806-814, November 1976.

Jakeman, E., and P.N. Pusey, "Statistics of Non-Rayleigh Microwave Sea Echo," *Proceedings of Radar 77*, Institution of Electrical Engineers, London, pp. 105-114, October 1977.

Jones, W.L., W.L. Grantham, L.C. Schroeder, J.W. Johnson, C.T. Swift, and J.L. Mitchell, "Microwave Scattering from the Ocean Surface," *IEEE Transactions on Microwave Theory and Techniques*, vol. MTT-23, pp. 1053-1058, 1975.

Jones, W.L., L.C. Schroeder, and J.L. Mitchell, "Aircraft Measurements of the Microwave Scattering Signature of the Ocean," *IEEE Transactions on Antennas and Propagation*, vol. AP-25, pp. 52-71, January 1977.

Kalmykov, A.I., and V.V. Pustovoyesko, "On Polarization Features of Radio Signals Scattered from the Sea Surface at Small Grazing Angles," *Journal of Geophysical Research*, vol. 81, pp. 1960-1964, 20 April 1976.

Kell, R.E., "On the Derivation of Bistatic RCS from Monostatic Measurements," *Proceedings of the IEEE*, vol. 53, pp. 983-988, August 1965.

Lammers, U.H.W and D.T. Hayes, "Multipath Propagation Over Snow at Millimeter Wavelengths," Rome Air Development Center Report RADC-TR-80-54, February 1980.

Larson, R.W., A.L. Maffett, R.C. Heimiller, A.F. Fromm, E.L. Johansen, R.F. Rawson, and F.L. Smith, "Bistatic Clutter Measurements," *IEEE Transactions on Antennas and Propagation*, vol. AP-26, pp. 801-804, November 1978.

Lee, P.H.Y, "Laboratory Measurements of Polarization Ratios of Wind Wave Surfaces," *IEEE Transactions on Antennas and Propagation*, vol. AP-26, pp. 302, March 1978.

Lennon, J.F., and R.J. Papa, "Statistical Characterization of Rough Terrain," Rome Air Development Center Report RADC-TR-80-9, February 1980.

Lewis, B.L. and I.D. Olin, "Some Recent Observations of Sea Spikes," *Proceedings of Radar 77*, IEE Conference Publication 155, London, pp. 115-119, October 1977.

Lind, G., "Frequency Agility Radar Range Calculation Using Number of Independent Pulses," *IEEE Transactions on Aerospace and Electronic Systems*, vol. AES-12, pp. 811-815, November 1976. Included in Barton (1977b).

Lind, G., and B. Bergkvist, "A Method to Calculate the Density of Sea Clutter at Low Grazing Angles," *IEEE International Radar Conference*, pp. 230-234, 1975.

Long, M.W., "A Radar Model for Land and Sea," *Proceedings of the Open Symposium of URSI*, La Baule, France, 28 April - 6 May, 1977.

Maalφe, J., "Sea Clutter Statistics," *Proceedings of Radar 82*, IEE Conference Publication 216, London, 18-20 October 1982.

Maresca, J.W., Jr. and J.R. Barnum," Theoretical Limitation of the Sea on the Detection of Low Doppler Targets by Over-the-Horizon Radar," *IEEE Transactions on Antennas and Propagation*, vol. AP-30, pp. 837-845, September 1982a.

Maresca, J.W., Jr. and J.R. Barnum, "Estimating Wind Speed from HF Skywave Radar Sea Backscatter," *IEEE Transactions on Antennas and Propagation*, vol. AP-30, pp. 846-852, September 1982b.

McCauley, J.F., *et al*, "Subsurface Valleys and Geoarcheology of the Eastern Sahara Revealed by Shuttle Radar," *Science*, vol. 218, pp. 1004-1020, 3 December 1982.

McEvoy, W.J., "Discrete Clutter Measurements Program: Operations in Western Massachusetts," Mitre Report MTR-2084, March 1972a.

McEvoy, W.J., "Discrete Clutter Measurements Program: Operations in the Metropolitan Boston Area," Mitre Report MTR-2085, March 1972b.

McGoogan, J.T., "Satellite Altimetry Applications," *IEEE Transactions on Microwave Theory and Techniques*, vol. MTT-23, pp. 970-978, December 1975.

Moore, R.K., "Active Microwave Sensing of the Earth's Surface — A Mini Review," *IEEE Transactions on Antennas and Propagation*, vol AP-26, November 1978.

Moore, R.K. and J.D. Young, "Active Measurement from Space of Sea Surface Winds," *IEEE Journal of Oceanic Engineering*, vol. OE-2, pp. 309-317, 1977.

Moore, R.K., K.A. Soofi, and S.M. Purduski, "A Radar Clutter Model: Average Scattering Coefficients of Land, Snow, and Ice," *IEEE Transactions on Aerospace and Electronic Systems*, vol. 16, pp. 783-799, November 1980.

Mrstik, A.V., and P.G. Smith, "Multipath Limitations on Low-Angle Radar Tracking," *IEEE Transactions on Aerospace and Electronic Systems*, vol. AES-14, pp. 85-102, January 1978.

NASA, "The SIR-B Science Plan," Jet Propulsion Laboratory Publication 82-78, December 1, 1982.

Olin, I.D., "Amplitude and Temporal Statistics of Sea Spike Clutter," *Proceedings of Radar 82*, IEE Conference Publication 216, London, 18-20 October 1982.

Onstott, R.G., R.K. Moore, S. Gogineni, and C. Delker, "Four Years of Low Altitude Sea Ice Broadband Backscatter Measurements," *IEEE Journal of Oceanic Engineering*, vol. OE-7, pp. 44-50, 1982.

Papa, R.J., J.F. Lennon, and R.L. Taylor, "Predictions of Electromagnetic Scattering for Rough Terrain Using Statistical Parameters Derived from Digitized Topographic Maps," Rome Air Development Center Report RADC TR-80-289, September 1980a.

Papa, R.J., J.F. Lennon, and R.L. Taylor, "Electomagnetic Wave Scattering from Rough Terrain," Rome Air Development Center Report RADC TR-80-300, September 1980b.

Pidgeon, V.W., "Bistatic Cross Section of the Sea," *IEEE Transactions on Antennas and Propagation*, vol. AP-14, pp. 405, 406, May 1966.

Reeves, R.G. (Ed.), *Manual of Remote Sensing*, American Society of Photogrammetry, Falls Church, Virginia, 1975.

Rigden, C.J., "High Resolution Land Clutter Characteristics," *Radar — Present and Future*, IEE Conference Publication No. 105, pp. 227-232, London, 23-25 October 1973. Reprinted in Barton (1975b).

Riley, J.H., "An Investigation into the Spatial Characteristics of Land Clutter at C-Band," Admiralty Surface Weapons Establishment Technical Report TR-71-6, Great Britain, 1971.

Rivers, W.K., and M.T. Tuley, "Radar Sea Clutter Statistics," *IEEE Symposium on Antennas and Propagation*, June 1977.

Rosenbaum, S., and L.W. Bowles, 'Clutter Return from Vegetated Areas," *IEEE Transactions on Antennas and Propagation*, vol. AP-22, pp. 227-236, March 1974.

Sackinger, W.M., and R.C. Byrd, "Backscatter of Millimeter Waves from Snow, Ice and Sea Ice," *Conference on Propagation of Radio Waves at Frequencies above 10 GHz*, Institution on Electrical Engineers, London, 1973.

Schleher, D.C., "Radar Detection in Log-Normal Clutter," *IEEE International Radar Conference*, pp. 262-267, Washington DC, April 1975.

Schleher, D.C., "Radar Detection in Weibull Clutter," *IEEE Transactions on Aerospace and Electronic Systems*, vol. AES-12, pp. 736-743, November 1976; see corrections, vol. AES-13, p. 435, July 1977.

Schleher, D.C., "Harbor Surveillance Radar Detection Performance," *IEEE Journal of Oceanic Engineering*, vol. OE-2, October 1977.

Schleher, D.C., *MTI Radar*, Artech House, Inc., Dedham, Massachusetts, 1978.

Schleher, D.C., *Automatic Detection and Radar Data Processing*, Artech House, Inc., Dedham, Massachusetts, 1980.

Sekine, M., S. Ohtani, T. Irabu, E. Kiuchi, T. Hagisawa, and Y. Tomita, "Weibull-Distributed Ground Clutter," *IEEE Transactions on Aerospace and Electronic Systems*, vol. AES-17, pp. 596-598, July 1981.

Sekine, M., S. Ohtani, T. Irabu, E. Kiuchi, T. Hagisawa, and Y. Tomita, "MTI Processing and Weibull-Distributed Ground Clutter," *IEEE Transactions on Aerospace and Electronic Systems*, vol. 18, pp. 729-730, November 1982.

Shotland, E. and R. Rollin, "The Complex Reflection Coefficient Over a Smooth Sea in the Micro- and Millimeter-Wave Bands for Linear and Circular Polarizations," Applied Physics Laboratory Report FP9-76-029, The Johns Hopkins University, March 1976.

Sittrop, H., "On the Sea Clutter Dependency on Windspeed," *Proceedings of Radar 77*, Institution of Electrical Engineers, London, pp. 110-114, October 1977.

Skolnik, M.I., *Introduction to Radar Systems*, McGraw-Hill Book Company, New York, New York, 1980.

Stiles, W.H., and F.T. Ulaby, "The Active and Passive Microwave Response to Snow Parameters: Part 1 — Wetness." *Journal of Geophysical Research*, vol. 85, pp. 1037-1044, February 20, 1980a.

Stiles, W.H., and F.T. Ulaby, "Microwave Remote Sensing of Snowpacks," Final Technical Report on Contract NAS5-23777, University of Kansas Center for Research, Inc., June 1980b.

Stiles, W.H., and F.T. Ulaby, "Dielectric Properties of Snow," University of Kansas Remote Sensing Laboratory Technical Report 527-1, December 1981.

Taylor, R.G., "Stationary Patterns in Radar Sea Clutter," *The Radio and Electonic Engineer*, vol. 46, pp. 103-108, March 1975.

Townsend, W.F., "An Initial Assessment of the Performance Achieved by the SEASAT-1 Radar Altimeter," *IEEE Transactions on Ocean Engineering*, vol. OE-5, pp. 80-92, 1980.

Trebits, R.N., N.C. Currie, and F.B. Dyer, "Multifrequency Millimeter Radar Sea Clutter Measurements," *Proceedings of EASCON-79*, October 1979.

Trebits, R.N., R.D. Hayes, and L.C. Bomar, "Millimeter Wave Reflectivity of Land and Sea," *Microwave Journal*, pp. 49-53, August 1978.

Trunk, G.V., "Non-Rayleigh Sea Clutter," Naval Research Laboratory Report 7986, June 1976. Reprinted in Schleher (1980).

Ulaby, F.T. "Vegetation Clutter Model," *IEEE Transactions on Antennas and Propagation*, vol. AP-28, pp. 538-545, July 1980.

Ulaby, F.T., and P.P. Batlivala, "Diurnal Variations of Radar Backscatter from a Vegetation Canopy," *IEEE Transactions on Antennas and Propagation*, vol. AP-24, pp. 11-17, January 1976.

Ulaby, F.T., and P.P. Batlivala, "Optimum Radar Paramteters for Mapping Soil Moisture," *IEEE Transactions on Geoscience Electronics*, vol. GE-14, pp. 81-93, April 1976b.

Ulaby, F.T., and T.F. Bush, "Corn Growth as Monitored by Radar," *IEEE Transactions on Antennas and Propagation*, vol. AP-24, pp. 819-828, November 1976.

Ulaby, F.T., and W.H. Stiles, "The Active and Passive Microwave Response to Snow Parameters: Part 2 — Water Equivalent of Dry Snow," *Journal of Geophysical Research*, vol. 85, pp. 1045-1049, February 20, 1980.

Ulaby, F.T., P.P. Batlivala, and M.C. Dobson, "Microwave Backscatter Dependence on Surface Roughness, Soil Moisture and Soil Texture, Part I: Bare Soil," *IEEE Transactions on Geoscience Electronics*, vol. GE-16, pp. 286-295, October 1978.

Ulaby, F.T., G.A. Bradley, and M.C. Dobson, "Microwave Backscatter Dependence on Surface Roughness, Soil Moisture, and Soil Texture: Part 2 — Vegetation-Covered Soil," *IEEE Transactions on Geoscience Electronics*, vol. GE-17, pp. 33-40, 1979.

Ulaby, F.T., T.F. Bush, and P.P. Batlivala, "Radar Response to Vegetation II: 8-18 GHz Band," *IEEE Transactions on Antennas and Propagation*, vol. AP-23, pp. 608-618, September 1975.

Ulaby, F.T., R.K. Moore, and A.K. Fung, *Radar Remote Sensing and Surface Scattering and Emission Theory*, vol. 2 of Remote Sensing Series, Addison-Wesley Publishing Company, Reading, Massachusetts, 1982.

Ward, H.R., "A Model Environment for Search Radar Evaluation," *EASCON 71 Record*, pp. 164-171, October 1971. Reprinted in Barton (1975b).

Ward, K.D., "A Radar Sea Clutter Model and Its Application to Performance Assessment," *Proceedings of Radar 82*, IEE Conference Publication 216, London, 18-20 October 1982.

Zehner, S.P., and M.T. Tuley, "Development and Validation of Multipath and Clutter Models for TAC ZINGER in Low Altitude Scenarios," Engineering Experiment Station, Georgia Institute of Technology, Final Technical Report, Contract F49620-78-C-0121, March 1979.

1 Remote Sensing by Radar

State of the Art

1.1 Overview

Radar was used extensively by the military during World War II and its applications have spread to many purposes on land, at sea, and in the air. During the Apollo 17 flight in December 1972, astronauts gathered data with radar to develop maps of the moon. Applications for which there will be greater use of radar in the years ahead include earth mapping; crop, forest, and flood disaster assessments; topographic studies for water resources and minerals; and iceberg and ocean surface (wave heights, wind speeds, direction and speed of surface currents, etc.) surveillance.

Radar is now being used to explore vast areas of the earth. More than 2 million square miles of South America, usually concealed by a heavy cloud cover, have been mapped by radar. Radar can monitor iceberg position, movement, and age to improve safety at sea. It is useful in studying oceans and their currents and for applications in geophysics, hydrology, mineral exploration, topographic mapping, land-use monitoring, agriculture, and forestry. Differences in the radar cross sections of vegetation produce images with variations in tone and texture. These variations can be used to generate land-use maps of crops and forests and to show forest-fire damage.

Radar senses the environment with short-wavelength electromagnetic (radio) waves. The frequency region of 300 megahertz (MHz) to 30 gigahertz (GHz) is called the microwave region; the corresponding wavelengths are one meter (m) and one centimeter (cm), respectively. The region 30 GHz (10 mm) to 300 GHz (1 mm) is called the millimeter wavelength region. Although millimeter-wave radars are becoming more widely used, most radars operate within the microwave region. Some operate at frequencies of a few megahertz and others use optical wavelengths. The commonly used radar bands are listed in table 1-1.

The band designations *K, X, S, L,* and *P* were secret code letters used during World War II, and the order "*K-X-S-L-P*" was sometimes recalled by the phrase "King Xerxes Seduced Lovely Princesses." The first radars used wavelengths of several meters. Microwaves were used by 1940 and by

Table 1-1
Radar Band Designations

Band	*Frequency*	*Wavelength*
P	300-1,000 MHz	30-100 cm
L	1,000-2,000 MHz	15-30 cm
S	2,000-4,000 MHz	7.5-15 cm
C	4,000-8,000 MHz	3.75-7.5 cm
X	8,000-12,500 MHz	2.4-3.75 cm
K_u	12,500-18,000 MHz (12.5-18.0 GHz)	1.67-2.4 cm
K	18.0-26.5 GHz	1.1 cm-1.67 cm
K_a	26.5-40.0 GHz	0.75 cm-1.1 cm

1945 operating frequencies had been extended into K band; by 1950 K_a band was being used and by 1960 experimental radars were operating at 70 GHz. Since then there have been a number of millimeter systems developed for wavelengths extending to about one millimeter.

The advantage of the higher frequencies (shorter wavelengths) is that narrower beamwidths are obtainable with a given overall antenna size. Disadvantages include a greater loss of signal strength associated with atmospheric absorption (and scattering), greater internal system losses, and less transmitter power.

For purposes of achieving narrow beamwidths (and simply exploring the unknown), radars using microwave techniques have been developed for wavelengths as short as 1 mm. Of course, radars have also been developed for the infrared and optical wavelength regions. Radars originally operated at decimeter wavelengths because the technology of microwaves had not yet been developed. There are renewed interests in these wavelengths for ''penetrating'' vegetation and the earth's surface. Moreover, decimeter wavelengths permit target detection at very great distances (over-the-horizon) via refraction from the ionosphere (sec. 2.9); for example, doppler radars operating at decimeter wavelengths are being used to sense sea surface conditions over large ocean areas (sec. 5.13). Combinations of long and short wavelengths have been used to simultaneously obtain improved penetration, directivity, and target classification.

Radar data are complementary with those from photographs obtained with passive systems using the infrared, visible, and ultraviolet regions. For microwaves, the atmosphere is essentially transparent.[a] Therefore, the all-weather, day or night capability of radar permits acquisition of informa-

[a]If all-weather performance is desired, the middle of K band should not be used (see sec. 2.13).

tion when it is impossible with aerial photography. Although a radar map is in many respects similar to a photograph, radar highlights different features. For example, radar is sensitive to vertical dimensions and consequently emphasizes topographic features.

Radar maps can be essentially free of difficulties with perspective. By suitable equipment design and use of the plan position indicator (PPI) or the more modern side-looking presentation of airborne radar (fig. 1-1), the picture obtained can be presented as a true plan view regardless of the depression angle from which the target area is viewed. Figure 1-2 compares a radar picture with a chart of the tip of Cape Cod and a direct photograph taken simultaneously from the airplane flying over Cape Cod Bay. There is little question of the superiority of the radar picture. Had it been made at night or had the weather been foggy, the radar picture would have been unaffected while photography or ordinary vision would have been useless.

The coordinants of the PPI presentation lend themselves to comparison with maps, but the correlation between the brightness patterns and the topographic features of the ground depends on reflectivity characteristics. Identification and correlation of the various features of a radar map are based upon intensity contrasts that fall into several categories such as the contrast between land and water, between hill and valley, and between built-up areas and open countryside. This, in essence, was the state of the art at the end of World War II.

Since World War II there have been new thrusts because of widespread interests and major extensions of aerial photography employing sophisticated microwave, infrared, ultraviolet, and signal processing techniques. The sensory techniques, in combination with conventional photographic techniques, are making significant economic impacts on various countries. Today, the general public is familiar with large-scale photographs obtained from high-altitude aircraft and satellites. These data have applications for weather prediction and analyses, and for general studies of land use.

Although radar mapping techniques have been developed to a high degree of sophistication, the state of the art for terrain mapping is still advancing at a rapid rate. These advances are permitting new opportunities for improving the quality of life in various countries. For example, radar data obtained at an aircraft elevation of 40,000 feet are being used to help integrate the Amazon region with other parts of Brazil. Figure 1-1 is a radar map of the type taken over dense forest areas of Brazil that are virtually unreachable by land and undetectable by other surveillance techniques because of a persistent cloud cover. Radar maps are providing an inventory of potential mineral resources, forestry and range resources, fresh water supplies, new transportation routes, and sites for agriculture. The possible applications of radar for major economic and social improvements seem almost limitless.

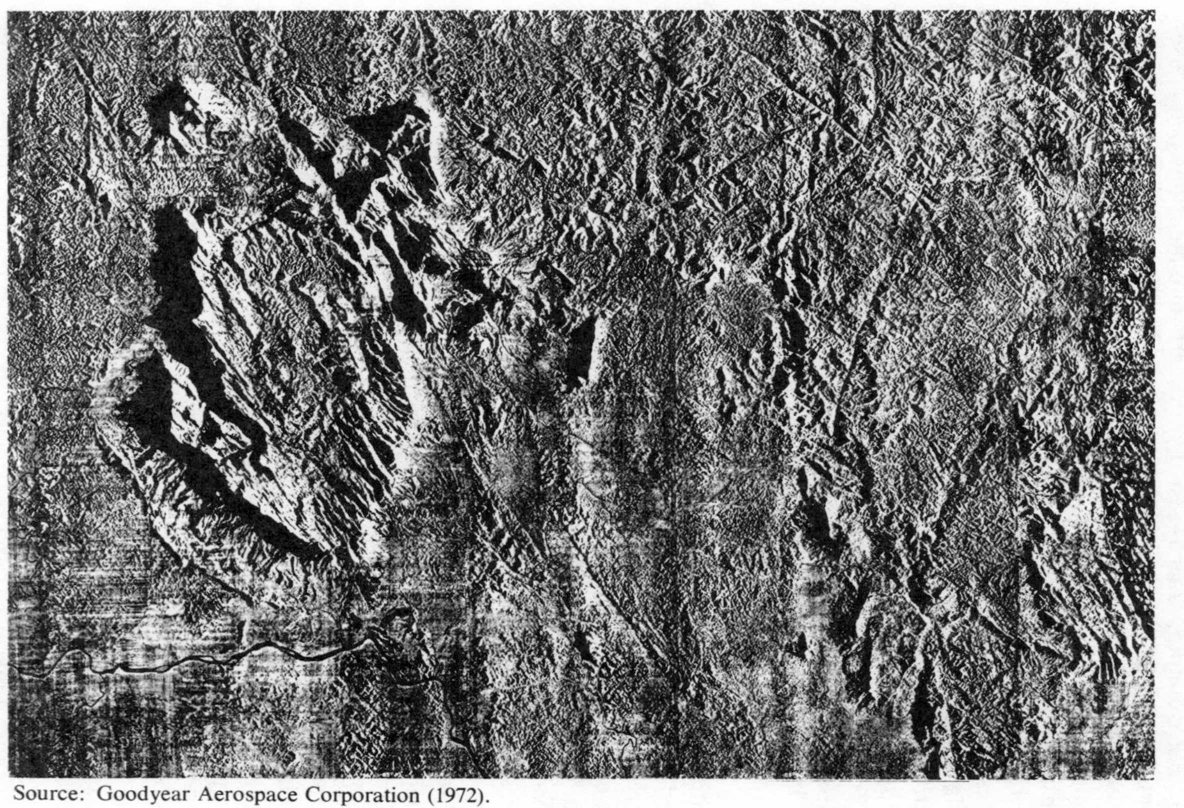

Source: Goodyear Aerospace Corporation (1972).

Figure 1-1. A 100-mile Swath of Venezuela.

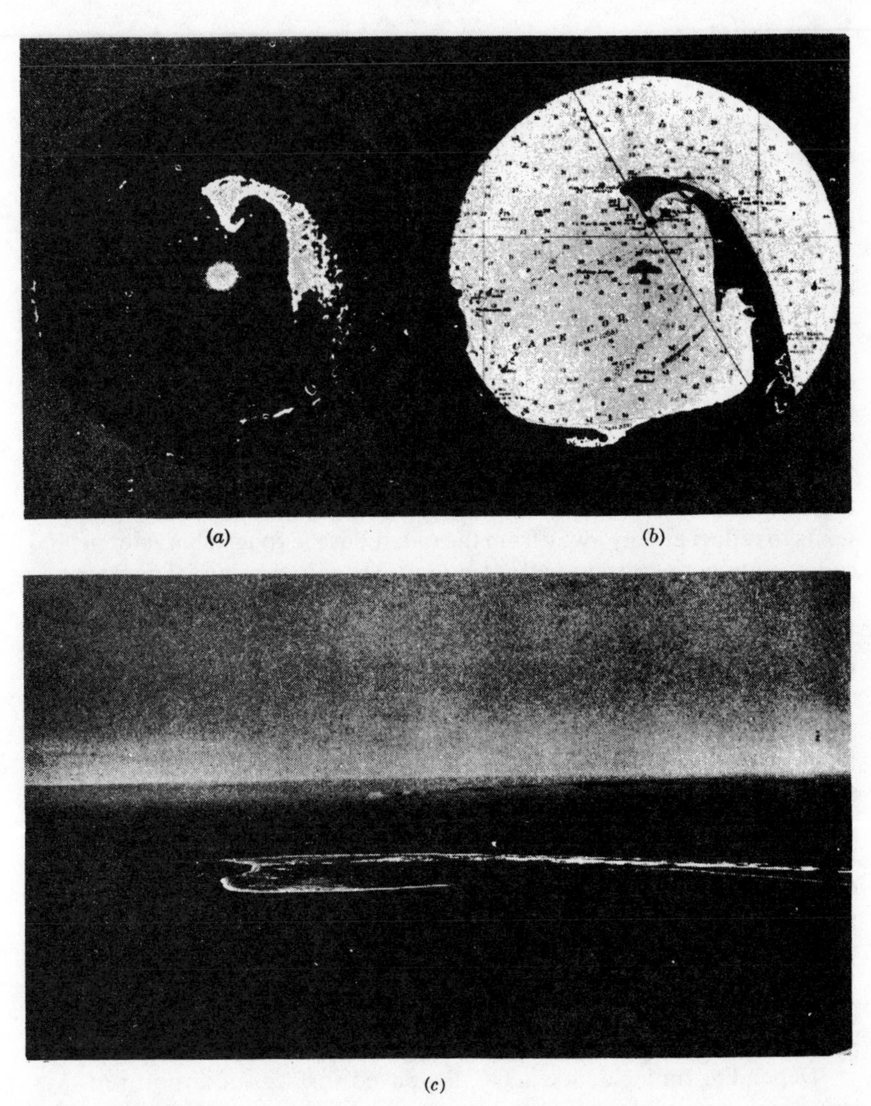

Source: L.N. Ridenour, *Radar System Engineering*. Copyright 1947, McGraw-Hill Book Company, Inc. Used with permission of McGraw-Hill Book Company.

Figure 1-2. A Radar Display, a Map, and an Optical Photograph of Cape Cod, Massachusetts.

1.2 Radar Capabilities at End of World War II

In this section a review of the capabilities of radar at the end of World War II is used to identify and explain pictorially the effects of the reflecting properties of land and sea. Much of this material was taken from L.N. Ridenour (1947) and J.S. Hall (1947).

By viewing the figures in this chapter, it is obvious that radar highlights different features than does conventional aerial photography. Details highlighted are, of course, dependent on the incidence angles for both the aerial photographs and the radar maps.

During World War II the basic range, azimuth, and signal strength information obtained by radar was used to form PPI maps. With radar, the individual elements of terrain tend to have a direct resemblance to their topographic characteristics; but the correlation between the radar presentation and the actual terrain varies with the radar parameters, the type of terrain, the altitude (depression angle), and the direction from which the target areas are viewed.

The most striking and readily identifiable terrestrial feature is usually the boundary between land and water. The smooth surface of calm water tends to reflect energy away from the radar, but the rough character of land causes energy to scatter in all directions. There is substantial backscatter from a rough sea, and such a surface can obliterate the demarcation between land and sea. Sea echo is clearly visible in figure 1-3, but here it does not interfere with identifying the land areas.

Built-up city areas are particularly effective in redirecting radiation toward the radar, thereby causing a strong radar signal. At small grazing angles of incidence (small depression angles for an airborne radar), vertical surfaces of built-up areas cause reflections back toward the radar that are much stronger than those from surrounding flat terrain. Therefore, cities are usually seen as strong targets. Figure 1-4 shows that cities yield characteristic shapes and collections of bright signals. From maps made at short ranges, such as figure 1-4, it was known that much picture detail would be ultimately obtainable with long-range radar of sufficiently high resolution. Since the end of World War II much effort has been directed toward obtaining improved picture detail for greater radar ranges, more rapid data processing, and higher reliability of target recognition.

Depending on incidence angle, the paved surfaces of airport runways usually backscatter less energy than the surrounding ground. Runways can be seen in figure 1-5. During World War II four-lane highways had been observed as dark lines on the available higher resolution radars, but most highways were too narrow to be resolved. Highways were often identifiable, however, because roads usually are bounded by telephone wires, fences, embankments, trees, billboards, and small buildings. The reflections from these objects would, collectively, produce a bright line on the PPI corresponding to the route of a highway. For the same reason, railroads were often recognized by bright lines. As might be expected, because of relative echo magnitudes it was possible to identify railroads across swamps but not across wooded country.

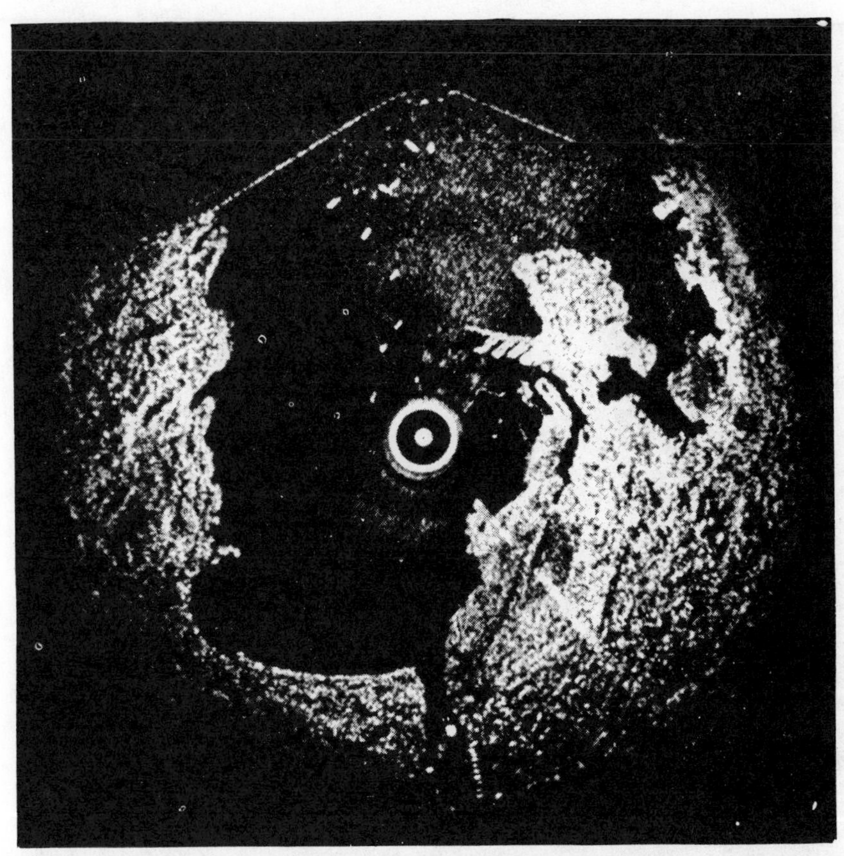

Source: L.N. Ridenour, *Radar System Engineering*. Copyright 1947, McGraw-Hill Book Company, Inc. Used with permission of McGraw-Hill Book Company.

Figure 1-3. Radar Map of Cristobal, Panama Canal Zone.

Mountains are indicated chiefly by dark areas because the shadowed areas (not illuminated by the radar) appear dark. The result is generally brighter signals from the near slopes and absence of signal from the far slopes—this gives a presentation that resembles a relief map (see fig. 1-1). The shapes of the bright and dark areas caused by mountainous terrain vary greatly with the angle of incidence, which is a function of the position and altitude of the radar; consequently, identification of a particular mountain in an extensive range is difficult. However, even during 1944 and 1945 the United States Army Air Force used radar displays as an aid to navigation through passes in the Alps Mountains. Isolated mountains in flat country were easily identified and they provided useful landmarks for air navigation (Hall 1947, p. 104).

Radar has been used for many years in the study of weather (Smith, Hardy, and Glover 1974). During World War II echoes from storms were

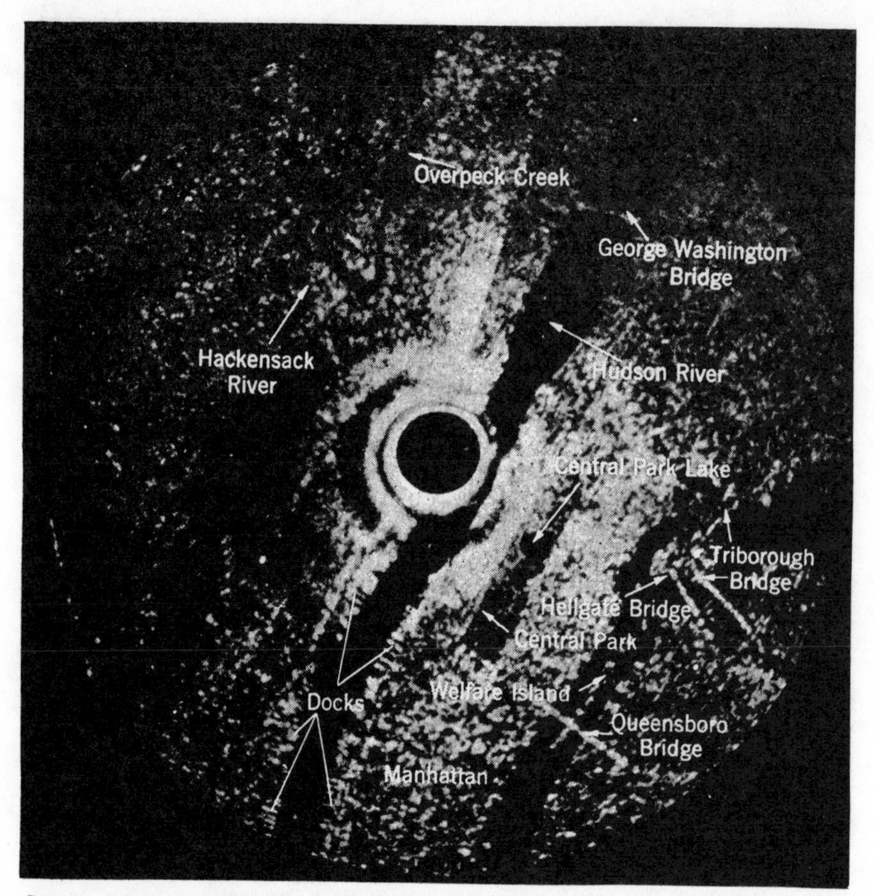

Source: J.S. Hall, *Radar Aids to Navigation*. Copyright 1947 by McGraw-Hill Book Company, Inc. Used with permission of McGraw-Hill Book Company.

Figure 1-4. Radar Map of New York City.

observed by airbone radar at ranges as great as 50 miles. They were recognizable because they had less distinct boundaries than other objects. Storms usually changed in shape and size and would persist even when the antenna was tilted upward so that ground signals disappeared.

Pilots in harbors, rivers, and lakes find radar extremely useful. The radars used for shipborne applications usually have narrow beamwidths, short pulse lengths, and fast indicator sweeps. Therefore, the radars used for ship navigation during World War II did provide higher resolution than was available on aircraft. Figure 1-6 is a radar picture taken from a ship as it passed through a canal. The range circle has a diameter of 2,000 yards. It is easy to see that the ship was centered in the canal; bridges are discernible. With the radars available during the war, individual wharves and piers were

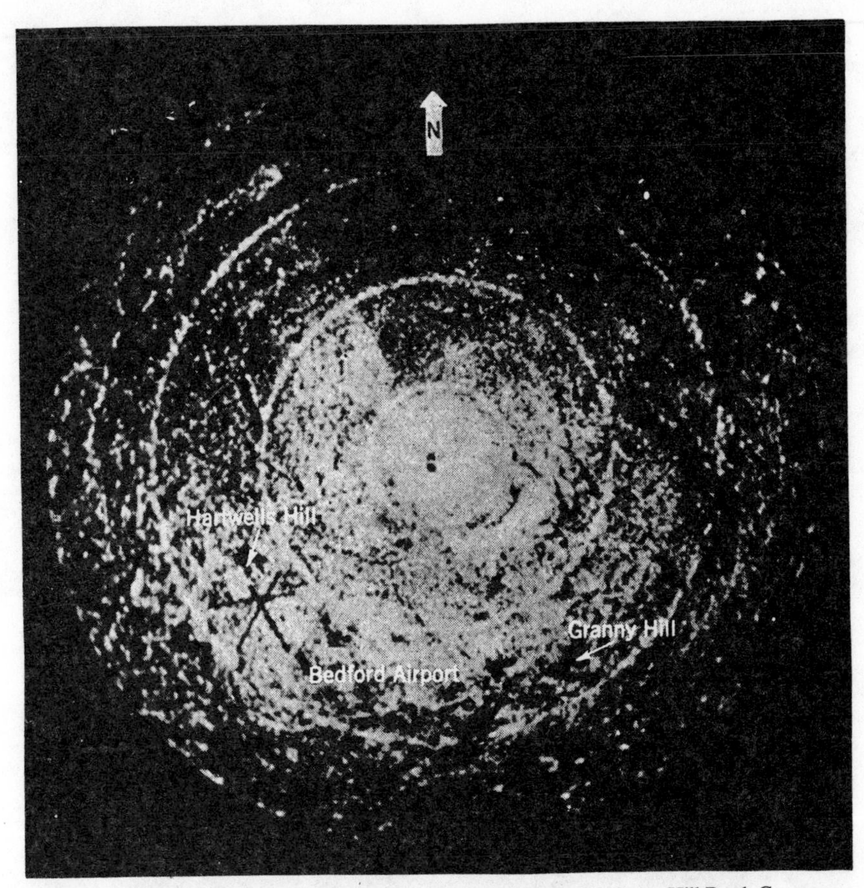

Source: J.S. Hall, *Radar Aids to Navigation*. Copyright 1947, McGraw-Hill Book Company, Inc. Used with permission of McGraw-Hill Book Company.

Figure 1-5. Radar Map of Runways at Army Airfield, Bedford, Massachusetts.

reliably identified. Ships in harbors are slow enough that even inexperienced pilots could identify various objects on the PPI by reference to a navigation chart.

At the end of the war, 10-cm and 3-cm radars had detected icebergs at ranges of many miles. Experience showed that as icebergs melted and disintegrated, detection became more difficult. Under rough sea conditions when echo from the sea masked that of smaller icebergs, radar was not a reliable method of detecting ''growlers'' that might be only 5 or 10 feet above the water, with the largest portions submerged.

During the war the main ways of distinguishing one target from another were by comparing shape, relative location, and echo (contrast) strength.

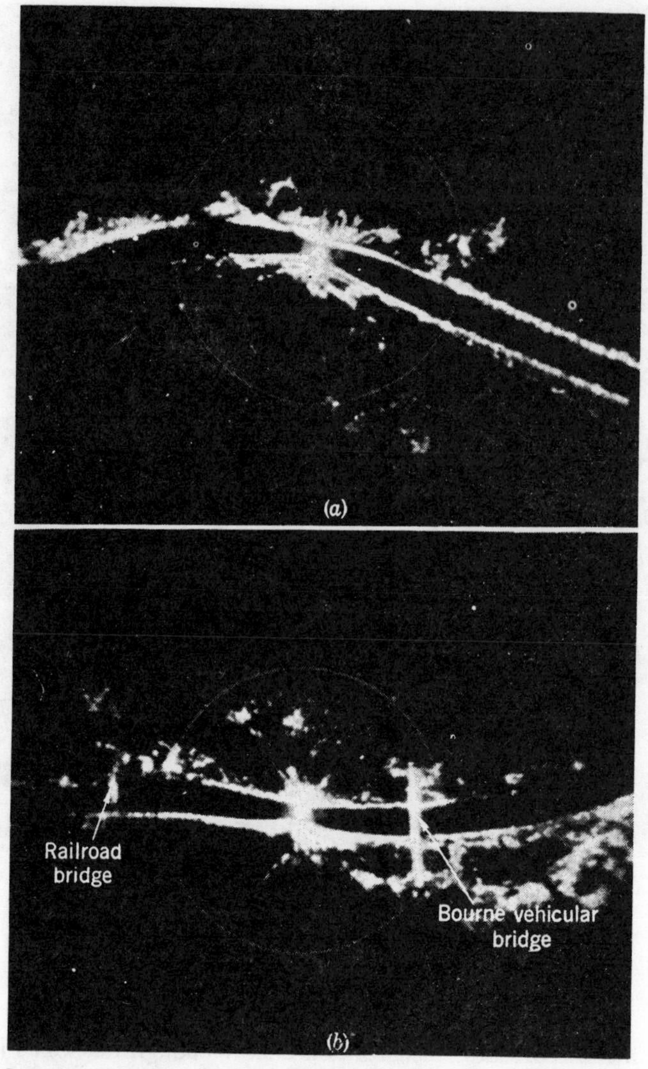

Source: J.S. Hall, *Radar Aids to Navigation*. Copyright 1947, McGraw-Hill Book Company, Inc. Used with permission of McGraw-Hill Book Company.

Figure 1-6. Radar Display on a Ship in Cape Cod Canal.

Target persistence provided another dimension. For example, small signals that appear for only a short time and then disappear might come from waves, birds, or whales and porpoises at play. Persistent signals might come from a rough patch of water, a submerged reef, a rain cloud, an iceberg, a ship, or some other floating object. Improved resolution was one of the early methods for obtaining increased radar-detection reliability.

Figure 1-7 shows PPI photographs obtained in the Boston Harbor. The improvement in recognition obtainable with the 1° beamwidth compared to the 12° beamwidth is profound.

As seen from the above discussions, even during World War II radar enabled certain classes of objects to be detected and located at distances far beyond those distinguishable by the unaided eye. Various improved signal-processing techniques have since been developed. Greater reliability in choosing targets of interest and ignoring the others is becoming a reality as more is learned about the reflectivity characteristics of targets.

1.3 Strip Maps and Side-looking Radar

The strip map represents a major improvement in data presentation that resulted from the development of side-looking radar. During flight the radar radiates a microwave beam at right angles to the aircraft, then detects and records the reflections received from the earth's surface. As it travels forward, successive strips of terrain are exposed to the radar beam and are detected at the aircraft. The reflections are used to produce images of the terrain covered.

As with conventional radar, side-looking radar uses pulse-ranging to determine distance to objects. The pulses are sent out and the distance to the object is determined by measuring the time they take to return. Side-looking radar makes use of the echo doppler frequency caused by aircraft motion to improve along-track or azimuthal resolution (Harger 1970). The radar is said to have a "synthetic aperture" because the apparent azimuthal resolution is greater than is obtainable from the antenna pattern. Therefore, signal processing is used to create the effect of antenna beam-sharpening.

Side-looking radar has proved to be particularly useful in areas that are unmapped, poorly known, sparsely populated, intensively cloud-covered, and in areas that otherwise have not lent themselves to traditional aerial mapping methods. Much of the interior of South America is of this nature. Figure 1-1 shows a 100-mile-wide swath of Venezuela obtained with a radar in less than 60 minutes. Heights of mountains can be estimated by knowing the radar incidence angle (depression angle) and the length of the shadow. The Orinoco River can be seen in the lower left of the figure, and the Marahuaca Mountains can be seen in the area above the river.

Figure 1-8 shows an area that includes the Willow Run Airport and Belleville, Michigan, as seen with a high-resolution radar. It shows that a wide variety of natural and man-made features can be identified. The airport, railroad lines and yards, highways, factories, housing patterns, and water bodies are plainly visible. Vertical objects such as poles and buildings appear bright; but the airport runway (smooth relative to the 3-cm radar wavelength) appears dark.

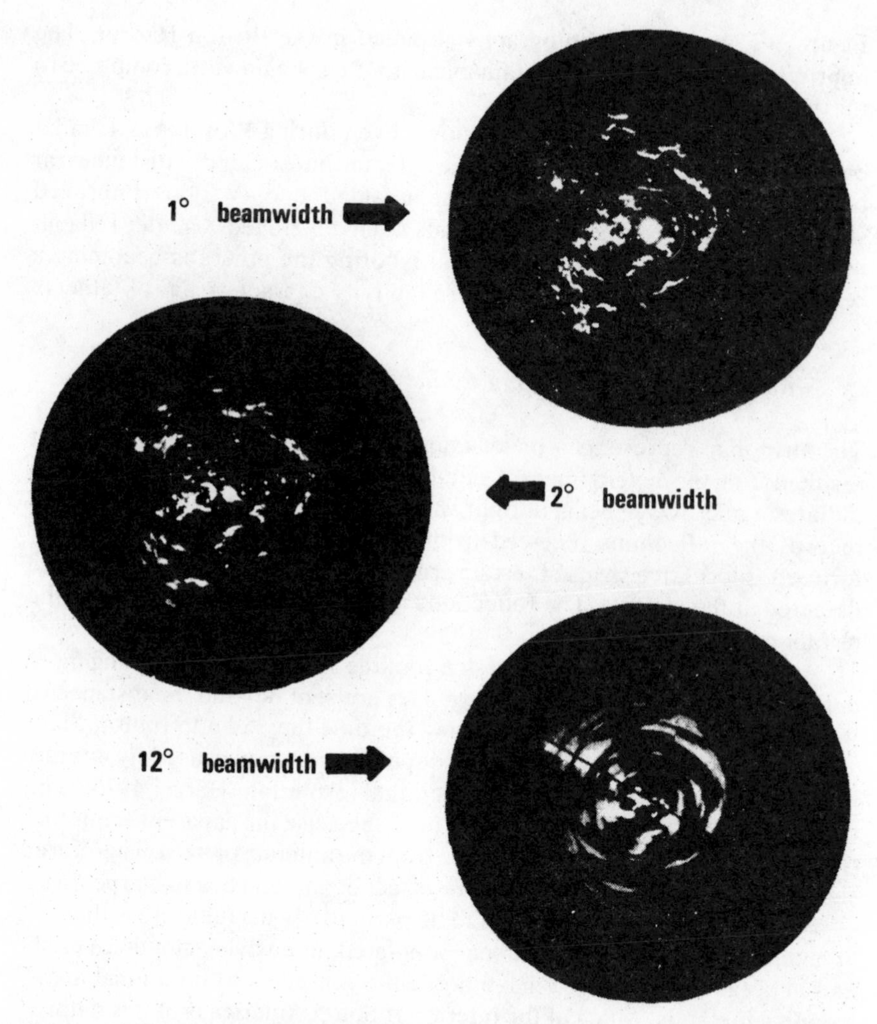

Source: J.S. Hall, *Radar Aids to Navigation*. Copyright 1947, McGraw-Hill Book Company, Inc. Used with permission of McGraw-Hill Book Company.

Figure 1-7. Pictures of the Same Area Taken from Three Radars with Different Resolutions.

Radar maps of vegetation contain variations in tone and texture, and these variations define boundaries of various types of natural growth. Soil moisture differences, potentially important to irrigated agriculture, can cause a significant influence on radar return. Figure 1-9 shows mapping of vegetation by radar of a sparsely populated agricultural area in northern Michigan. Agricultural fields, woodlands, roads, and stream drainage patterns are the most visible features. The fields differ in brightness, caused by

Courtesy of Environmental Research Institute of Michigan.

Figure 1-8. Radar Map of Area Including Willow Run Airport and Belleville, Michigan.

the presence or lack of vegetation, type of crop, orientation of rows, soil type, and moisture content.

The broad aerial coverage possible with side-looking radar makes it extremely valuable for studies of water drainage and erosion. Effects of erosion are underscored in the strip map of figure 1-10. This map shows the grand pattern of sweeping ridge and valley curves in eroded sandstone and shale near Conway, Arkansas.

Side-looking radar has the ability to display hundreds of square miles of terrain onto a single map. To the trained geologist, the radar imagery shows features of the earth's crust that appear faintly or not at all in aerial photographs. This is particularly true of relief variations such as hills, valleys, domes, and mesas.

Surface Effects and Emerging Techniques

1.4 Effects of Surface Characteristics

A smooth surface such as a paved street appears as a black area in a radar picture (small echo strength), while a rough surface appears bright in contrast (large echo strength). In the case of a smooth surface, the reradiated energy is directed primarily away from the radar (angle of reflection equal to the angle of incidence). When the ground or sea is rough, considerable radiation is scattered in all directions; therefore a significant part of it is returned to be displayed on the radar. The echo from land areas is usually much larger than that from water. Thus, the definition of a land-water boundary is usually one of the simplest tasks in target recognition.

A quiet-water surface is as near perfect a specular reflector for microwaves as nature provides, but a water surface agitated by wind backscatters a strong signal known as sea return or sea clutter. Sea return is clearly visible in the PPI photograph of figure 1-3, and it is easily distinguishable from land.[b] Sea return is strongest in the direction from which the wind is blowing (upwind) and this effect can be seen in the figure. The figure shows not only substantial sea return in the upwind and downwind directions (north and south), but also there is less (none visible) return in the crosswind directions. From the photograph, it is apparent that the sea return is weaker in the protected inner harbor than it is in the outer harbor.

Oil smooths rough water by modifying the effects of breaking waves and by damping the small gravity and capillary waves; this smoothing effect

[b]For rough seas, average echo strength can exceed that of land. Therefore, sea echo can destroy the land-sea interface on a radar map (for an example, see fig. 3-1 of Hall [1947]).

Courtesy of Environmental Research Institute of Michigan.

Figure 1-9. Radar Map of a Sparsely Populated Agricultural Area.

Source: Goodyear Aerospace Corporation [1971].

Figure 1-10. Radar Map of Eroded Sandstone and Shale.

permits oil slicks to be detected and monitored by radar. Slicks have been observed (Guinard 1971; Pilon and Purves 1973) with airborne radars at four operating frequencies (428 MHz, 1,228 MHz, 4,455 MHz, and 8,910 MHz). Oil films reduce sea echo more for vertical than for horizontal polarization. For vertical polarization and low-sea states, the radars could reliably detect and map oil spills from the initial thickness of a polluting spill to thicknesses of 1 micron or less. K. Krishen (1972) reported that for depression angles between 40° and 75°, the vertically polarized 13.3-GHz sea echo strength was reduced 5 to 10 dB by the presence of oil.

Whether a particular surface appears rough or smooth depends on the wavelength of the radar. For example, a paved airport runway will stand out clearly against grass-covered ground between the runways. At a wavelength of 1.25 cm the grass is thoroughly rough and the runways quite smooth, resulting in a strong contrast. The contrast is reduced at shorter wavelengths; for example, for ordinary light runways and grass appear rough. At much longer wavelengths both the grass and the runways appear smooth in comparison with wooded areas near an airport. In general, the radar picture of a surface that is smooth with respect to the wavelength appears darker than a rough surface, except when the flat surface is viewed at near-normal incidence. In that case, the flat surface stands out as a very strong radar target. Therefore, when looking straight down with radar, a smooth sea can be a much stronger target than land.

Strong reflection will occur if a flat surface happens to be oriented normal to the line-of-sight. However, the mere presence of flat surfaces is not enough to assure strong echo. For example, if flat surfaces are oriented in random directions, the probability of one being at a favorable (strong reflection) orientation can be so low that the average echo intensity also might be low. For a group of buildings, many of the flat surfaces are vertical walls and others are smooth pavements and flat roofs. Therefore, there are many possibilities for combinations of three flat surfaces at right angles to one another; such combinations form targets (corner reflectors) that are highly retrodirective.

Vertical and horizontal surfaces can also combine to form highly retrodirective surfaces that are two-sided corner reflectors. The combination of two surfaces at right angles, called a diplane, is highly retrodirective if the intersection (seam) of the two surfaces is oriented nearly normal to the line-of-sight. Rough surfaces, such as the earth, can serve as a reflecting plane for distant targets when the angle of incidence is sufficiently small (grazing). Buildings, or groups of buildings, are known to provide a strong echo at long ranges. At shorter ranges the echo from buildings might be substantially less than that from other targets, and in fact the echo may tend to fade at shorter ranges when the larger grazing angle reduces the retrodirectivity. Strong signals may sometimes be observed even without a corner-reflector effect because of direct reflection (line-of-sight perpendicular to surface) at long ranges when the line-of-sight is almost horizontal.

Hills and mountains are indicated by bright echoes from the slopes facing the radar and by the shadowed regions on the far sides of the crests. These conditions follow directly from the geometry of the illumination and produce the very realistic effect of land forms in relief. As an aircraft approaches a hill or mountain, if its altitude is greater than the height of the hill or mountain, it will eventually reach a position where the farther slope is no longer hidden behind the crests. Although the shadow will disappear, it is still possible to distinguish a mountain on the radar picture by the contrast between the brighter signal from the near slope and the weaker signal from the far slope.

In detailed geologic studies, the most significant contribution of radar is probably in detecting subtle changes in terrain configuration. This capability has been demonstrated at a number of locations where both ground investigations and aerial photographic studies have been conducted. H.C. MacDonald, P.A. Brennan, and L.F. Dellwig (1967) give data for the Cane Springs, Arizona, region which reveal fundamental differences between aerial photography and radar maps. The *K*-band radar revealed terrain features not readily apparent in the aerial photographs, even though the radar resolution was somewhat less than that of the optical photographs. In figure 1-11, the fault (*d-d'*) that bisects the area is more pronounced on the radar presentation than on the optical one. For the air photograph, vegetation serves as a camouflage by contributing variable gray-scale values along the fault. In other words, the enhancement of this feature on the radar imagery is related to the absence of scattering by the dry sparse vegetation and by the consequent domination of the radar signal of terrain surface characteristics.

The radar intensities of areas *a* and *e* contrast with one another and with the more extensive alluvial surface that borders these areas, presumably because of differences in surface roughness and dielectric properties. Area *a* consists of compact, dry, fine-grained dune sand that yields a very weak echo; area *e* is characterized by an admixture of fine-grained sand and sandstone fragments that yields an intermediate echo strength. In contrast, the surrounding alluvial surface is compact and consists of numerous rock fragments—presumably the geometric irregularies cause the comparatively strong echo.

There is tertiary lava in the area near *b* that is somewhat uniformly gray on the radar imagery, but it has a variable gray tone in the air photograph. Much of the lava surface is reportedly covered by a thin carbonate residuum varying up to 1 cm in thickness, producing the wide gray-scale range on the air photograph. The radar does not record the carbonate

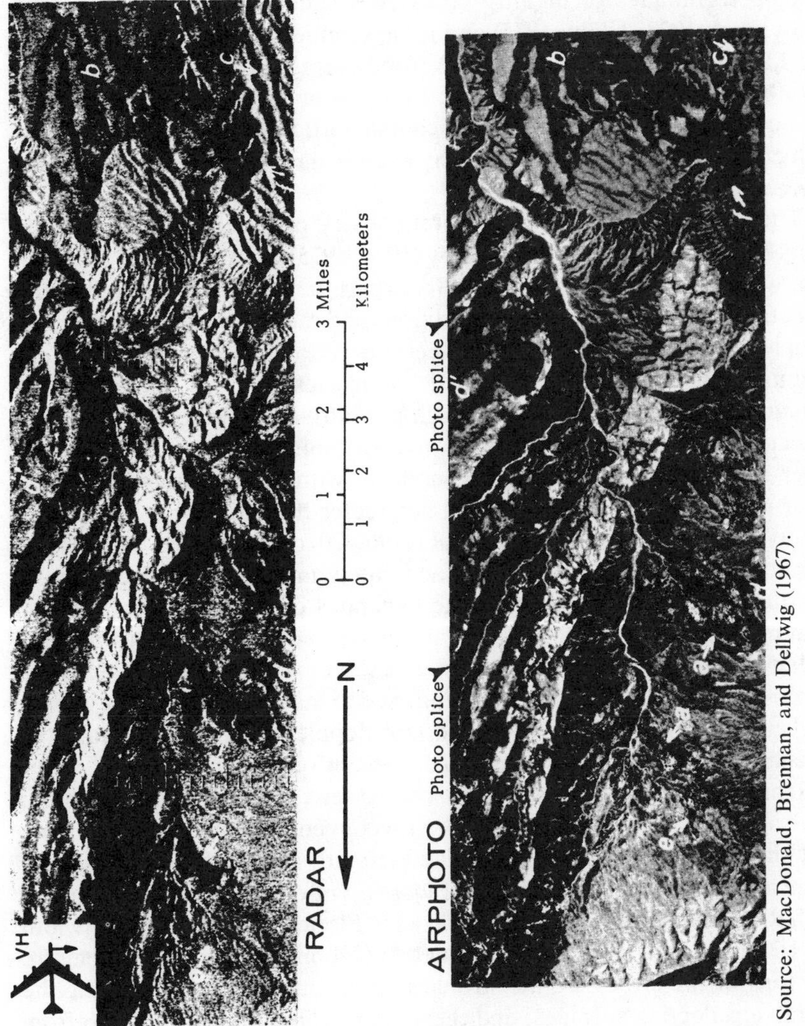

Source: MacDonald, Brennan, and Dellwig (1967).

Figure 1-11. Comparison of Radar Imagery and Aerial Photography.

residuum. Consequently, the boundaries and aerial extent of the lava are displayed with radar imagery as a distinct lithologic unit that is not as apparent in the air photograph. In this case, both sensors contribute information about the lava surface.

Research has been underway for a number of years in an effort to develop techniques to identify crop types (Schwarz and Caspall 1968; Ulaby et al. 1972; Ulaby 1975) and to determine moisture content (Waite and MacDonald 1971; Ulaby 1974) from averages of σ°. Although there have been some promising results, these radar problems have not been solved. Such determinations are, of course, difficult or impossible to make with conventional aerial photography even in daylight with a clear atmosphere.

The effect of soil moisture on reflectivity of bare and crop-covered fields has been studied by F.T. Ulaby (1974, 1975) as a function of polarization, wavelength, and look angle. His experimental data indicate that the reflectivity is dependent on soil conditions and radar parameters in various complex ways. However, within certain terrain and operational constraints, radar images have defined boundaries that are related to soil moisture content. Because of the high permittivity of water, moist soil is expected to have a higher reflectivity than dry soil. Also, the permittivity of water is larger than that of ice. Therefore, with other factors equal, the radar reflectivity for moist soil will be greater than for dry soil or frozen soil. W.P. Waite and H.C. MacDonald (1969-70) observed that the contrast on radar imagery was sufficient to accurately map areas of glacier ice.

Much is being learned about the dynamics of the ocean surface with radar and about what it is that the various types of systems measure best. Radar imagery is being used to investigate waves, swells, current edges and roughness fronts. Altimeters are being used to measure local mean surface level and sea roughness. Over-the-horizon doppler radars are being used to measure waves and winds over large areas of ocean surface.

It has long been known that well-ordered sea waves are clearly visible at small and medium depression angles with conventional pulse radar. Figure 1-12 shows sea waves on a B-display (azimuth angle along abscissa and range along ordinate) of an X-band radar (Flynt et al. 1967) at the Georgia Tech Radar Site along the Atlantic Coast in Florida. Azimuthal beamwidth is 1 1/2° and range resolution is 38 meters (1/4 microsecond pulse length). The tops of the crests are the strongest contributors toward defining the effective reflecting surfaces, and these tops prohibit (shadow) echo from the far sides of the waves. The result is bright signals from the near slope and absence of signals from the far slope. Clearly, radar provides an easy method for measuring the distance between waves (wavelength). Figure 5-23 is a graphical presentation of the theoretical relationship between wavelengths, velocities, and periods in deep water. Therefore a measurement of wavelength will also provide estimates of wave velocity and period for deep water.

One can determine wave period and wave velocity simply with a radar by counting the waves passing a point and by measuring the time for waves to transverse a given distance. Sea waves are not perfectly periodic, but an average of ten periods as automatically measured by radar seems to provide a useful way of remotely sensing the periods of gravity waves (sec. 5.9). Alternatively, an average wave period can be obtained from the "periodic" component of the autocorrelation function (sec. 5.10).

Reflectivity for microwaves is caused principally by small wind ripples and capillary waves. These scatterers have dimensions of a few centimeters that are about equal to wavelengths at microwave frequencies. Even though echo strength is strongly influenced by scatterers with dimensions of the order of a radar wavelength, the effect of other scatterers is observable. For example, the shape of the *microwave* reflectivity versus incidence angle curve is strongly influenced by *wave height* (sec. 6.15). For frequencies of a few megahertz (corresponding to decimeter wavelengths), sea wave height is usually small with respect to radar wavelength. For these frequencies, the gross wave structure is more susceptible to detection because the fine structure (ripples and wavelets) is minuscule compared to the radar wavelength. However, the doppler spectrum for these frequencies is perceptibly influenced by the wind-generated *fine structure* of the sea (sec. 5.13).

1.5 Emerging Techniques

Terrain mapping and target classification have been based primarily on differences in echo intensity, shape, and location. This section describes improvements in target detection and classification that are being made by using new techniques involving choices and combinations of incidence angle, polarization, wavelength, and satellite platforms. A past thrust has been toward making radar resolution finer in order to improve target classification. Now, the broad overview of extremely large areas possible with satellite-borne radar adds another dimension. The resulting information brings new insights into interpreting data. For example, the present approach of determining the water content of soil is to extrapolate on-site point measurements. However, such extrapolation is questionable because precipitation and soil properties vary spatially. Possibly an alternative approach is to estimate (average) soil moisture content with satellite radar that can directly cover large areas within a short time period. Another example of the potential of satellite platforms is that of measuring gradual

Courtesy F.B. Dyer, Engineering Experiment Station, Georgia Institute of Technology.

Figure 1-12. B-Display of Atlantic Ocean Near Boca Raton, Florida. Range interval 5 nautical miles. Azimuthal sector 70°.

deviations in the earth's shape, that is, bulges and trenches that extend over large areas.

As used here, scatterometry refers to the study of the magnitude of scattering of terrain as a function of incidence angle. Terrain surface characteristics (moisture content, dielectric constant, fine-scale surface roughness, shape, etc.) affect the average intensity for a given incidence angle and the curve shape of average-intensity-versus-incidence-angle; intensity and intensity versus incidence angle are also affected by polarization and radar frequency. The possible applications of scatterometry are numerous. For example, J.W. Rouse, H.C. MacDonald, and W.P. Waite (1969) reported variations in echo from Arctic ice as a function of incidence angle, and J.W. Rouse et al. (1973) reported the development of a radar to automatically classify ice into various types. The classification is accomplished through a determination of the incidence angle variation (slope) and mean value of echo from ice illuminated by the radar.

S.K. Parashar et al. (1974) have described a study on classifying ice by thickness, age, surface roughness, and texture. The ice was ground-truthed according to visual interpretation of aerial photographs; only radar data corresponding to large homogeneous patches of ice were considered. By comparing images from a 16.5 GHz radar, it was possible to identify ice and classify it into the following four categories:

Open water
Thin first-year ice (less than 18 cm thick)
Thick first-year ice (18 to 90 cm thick)
Multiyear ice and very thick first-year ice (greater than 90 cm thick)

Radar data were taken for 12 forward angles to produce intensity versus angle curves. Two radars (400 MHz and 13.3 GHz) were used. By using intensity versus angle to classify according to the four image analysis categories, a correct classification of 85 percent was reported possible from the scatterometry data. It seems that radar will become a useful tool for automatically classifying sea ice.

Recently a synthetic aperture system for 60 meters, 20 meters, and 2 meters (5, 15, 150 MHz) has been developed (Porcello et al. 1974). This system was used for the Apollo 17 lunar experiment. From an electromagnetic point of view, subsurface detections are easier for the moon than for the earth because lunar rock has very low attenuation. However, the ultimate availability of decimeter systems will provide improved capabilities for penetrating vegetation and soil.

R.D. Ellermeier, D.S. Simonett, and L.F. Dellwig (1967) observed that comparative data for two wavelengths can be of value. They described geologic areas with relatively (compared with surrounding areas) small reflectivity at *K* band that have relatively large reflectivity at 428 MHz. The authors conjectured that at the longest wavelength loose sediment was being penetrated and the echo was dominated by scattering from rough underlaying lava surfaces. The surface penetrated in this case was fine-grained, unconsolidated, and without much moisture.

Dual polarization also provides improved sensory capability for radar. R.K. Moore (1969; Skolnik 1970, chap. 25) has described results for transmitting horizontal polarization and receiving that polarization (*HH*) and the cross-polarized component (*HV*). The contrast for *HH* imagery and *HV* imagery is sometimes different in very important ways. For example, L.F. Dellwig and R.K. Moore (1966) reported that the cross-polarized imagery enhanced a number of lava-alluvium boundaries relative to those displayed on like-polarized imagery. Similarly, R.D. Ellermeier, A.K. Fung, and

D.S. Simonett (1966)noted that cross-polarized imagery was better defined. and that the differentiation allowed separation of lava flows of different ages, weathering, and roughness. There have even been geological findings resulting directly from the availability of dual-polarization imagery (see fig. 7-1) that had not been detected by previous ground or aerial surveys.

Dual polarization (*HH*, *HV*) imaging radar has also been used (Moore 1969; Skolnik 1970, chap. 25) to accomplish separation of crops and natural vegetation into certain general classes. The separation was obtained with a single frequency (*K* band) radar, but Moore expressed the view that multi-frequency plus dual polarization systems are needed for full identification of crops. A number of papers (Ulaby 1974; Ulaby, Cihlar, and Moore 1974; Ulaby 1975) have been published on the sensitivity of echo from vegetation and soil to wavelength, polarization, and incidence angle. The data are presented with a view toward developing improved microwave remote sensors (sec. 6.12).

Applications of high resolution (synthetic aperture) side-looking air-borne radar (SLAR) imagery are continuing to be extended into new areas. Ben Drake et al. (1974) describe experiments using *X*- and *L*-band radars with a choice of transmit and receive polarizations. During flight each radar simultaneously received horizontal and vertical polarizations. Either horizontal or vertical polarization could be transmitted, but only one transmitted polarization was available during a given flight. By obtaining four images simultaneously, comparisons could be made for data from the same imaged swath, the same imaging parameters (radar-look direction, depression angles, etc.), the same motion errors, and the same terrain conditions (surface texture, slope orientations, vegetation, etc.). Tone and texture were the primary identification parameters used; shape, pattern, and location were secondary identification parameters. The authors referred to their system as a multiplex SLAR, and they found that the system was an excellent indicator of the relative heights of vegetation.

The studies were conducted in flat agricultural regions of Michigan and along Florida river, lake, and pond areas. By using both the *X*- and *L*-band imagery, the relative heights of crops could be discerned to within 60 cm (sometimes as small as 30 cm). Several types of vegetation, both in water and on land, were differentiated with the multiplex imagery by their relative heights, densities, and surface roughnesses. The vegetation type, per se, was not sensed.

Most of the differences in the radar return observed in the imagery was due to wavelength—the like- and cross-polarized image of the same wavelength usually looked the same. As expected, there was no significant penetration of vegetation observed at *X* band; however, at *L* band the radiation penetrated low grasses and reeds standing 0.9 to 1.5 meters out of

the water. It seems that use of multiplex imagery may make it possible to determine the particular growth stage (e.g., from emergent to full grown) of some crops.

Although tone and texture are primary features for target identification, parameters such as shape, size, pattern, and location are also very important in the identification process. For a given size of antenna aperture, beamwidths obtainable are directly proportional to radar wavelength. For example, at 70 GHz (4.3 mm) a 1° beamwidth can be obtained with an antenna approximately one foot in diameter. At 700 MHz, an antenna diameter of 100 feet would be necessary to achieve the same beamwidth. Therefore, there is continuing interest for using the shorter wavelengths for providing improved target recognition.

Figure 1-13 is a radar picture (Long, Rivers, and Butterworth 1960) obtained with a ground-based mobile system operating at 70 GHz. The figure is of a *B*-display (azimuth angle along the abscissa and range along the ordinate). The azimuthal beamwidth was 0.2°, the range resolution was 7.5 meters (0.05 microsecond pulse length), and the polarization was vertical.

Figure 1-13 shows the town of Sierra Vista, Arizona, as seen from a test area at Fort Huachuca. The lowest range marker visible represents a distance of 4 kilometers from the radar, and the range markers are spaced at intervals of 1 kilometer. Some of the principal targets are identified in the figure. It should be noted that the display in many respects resembles those for high resolution radars obtained at longer wavelengths. Therefore figure 1-13 illustrates that it is possible to obtain improved target discrimination with the higher angular resolution available by using millimeter wavelengths.

The early radars operated in the decimeter wavelength region because equipment had not been developed for shorter wavelengths. There is a renewed interest in the frequency region of 1 to 40 MHz because these waves will propagate beyond the line-of-sight by either diffraction along the curvature of the earth or by sky waves refracted by the ionosphere. The radars are called HF OTH (high frequency over-the-horizon) systems even though strictly speaking the HF band is defined as 3 to 30 MHz. Modern radar experiments with these frequencies began at the Naval Research Laboratory in the 1950s, and HF OTH radars have detected targets at distances of hundreds to thousands of miles. Therefore, OTH radar can be used to achieve detection ranges that are an order of magnitude greater than can be accomplished with microwaves. J.M. Headrick and M.I. Skolnik (1974) review some of the characteristics, capabilities, and limitations of OTH radar and discuss applications for air-traffic control and for sensing sea conditions (sec. 5.13) over vast ocean areas. Although

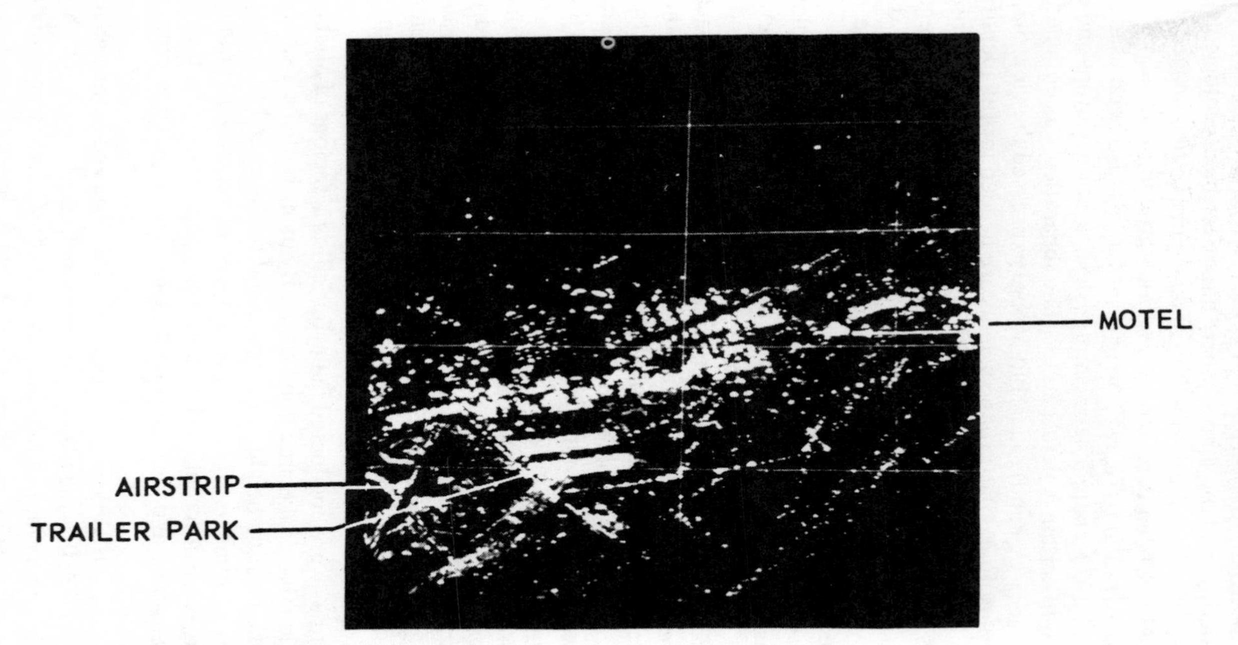

Source: Long, Rivers, and Butterworth (1960).

Figure 1-13. B-Display of Sierra Vista, Arizona. Range interval 5 kilometers. Azimuthal sector 30°.

HF OTH radars require very large antennas, this disadvantage is sometimes offset because these radars will cover much larger regions than is possible with microwaves.

P.J. Cannon (1974) has reported on the application of radar imagery to extensive environmental geologic mapping in Texas. In his work, radar imagery is considered an important aid for preliminary compilation of environmental maps and for periodic updating of completed maps. According to Cannon, when mapping large regions radar imagery is less expensive than photographic coverage of the same area, and it should take less time to map with radar imagery than with low- or intermediate-altitude aerial photographs. Less map-compilation time is required with radar imagery because aerial photographs must be viewed stereoscopically in order to observe what can be seen with the unaided eye with radar imagery. According to Cannon, radar imagery is the least expensive method for monitoring the growth and development of environmentally significant man-made features over areas of several hundred square miles. As already mentioned, radar imagery is also being used to investigate sea waves, swell, current edges, and roughness fronts. We therefore foresee an upsurge in the use of radar imagery for a wide range of applications.

Satellites provide a promising new platform for making large-scale measurements of land and sea. The mounds and trenches of the sea surface can be measured from space. For example, from space radar measurements the Puerto Rican trench has been reported to have a depth of approximately 10 meters (Vonbun, Marsh, and Lerch 1974). The radar used was the S-193 altimeter (McGoogan et al. 1974) aboard the Skylab satellite that operates at a frequency of 13.9 GHz. The topographic capability of the Skylab altimeter is reported to be about 1 meter, and the objective for the measuring capability of the SEASAT-A altimeter (proposed for launch in 1978 by NASA [Apel 1974]) is 10 centimeters. Altimeters, in addition to having a geodetic profiling capability, will become useful for measuring sea roughness and large-scale currents (Barrick 1974).

It ultimately should be possible with space radar to classify land use into various broad categories and to determine soil moisture content. The National Aeronautics and Space Administration SEASAT-A ocean dynamics satellite will include a short pulse radar altimeter, a synthetic aperture imaging radar, a microwave scatterometer, a scanning microwave radiometer, and a scanning infrared radiometer. These instruments will measure information on waves, surface winds, sea surface temperatures, the geoid, currents, tides, and features of the ocean, land, and the atmosphere. Thus, satellites are beginning to provide data on a global basis that will permit geographers, oceanographers, geologists, and meteorologists to analyze large-scale phenomena on a timely basis.

References

Apel, J.R., "SEASAT: An Integrated Spacecraft Observatory for the Oceans," *Program and Abstracts for Annual Meeting of the International Union of Radio Science,* Boulder, Colorado, p. 33, October 14-17, 1974.

Barrick, D.E., "Wind Dependence of Quasi-Specular Microwave Sea Scatter," *IEEE Transactions on Antennas and Propagation,* vol. AP-22, pp. 135-6, January 1974.

Cannon, P.J., "Application of Radar Imagery to Environmental Geologic Mapping of Texas," *Ninth International Symposium on Remote Sensing of Environment,* Environmental Research Institute of Michigan, April 15-19, 1974.

Dellwig, L.F. and R.K. Moore, "The Geologic Value of Simultaneously Produced Like- and Cross-Polarized Radar Imagery," *Journal of Geophysical Research,* vol. 71, pp. 4995-98, October 1966.

Drake, B., R.A. Shuchman, R.F. Rawson, F.L. Smith, and R.W. Larson, "Feasibility of Using Multiplex Slar Imagery for Water Resource Management and Mapping Vegetation Communities," *Ninth International Symposium on Remote Sensing of Environment,* Environmental Research Institute of Michigan, April 15-19, 1974.

Ellermeier, R.D., A.K. Fung, and D.S. Simonett, "Some Empirical and Theoretical Interpretations of Multiple Polarization Radar Data," *Proceedings of the 4th Symposium on Remote Sensing of the Environment,* University of Michigan, Ann Arbor, pp. 657-70, 1966.

Ellermeier, R.D., D.S. Simonett, and L.F. Dellwig, "The Use of Multiparameter Radar Imagery for the Discrimination of Terrain Characteristics," *1967 IEEE International Convention Record,* vol. 15, pp. 127-35.

Flynt, E.R., F.B. Dyer, R.C. Johnson, M.W. Long, and R.P. Zimmer, "Clutter Reduction Radar," Final Report on Contract NObsr-91024, Engineering Experiment Station, Georgia Institute of Technology, December 1967.

Goodyear Aerospace Corporation, *Profile,* vol. 9, Summer 1971.

Goodyear Aerospace Corporation, *Profile,* vol. 10, Fall 1972.

Guinard, N.W., "The Remote Sensing of Oil Slicks," *Proceedings of the 7th International Symposium on Remote Sensing of the Environment,* University of Michigan, Ann Arbor, May 1971.

Hall, J.S., *Radar Aids to Navigation,* Massachusetts Institute of Technology Radiation Laboratory Series, vol. 2, McGraw-Hill Book Company, Inc., New York, New York, 1947.

Harger, R.O., *Synthetic Aperture Radar Systems,* Academic Press, New York, New York, 1970.

Headrick, J.M. and M.I. Skolnik, "Over-the-Horizon Radar in the HF Band," *Proceedings of the IEEE,* vol. 62, pp. 664-73, June 1974.

Krishen, K., "Detection of Oil Spills Using a 13.3 GHz Radar Scatterometer," *Proceedings of the 8th International Symposium on Remote Sensing of the Environment,* Environmental Research Institute of Michigan, October 1972.

Long, M.W., W.K. Rivers, and J.C. Butterworth, "Combat Surveillance Radar AN/MPS-29 (XE-1)," *Record of the Sixth Annual Radar Symposium,* University of Michigan, Ann Arbor, 1960.

MacDonald, H.C., P.A. Brennan, and L.F. Dellwig, "Geologic Evaluation by Radar of NASA Sedimentary Test Site," *IEEE Transactions on Geoscience Electronics,* vol. GE-5, pp. 72-78, December 1967.

McGoogan, J.T., L.S. Miller, G.S. Brown, and G.S. Hayne, "The S-193 Radar Altimeter Experiment," *Proceedings of the IEEE,* vol. 62, pp. 793-803, June 1974.

Moore, R.K., "Radar Return from the Ground," Bulletin of Engineering No. 59, University of Kansas, 1969.

Parashar, S.K., A.W. Biggs, A.K. Fung, and R.K. Moore, "Investigation of Radar Discrimination of Sea Ice," *Ninth International Symposium on Remote Sensing of Environment,* Environmental Research Institute of Michigan, April 15-19, 1974.

Pilon, R.O. and C.G. Purves, "Radar Imagery of Oil Slicks," *IEEE Transactions on Aerospace and Electronic Systems,* vol. AE2-9, pp. 630-36, September 1973.

Porcello, L.J., R.L. Jordan, J.S. Zelenka, G.F. Adams, R.J. Phillips, W.E. Brown, Jr., S.H. Ward, and P.L. Jackson, "The Apollo Lunar Sounder Radar System," *Proceedings of the IEEE,* vol. 62, pp. 769-83, June 1974.

Ridenour, L.N., *Radar System Engineering,* Massachusetts Institute of Technology Radiation Laboratory Series, vol. 1, McGraw-Hill Book Company, Inc., New York, New York, 1947.

Rouse, J.W., Jr., H.C. MacDonald, and W.P. Waite, "Geoscience Applications of Radar Sensors," *IEEE Transactions on Geoscience Electronics,* vol. GE-7, pp. 2-19, January 1969.

Rouse, J.W., J.A. Schell, W.D. Nordhaus, and J.A. Permenter, "On Development of an Arctic Ice Classification System," *Texas Engineering Experiment Station Bulletin,* pp. 16-24, April 1973.

Schwarz, D.E. and F. Caspall, "The Use of Radar in the Discrimination and Identification of Agricultural Land Use," *Proceedings of the Fifth Symposium on Remote Sensing of the Environment,* University of Michigan, Ann Arbor, April 1968.

Skolnik, M.I., *Radar Handbook,* McGraw-Hill Book Company, Inc., New York, New York, 1970.

Smith, P.L., Jr., K.R. Hardy, and K.M. Glover, "Applications of Radar to Meteorological Operations and Research," *Proceedings of the IEEE,* vol. 62, pp. 724-45, 1974.

Ulaby, F.T., "Radar Measurement of Soil Moisture Content," *IEEE Transactions on Antennas and Propagation,* vol. AP-22, pp. 257-65, March 1974.

Ulaby, F.T., "Radar Response to Vegetation," *IEEE Transactions on Antennas and Propagation,* vol. AP-23, pp. 36-45, January 1975.

Ulaby, F.T., R.K. Moore, R. Moe, and J. Holtzman, "On Microwave Remote Sensing of Vegetation," *Proceedings of the Eighth Symposium on Remote Sensing of the Environment,* Environmental Research Institute of Michigan, October 1972.

Ulaby, F.T., Josef Cihlar, and R.K. Moore, "Active Microwave Measurement of Soil Water Content," *Remote Sensing of the Environment*, vol. 3, pp. 185-203, 1974.

Vonbun, F.O., J. Marsh, and F. Lerch, "Sea Surface Topography from Space—GSFC Geoid," *Program and Abstracts for Annual Meeting of the International Union of Radio Science,* Boulder, Colorado, p. 34, October 14-17, 1974.

Waite, W.P. and H.C. MacDonald, "Snowfield Mapping with *K*-Band Radar," *Remote Sensing of Environment,* vol. 1, pp. 143-50, 1969-70.

Waite, W.P. and H.C. MacDonald, "Vegetation Penetration with *K*-Band Imaging Radars," *IEEE Transactions on Geoscience Electronics*, vol. GE-9, pp. 147-55, July 1971.

2 Basic Concepts and Definitions

Radar Reflectivity

Radar echo is directly related to the nature of the terrain illuminated by the transmitted electromagnetic wave. The relationship is complicated because there are many involved terrain parameters that affect the echo. Radar echo is affected by the detailed composition of the illuminated area and even by its moisture content. The relationship between echo and terrain is further complicated by the fact that it is a function of the angle of incidence of the transmitted wave, the transmitted and received polarizations, and radar frequency.

The radar cross section σ is a frequently used defining parameter that is directly proportional to echo power. A radar receives signals related to the radar cross section (RCS) of targets in accordance with the radar equation

$$P_r = \frac{P_t G^2 \lambda^2}{(4\pi)^3 R^4} \sigma \tag{2.1}$$

where P_r = received power

P_t = transmitted power

G = antenna gain

λ = wavelength of transmitted waves

R = range (distance) between antenna and target

This equation is for the conventional monostatic radar (transmit and receive antennas co-located) with each antenna having a one-way gain G. Equation (2.3) is a more general radar equation.

The quantity σ°, radar cross section per unit area of illuminated surface, is a normalized parameter usually used to describe radar cross section of extended target areas (sec. 2.3). It has been found that σ° depends on the angle θ (see fig. 2-1), the transmitter wavelength λ, and the polarizations of the incident and reflected waves. Depending on the nature of the surface, echo can be sensitive to wind speed, wind direction, moisture content, and surface roughness.

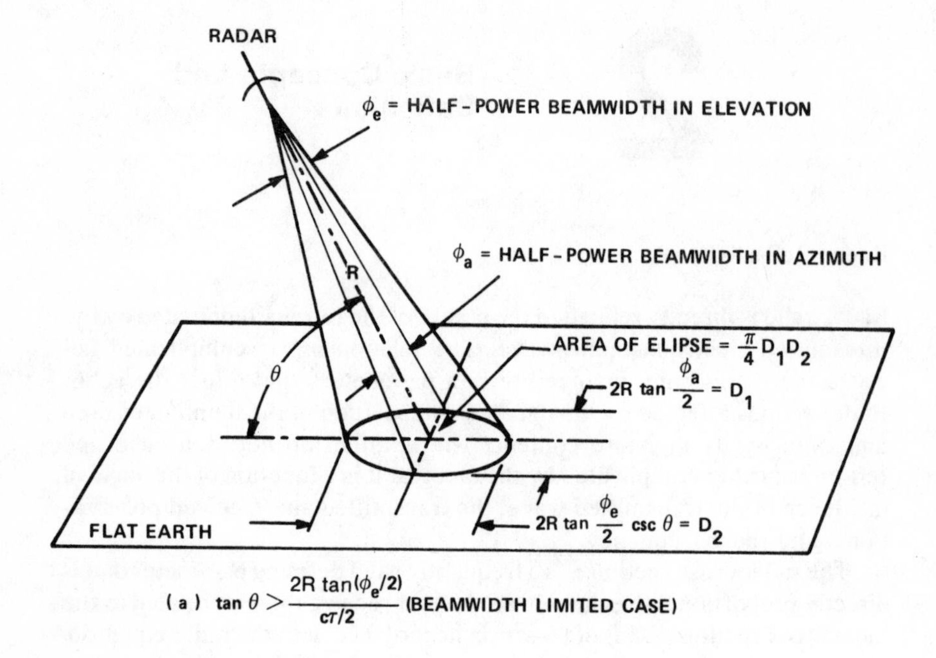

(a) $\tan\theta > \frac{2R\tan(\phi_e/2)}{c\tau/2}$ (BEAMWIDTH LIMITED CASE)

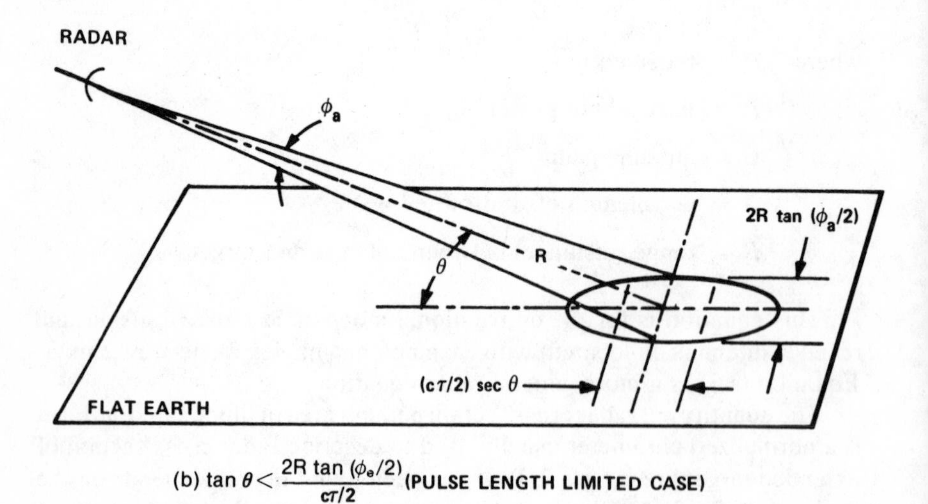

(b) $\tan\theta < \frac{2R\tan(\phi_e/2)}{c\tau/2}$ (PULSE LENGTH LIMITED CASE)

Source: Adapted from F.E. Nathanson, *Radar Design Principles*. Copyright 1969 by McGraw-Hill Book Company, Inc. Used with permission of McGraw-Hill Book Company.

Figure 2-1. Area Illumination for Pulse Radar.

2.1 The Radar Equation for Free Space

This section gives equations for scattering under the assumption that the target(s) and antenna(s) are located in a vacuum (free space). Although the concept is highly idealized, the resulting equations are useful for describing propagation within the earth's atmosphere.

If power P_t is radiated isotropically (equally in all directions), the power per unit area (power density) at a distance R_t from the source is $P_t/4\pi R_t^2$. It should be recognized that $4\pi R_t^2$ is the surface area of a sphere of radius R. Antennas do not radiate equally in all directions, and antenna gain G is defined such that power density is

$$\text{Power density from directive antenna} = P_t\, G/4\pi R_t^2$$

Antenna gain is also described as

$$G = 4\pi(A_e/\lambda^2)$$

where λ is wavelength and A_e is the "effective" aperture or collecting area of the antenna. Although most antennas are reciprocal in the sense that the gains for transmission and reception are equal, let G_r and G_t represent gain of receive and transmit antennas, respectively. Then power received by an antenna can be expressed as

$$\begin{aligned}\text{Power received} &= (\text{power density}) \cdot A_e \\ &= \frac{P_t G_t}{4\pi R_t^2} \cdot A_e = \frac{P_t G_t G_R \lambda^2}{(4\pi)^2 R_t^2}\end{aligned} \tag{2.2}$$

Equation (2.2) gives received power for one-way transmission.

The radar cross section σ of an object is defined as *the area intercepting that amount of power which, when scattered isotropically, produces an echo equal to that received from the object*. Therefore, the power density at a distance R_r from the scatterer is given by

$$\text{Power density from scatterer} = \frac{P_t G_t}{4\pi R_t^2}\,\frac{\sigma}{4\pi R_r^2}$$

If the receiving antenna is located at a distance R_r from the scatterer and has gain G_r, then

$$\begin{matrix}\text{Power received} \\ \text{from object}\end{matrix} = \frac{P_t G_t}{4\pi R_t^2}\,\frac{\sigma}{4\pi R_r^2} A_e = \frac{P_t G_t G_r \lambda^2}{(4\pi)^3 R_t^2 R_r^2}\sigma \tag{2.3}$$

A radar for which the transmitting and receiving antennas are not co-located is called a bistatic radar; therefore, equation (2.3) is applicable to bistatic radar.[a] Most radars use the same antenna for transmitting and

---

[a]For research reports on bistatic scattering, see A.R. Domville [1967, 1968].

receiving ($G_t = G_r$); such radars are said to be monostatic. For monostatic radar, equation (2.3) gives the following for received power P_r

$$P_r = \frac{P_t G^2 \lambda^2}{(4\pi)^3 R^4} \sigma \tag{2.4}$$

Equation (2.4) is known as "the radar equation." Obviously the equation must be modified to include propagation losses caused by the atmosphere and the earth (secs. 2.9, 2.10, 2.12, 2.13).

2.2 Radar Cross Section of Targets

Radar echo is caused by the reradiation of an object; in other words, an electromagnetic wave illuminates an object and this object acts as an antenna. Therefore, calculation techniques developed from antenna theory can be used to predict scattering by a target.

The radar cross section (RCS) of an object is, in general, a sensitive function of orientation and wavelength. From the definition of σ, the cross section of an object that scatters equally (isotropically) in all directions is equal to its projected area. A sphere with radius a that is large with respect to wavelength scatters isotropically; therefore, the cross section (if $a > \lambda$) of a metal (lossless) sphere is

$$\sigma = \pi a^2 \tag{2.5}$$

For a sphere with radius small with respect to λ, the σ will vary as λ^4. This fourth-power dependence of σ on λ is known as the Rayleigh law and is characteristic of scattering from objects that are small compared to λ. For linear dimensions of the same order as λ, a highly sensitive and oscillating dependence of σ on λ exists (resonance phenomenon).

It should be apparent that for an object that produces directive scattering tending to be oriented toward the radar, σ will be larger than its projected physical area. This can be thought of as being due to the elemental areas of the object reradiating so that the separate radiations add constructively.

For a flat metal sheet of area A perpendicular to the incident radar beam,

$$\sigma = 4\pi A^2/\lambda^2$$

For general angles of incidence, the cross section is a function of the angle of incidence and it varies very rapidly if the wavelength is small compared with the linear dimensions of the plate. In a diagram of return power versus angle of observation these variations appear as "lobes." The strong main lobe is normal to the plate and the peaks of the side lobes decrease rapidly with increasing angle. For small angles γ, but excluding the main lobe, the average cross section $\bar{\sigma}$ (averaged over several lobes) is given (Ridenour 1947) approximately by

$$\bar{\sigma} \approx 4\pi\lambda^2/(2\pi\gamma)^4$$

This result is independent of the size of the target, subject of course to the limitation that the linear dimensions of the target are large compared with the wavelength.

Equation (2.5) for σ of a sphere is a special case of more general equations for surfaces with linear dimensions and radii large compared with a wavelength. The reader is referred to G.T. Ruck et al. (1970) for information on cross section for objects of various shapes.

Most man-made targets have complex reradiation patterns. In general, average echo from buildings is larger than that from the ground, and the echo from ground is usually larger than that from the sea. Average values for motor vehicles at X band for various polarizations are given by R.D. Hayes et al. (1958). Comprehensive reports have been prepared on echo from ships (Corriher et al. 1967) and from aircraft (Corriher 1970). Sections 3.6 through 3.15 are on theories of RCS for rough, extended target including land and sea.

2.3 Normalized Radar Cross Section

Echo from land and sea is caused by the various scatterers within the resolution cell of a radar. Herbert Goldstein (1950) introduced the quantity σ°, radar cross section per unit area of surface, to provide a normalized parameter that can be used to describe radar cross section. Using the definition of σ°, the radar cross section σ equals $\sigma^\circ A$. Here A is the area of a smooth surface that corresponds to the mean for the land or sea surface contained within the radar's cell of resolution (fig. 2-1).

The illuminated area is a very complicated function of system parameters. For example, for a pulse radar the effective area actually depends on details in pulse and antenna beam shapes (Kerr 1951). However, for most practical purposes, a good approximation to the area can be obtained by using the nominal pulse width and the beamwidths in the elevation and azimuthal planes. Early references specified antenna pattern effects in terms of the one-way, 3-dB beamwidths (Kerr 1951; Skolnik 1962). More recently the two-way, 3-dB beamwidths (Barton 1964; Nathanson 1969) have been used. A discussion of this point is included in the last paragraph of this section. To the best of this author's knowledge, all data included within this book are based on using the one-way, 3-dB beamwidths as defining quantities.

Goldstein introduced the concept of σ° while working with sea clutter in narrow-beam pulse radars for near grazing incidence angles. In this case, range resolution is $(c\tau/2)\sec\theta$, where c is the velocity of propagation (3×10^8 meters per second for free space), τ is the nominal pulse length, and θ is the depression (grazing) angle. The azimuthal width of the resolution cell in the case of a narrow azimuthal beamwidth is $R\phi_a$, where R and ϕ_a are range and azimuthal beamwidth, respectively. Therefore, for a narrow azimuthal beamwidth and near grazing incidence angles

$$A = (R)(c\tau/2)(\phi_a \sec\theta) \tag{2.6}$$

For a near vertical incidence angle, pulse length usually does not affect the resolution cell. Then, for narrow beamwidths the surface area illuminated can be approximated by

$$A = (\pi/4)(R\phi_a)(R\phi_e \csc\theta) = (\pi/4)(R^2\,\phi_a\,\phi_e \csc\theta) \tag{2.7}$$

A represents the area of an ellipse with axes of lengths $R\phi_a$ and $R\phi_e \csc\theta$.

Equations (2.6) and (2.7) are valid providing antenna beamwidths are sufficiently narrow so that the tangent of the angle can be approximated by the angle (i.e., $\tan\phi_a \approx \phi_a$). By referring to the geometry in figure 2-1, for wider angles equations (2.6) and (2.7) must be replaced by

$$A \approx 2R\frac{c\tau}{2}\tan\frac{\phi_a}{2}\sec\theta \quad \text{if} \quad \tan\theta < \frac{2R\tan(\phi_e/2)}{c\tau/2} \tag{2.8}$$

and

$$A \approx \pi R^2 \tan\frac{\phi_a}{2}\tan\frac{\phi_e}{2}\csc\theta \quad \text{if} \quad \tan\theta > \frac{2R\tan(\phi_e/2)}{c\tau/2} \tag{2.9}$$

Whether ϕ_a and ϕ_e are best represented by one- or two-way 3-dB beamwidths has not been firmly established. That difference will amount to approximately 1.5 dB for equation (2.6) and about 3 dB for equation (2.7). Most papers consider ϕ_a and ϕ_e to be the one-way, 3-dB widths, but a number of more recent references consider ϕ_a and ϕ_e in equations (2.8) and (2.9) to be more accurately represented by the two-way beamwidths. The belief seems to be developing that a better estimate for effective area when using equations (2.8) and (2.9) is obtained if it is assumed that the widths are for two-way propagation; or, alternately, ϕ_a and ϕ_e are $\phi'_a/\sqrt{2}$ and $\phi'_e/\sqrt{2}$, respectively, where ϕ'_a and ϕ'_e are the one-way, 3-dB beamwidths (Daley 1973). As stated previously, the data within this book are to the best of this author's knowledge all based on ϕ_a and ϕ_e being the one-way, 3-dB beamwidths.

2.4 Rayleigh Roughness Criterion

If there is a height difference Δh between two points on a surface (see fig.

2-2), then the waves reflected at these two points will be shifted in phase with respect to each other by

$$\Delta\Phi = 4\pi\,\Delta h \sin\theta/\lambda \tag{2.10}$$

where λ is the wavelength and θ is the grazing angle. To give a qualitative indication of surface roughness, whether smooth or rough in the electromagnetic sense, Lord Rayleigh used equation (2.10). He considered a surface smooth if

$$\Delta h \sin\theta < \lambda/8$$

Obviously, most surfaces are neither perfectly smooth nor rough. More precise calculations can be made by following the material in section 4.9.

2.5 Coherence and Incoherence of a Scattered Field

The terms "coherent" and "incoherent" are sometimes used to help describe scattering phenomena. If the phase of a wave is constant or varies in a deterministic manner, it is called coherent. If the phase of a wave is random and uniformly distributed over an interval of phase 2π, it is called incoherent. The mean power density of the sum of incoherent waves is the algebraic sum of separate power densities; the total power density of coherent waves is obtained by summing the individual fields *vectorially* and determining the total power from the resulting total field. For example, for n waves of equal power, the total power is directly proportional to n if the waves are incoherent and to n^2 if the waves are all in phase.

The field scattered by terrain is the result of scattering from many separate objects, thus the field at a given point is the sum of many elementary waves. In general, the field scattered by a rough surface in the specular direction has a constant, coherent component and a random component. If the surface is very rough, there is, in principle, only one component—an incoherently scattered field.

In mathematical terms, if a target is composed of a number of scatterers with cross section for each denoted by σ_i, total cross section σ is expressed as follows:

$$\sigma = \left| \sum_{i=1}^{n} (\sqrt{\sigma_i}\, e^{j\psi_i}) \right|^2$$

The phase factor $e^{j\psi_i}$ includes phase change on reflection and phase delay because of distance from the radar. Therefore, this equation includes the addition of reradiated fields from all scatterers while appropriately accounting for the differences in the relative phases.

Certain simplifying assumptions are possible if only an average is

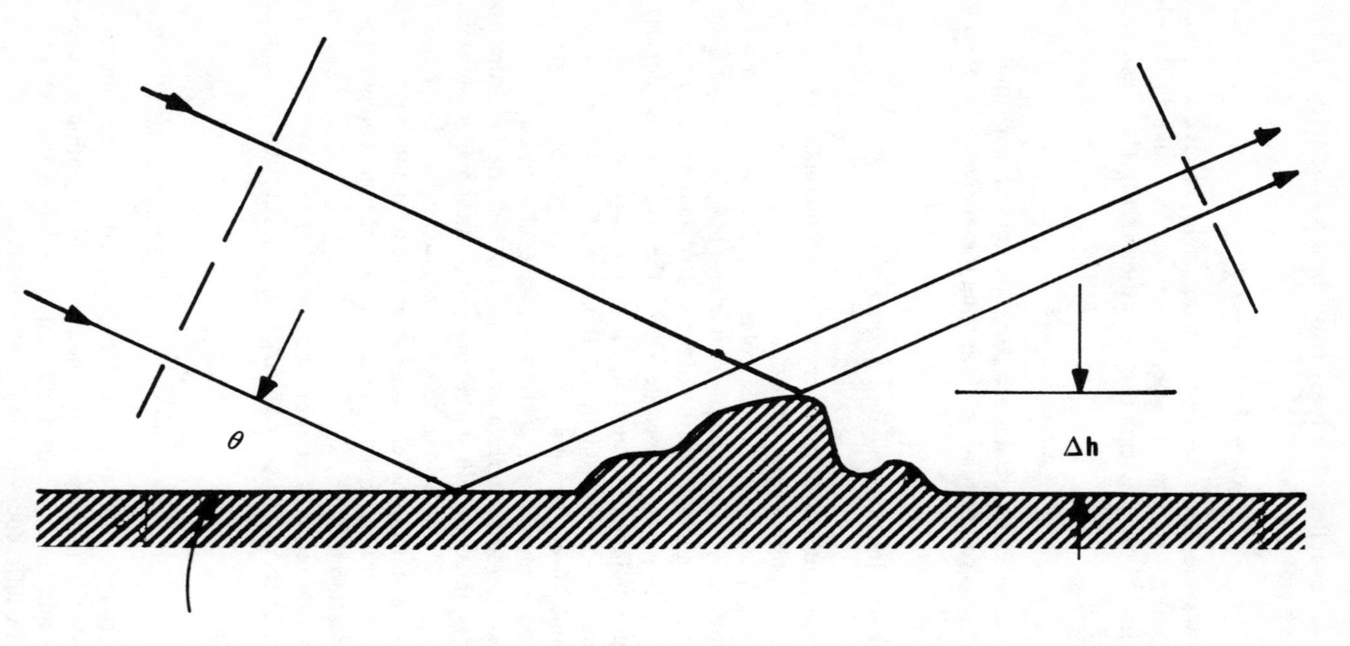

Figure 2-2. Forward Scattering from an Irregular Surface.

needed for a large number of randomly positioned scatterers. For example, the time average or average for many different positions of the scatterers is

$$\overline{\sigma} = \sum_{i=1}^{n} \sigma_i$$

To illustrate coherent and incoherent scattering, assume that there are a total of n scatterers each with the same RCS ($\sigma_i \equiv \sigma_1$). Then, if the various relative phases are randomly distributed, the expected average value of σ is

$$\overline{\sigma} = \sum_{i=1}^{n} \sigma_i = n\sigma_1$$

Furthermore, if the relative phases are zero (i.e., scattered fields all of the same phase at point of reception), the expected value of σ is

$$\sigma = \left| \sum_{i=1}^{n} \sqrt{\sigma_i} \right|^2 = \left| n \sqrt{\sigma_1} \right|^2 = n^2 \sigma_1$$

2.6 *Far Field of Radar Targets*

The RCS of a target is defined only for large distances from the radar. Since a target (like an antenna) is a radiator, the well-known concept of a "far-zone" as used in antenna measurements theory is applicable (Kouyoumjian and Peters 1965; Silver 1949). In the absence of target surface roughness, this requirement means that the distance R to the target must satisfy the criterion $R > 2(L_t + L_a)^2/\lambda$. L_t and L_a are the largest lateral dimensions of the target and the measurements antenna. Usually L_t and L_a differ sufficiently so that the far-zone criterion is customarily expressed simply as $R = 2L^2/\lambda$. E.F. Knott and T.B.A. Senior (1974) have observed that large measurement errors can exist for this separation, and reportedly it may be necessary to exceed this separation by a factor of 5 or more in order to achieve acceptable precision. There does, however, exist at Georgia Tech a novel "compact range" that permits targets to be accurately measured with small separations providing L_t is smaller than L_a (Johnson, Ecker, and Moore 1969; Johnson, Ecker, and Hollis 1973).

It has been shown (Barrick 1965) that if one is considering the power scattered incoherently from a roughened surface, the far-field requirement reduces to $R > 2\ell^2/\lambda$ where ℓ is the roughness correlation length. This means that equations defined for the far zone, such as those for σ, are applicable at shorter ranges for a rough surface than for a smooth one because ℓ is usually much smaller than L.

2.7 Relationship between the Autocorrelation Function and the Power Density Spectrum

The fluctuations of radar echoes are discussed in chapter 5. Sections 2.7 and 2.8 give some mathematical background that the reader may find useful as supplementary material for that chapter. Rates of fluctuations can be described by the power density spectrum and by the time autocorrelation function. The autocorrelation function $R(\tau)$ of a stationary (that is, temporally homogeneous) function $X(t)$ is defined as

$$R(\tau) = \lim_{T\to\infty} \frac{1}{2T} \int_{-T}^{T} X(t) \cdot X(t+\tau)\, dt \qquad (2.11)$$

The power density spectrum

$$P(f) = \lim_{T\to\infty} \frac{1}{2T} \left| \int_{-T}^{T} X(t)\, e^{-j2\pi ft}\, dt \right|^2 \qquad (2.12)$$

is the Fourier transform of the correlation function; that is,

$$P(f) = \int_{-\infty}^{\infty} R(\tau) \cdot e^{-j2\pi f\tau}\, d\tau$$

Hence, by inverse transformation,

$$R(\tau) = \int_{-\infty}^{\infty} P(f) \cdot e^{j2\pi f\tau}\, df$$

Therefore, the total power contained in the spectrum is equal to the autocorrelation coefficient at $\tau = 0$:

$$R(0) = \int_{-\infty}^{\infty} P(f)\, df \qquad (2.13)$$

$R(\tau)$ and $P(f)$ are even functions of their respective arguments. Hence, the relation between them can be expressed more simply as

$$P(f) = \int_{-\infty}^{\infty} R(\tau) \cdot \cos 2\pi f\tau \cdot d\tau = 2\int_{0}^{\infty} R(\tau) \cdot \cos 2\pi f\tau \cdot d\tau \qquad (2.14)$$

and

$$R(\tau) = \int_{-\infty}^{\infty} P(f) \cdot \cos 2\pi f\tau \cdot df = 2\int_{0}^{\infty} P(f) \cdot \cos 2\pi f\tau \cdot df \qquad (2.15)$$

From equations (2.14) and (2.15), it is clear that correlation functions and the corresponding power spectra contain the same information displayed in different forms. Therefore, the problem of determining the power spectrum corresponding to $R(\tau)$ is analogous to determining other Fourier transform related parameters; for example, a power spectrum from a pulse shape or a radiation pattern from an antenna aperture distribution. As a specific example, the problem of determining the power spectrum corresponding to $R(\tau)$ is mathematically the same as that of determining the linear antenna distribution for a symmetric aperture with constant phase (since the $R(\tau)$ is real and symmetric about τ).

Examples of useful transform pairs taken from J.L. Lawson and G.E. Uhlenbeck (1950, p. 42) follow. In these examples a normalization in multiplication factors is used with $R(\tau)$ and $P(f)$ so that $R(0)$ is always unity and the total area of the spectrum is

$$\int_0^\infty P(f)\,df = 1$$

The Gaussian function is unusual in that its Fourier transform is also Gaussian. Therefore, if the spectral shape is Gaussian the autocorrelation function is Gaussian, and conversely. Let

$$R(\tau) = e^{-\alpha^2\tau^2} \quad \text{then} \quad P(f) = (2\sqrt{\pi}/\alpha)(e^{-\pi^2 f^2/\alpha^2}) \tag{2.16}$$

Define $f_{\frac{1}{2}}$ as the width of the frequency spectrum between f equals zero and the point that the power is one-half its maximum value. Let τ_o be the half-amplitude point for the correlation function, that is, $R(\tau_o) = \frac{1}{2}R(0)$. Since $P(f_{\frac{1}{2}}) = \frac{1}{2}P(0)$ and $R(\tau_o) = \frac{1}{2}R(0)$, it can be seen from equation (2.16) that

$$f_{\frac{1}{2}} = \frac{\log_e 2}{\pi}\,\frac{1}{\tau_0} = 0.22/\tau_o \tag{2.17}$$

If

$$R(\tau) = e^{-\beta\tau}$$

then

$$P(f) = \frac{4\beta}{\beta^2 + (2\pi f)^2} \tag{2.18}$$

By using the definitions as before, that is, $R(\tau_o) = \frac{1}{2}R(0)$ and $P(f_{\frac{1}{2}}) = \frac{1}{2}P(0)$, it can be shown from above that

$$f_{\frac{1}{2}} = \frac{\log_e 2}{2\pi}\,\frac{1}{\tau_0} = \frac{0.11}{\tau_o} \quad \text{if} \quad R(\tau) = e^{-\beta\tau} \tag{2.19}$$

When $R(\tau)$ is a monotonically decreasing function of τ, $P(f)$ is also a monotonically decreasing function of f. The function $P(f)$ will become wider and flatter as $R(\tau)$ becomes narrower. For example, if $R(\tau)$ drops to zero in a short time Δ, then $P(f)$ will be essentially constant up to a high frequency of the order of magnitude $1/\Delta$.

2.8 Fluctuation Statistics

The probability of an event is defined as a frequency ratio υ/n, the number of times υ an event is observed to occur in a sequence of n identical experiments. Probability is a measure of likelihood of occurrence of an event.

Let $p(x)$ be called the probability density function. It is equal to or greater than zero. The probability of x being between x and $x + dx$ is $p(x)dx$; the probability that x is between all values of x is obviously unity

$$\int_{-\infty}^{\infty} p(x)\,dx = 1$$

If the probability of an event is zero, it does not mean the event is impossible. For example,

$$\int_{a}^{a} p(x)\,dx = 0$$

illustrates that the probability of x being equal to the point a is zero, and yet it is possible for x to be equal to a. The average or mean value of a random variable x is

$$\bar{x} = \int_{-\infty}^{\infty} xp(x)\,dx \tag{2.20}$$

The best known distribution is the Gaussian or normal distribution. Its probability density function can be written as

$$p(x) = \frac{1}{s\sqrt{2\pi}}\exp\left[-\frac{(x-a)^2}{2s^2}\right] \tag{2.21}$$

Here $1/s\sqrt{2\pi}$ is a measure of the height (frequency) scale, because $p(x) = 1/s\sqrt{2\pi}$ at the center of the distribution $(x = a)$. The curve is symmetric about x $= a$ and approaches zero asymmetrically; therefore, a is the median of x. The quantity $1/2s^2$ governs that width of the curve, because it is only a multiplier on the x scale. If $1/2s^2$ is large, the curve is narrow and high; if small, the curve is low and broad. The quantity s is called the standard deviation. For the Gaussian function, 0.68 of the data points are within plus or minus one standard deviation of the median, and 0.95 of the points are within plus or minus two standard deviations of the median.

Four frequently used probability density functions for describing the fluctuations in echo strength are the Rayleigh, Ricean, lognormal, and Weibull functions. The first three functions are described below and the Weibull is discussed in the Appendix.

The Rayleigh density function is

$$W(X) = \frac{2X}{\alpha}\ exp\left[-\frac{X^2}{\alpha}\right] \tag{2.22}$$

where $X \geq O$ and for a given $W(X)$ the variance α is a positive constant equalling the average of X^2. The density function describes the fluctuations in amplitude (e.g., magnitude of voltage) of thermal noise from a linear circuit or fluctuations in amplitude of echo due to Raleigh clutter.

From a physical viewpoint, Rayleigh clutter or scattering is due to many scatterers of about equal strength and that are independently positioned. The probability density function for power due to Raleigh scattering (or thermal noise in a linear circuit), where power P is proportional to X^2 of equation (2.22), is

$$W(\overline{P}) = (1/\overline{P})\ e^{-P/P} \tag{2.23}$$

where $\overline{P}$ is average power. Mathematically $W(\overline{P})$ is known as the exponential density function, but for obvious reasons it is sometimes called the Raleigh probability density function for power. Strictly speaking, equation (2.23) is valid only if the number of scatterers is very large. However, for certain special cases the number of scatterers can be as small as four or five (Kerr 1951, p. 554).

Average or mean power can be obtained as

$$\overline{P} = \int_0^\infty P\ W(P)\, dP$$

The probability that the power is between two finite levels P_1 and P_2 can be expressed as

$$Pr\{P_1 < P < P_2\} = \int_{P_1}^{P_2} W(P)\, dP$$

$$= (1/\overline{P})\int_{P_1}^{P_2} e^{-P/\overline{P}}\, dP = e^{-P_1/\overline{P}} - e^{-P_2/P}$$

The probability that P is less than P_2 is

$$Pr\{0 < P < P_2\} = \int_0^{P_2} W(P)\, dP = 1 - e^{-P_2/\overline{P}}$$

The level of power for which half the time power is either above or below that level is the median level P_m. Therefore, if $Pr\,\{0 < P < P_2\} = 0.5$; P_2 is equal to P_m. In this case

$$e^{-P_m/\overline{P}} = 0.5$$

and

$$P_m = 0.693\ \overline{P} \tag{2.24}$$

In decibels,

$$\overline{P}\text{ (in dB)} = P_m\text{ (in dB)} + 1.6\text{ dB} \tag{2.25}$$

If the received signal contains a constant component in addition to a Rayleigh distributed random component, the peak of the distribution is shifted so that the most probable value of received power is not zero and can be written as

$$W(P) = (1 + m^2)e^{-m^2} e^{-P(1+m^2)/\overline{P}}\ J_0\left(2im\sqrt{1 + m^2}\,\sqrt{P/\overline{P}}\right) dP/\overline{P} \tag{2.26}$$

where J_0 is the Bessel function described by the series

$$J_0(x) = 1 - \frac{x^2}{2^2} + \frac{x^4}{2^2 \cdot 4^2} - \frac{x^6}{2^2 \cdot 4^2 \cdot 6^2} + \ldots \tag{2.27}$$

The distribution $W(P)$ is called the Ricean distribution (Rice 1944) and is plotted in figure 5-3 for several values of m^2 (ratio of the constant power to the random power). Tables for this function have been published by K.A. Norton et al. (1955).

A random variable Y is said to be lognormally distributed if its logarithm is normally distributed. Let $X = \ln Y$ (natural logarithm) be normally distributed with mean μ and standard deviation s. Then Y is said to be lognormally distributed, and the distribution of Y is completely specified by the two parameters μ and s of the distribution X.

The relationship $Y = e^X$ is the mathematical inverse of $X = \ln Y$. Therefore, the normally distributed variable X can be positive or negative, but the lognormally distributed variable Y takes on only positive values. For a normal distribution the mode (most probable value), median, and mean are coincident; for a lognormal distribution the mean exceeds the median and the median exceeds the mode (see fig. 2-3). J. Aitchison and J.A.C. Brown (1969) give the positions of the mean, median, and mode; namely, $Y = e^{\mu + \frac{1}{2}s^2}$, $Y = e^{\mu}$, and $Y = e^{\mu - s^2}$, respectively. Figure 2-4 includes graphs of the Y distribution for various values of s^2.

The probability density function for the lognormal distribution can be obtained from the normal distribution by using the transformation $X = \ln Y$ (see, for example, Petr Beckmann [1967, p. 60]):

$$\begin{aligned} W(Y) &= \frac{1}{Ys\sqrt{2\pi}} \exp\left[-\frac{1}{2s^2}(\ln Y - \mu)^2\right] \\ &= \frac{1}{Ys\sqrt{2\pi}} \exp\left[-\frac{1}{2s^2}\left(\ln\frac{Y}{Y_m}\right)^2\right] \end{aligned} \tag{2.28}$$

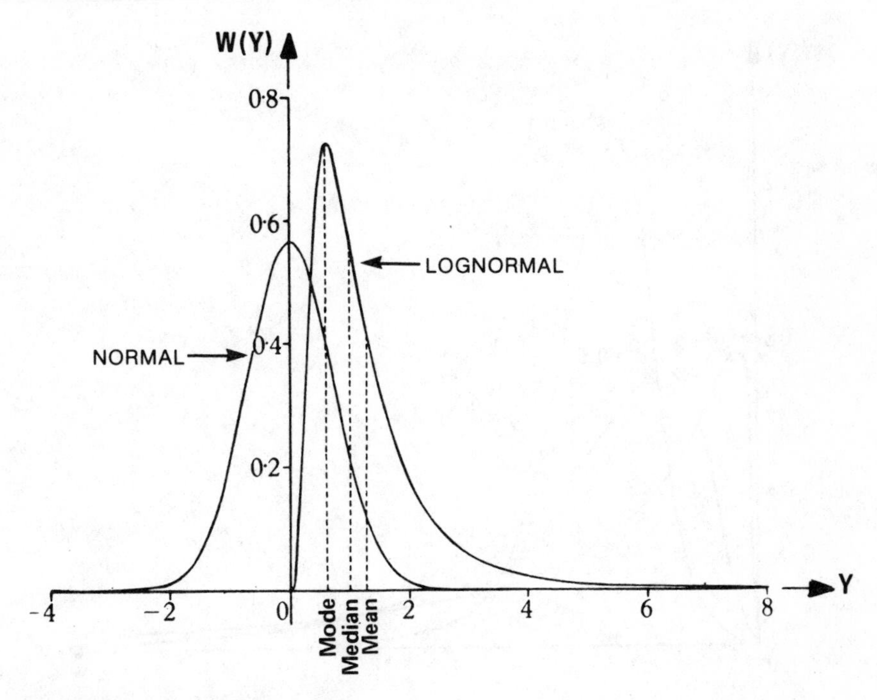

Source: Aitchison and Brown (1969, p. 9).

Figure 2-3. Probability Density Functions $W(Y)$ Versus Y for a Normal and a Lognormal Distribution.

where Y = the lognormally distributed variable

Y_m = the median value of Y

s = the standard deviation of $\ln (Y/Y_m)$

It is sometimes observed that distributions of radar cross section (RCS) when expressed in decibels can be approximated by a normal distribution. Then it is said that the RCS is lognormally distributed. The material that follows is included to illustrate how calculations of the average and most probable RCS can be determined from the median and standard deviation of lognormally distributed echo data.

Assume that the median and standard deviations, expressed in decibels, are given. Let σ be RCS as a ratio of linear units (normalized RCS), let σ(nat) be $\ln \sigma$ (natural logarithm) and let σ(dB) be $10 \log_{10} \sigma$. Further, let $\overline{\sigma}$ denote average or mean, σ_m denote median, and σ_{mode} denote mode.

From above, for the mean and mode

$$\overline{Y} = e^{\mu + \frac{1}{2}s^2} \qquad Y_{\text{mode}} = e^{\mu - s^2}$$

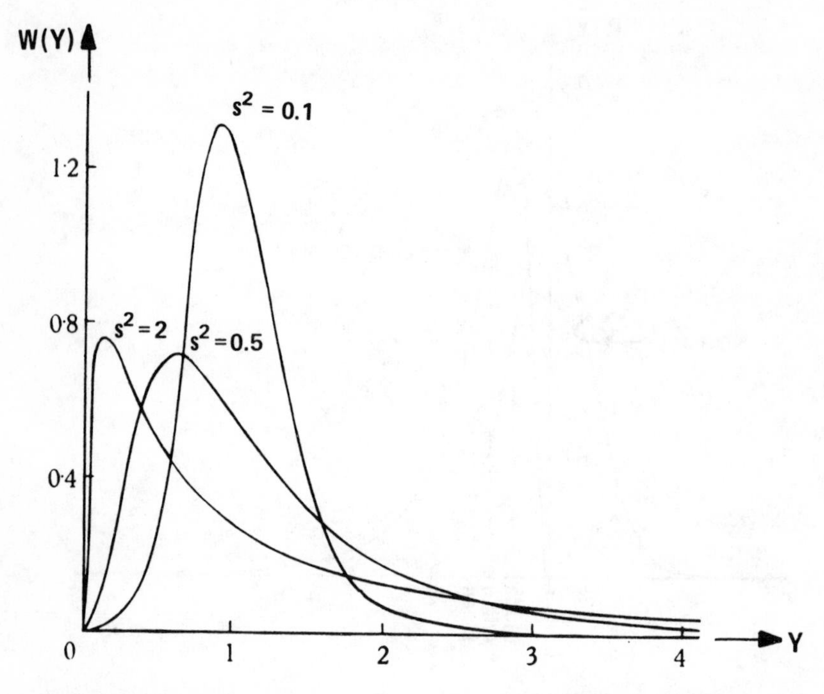

Source: Aitchison and Brown (1969, p. 10).

Figure 2-4. Probability Density Functions for Lognormal Distributions with Different Values of s^2.

where for RCS this becomes

$$\overline{\sigma} = e^{\overline{\ln \sigma}}\, e^{\frac{1}{2}s^2} \tag{2.29}$$

$$\sigma_{\text{mode}} = e^{\overline{\ln \sigma}}\, e^{-s^2} \tag{2.30}$$

For a normally distributed function, $\overline{X} = X_m = X_{\text{mode}}$. Therefore, since $\ln \sigma$ is normally distributed, mean of $\ln \sigma \equiv \overline{\ln \sigma}$ = mode of $\ln \sigma$ = median of $\ln \sigma$ $= \ln \sigma_m$. Equations (2.29) and (2.30) can thus be written

$$\overline{\sigma} = e^{\ln \sigma_m}\, e^{\frac{1}{2}s^2} \tag{2.31}$$

$$\sigma_{\text{mode}} = e^{\ln \sigma_m}\, e^{-s^2} \tag{2.32}$$

If a is positive, $e^{\ln a} = a$. Therefore, since all $\ln \sigma$ are positive,

$$\overline{\sigma}/\sigma_m = e^{\frac{1}{2}s^2} \tag{2.33}$$

and

$$\sigma_{\text{mode}}/\sigma_m = e^{-s^2} \tag{2.34}$$

To review, $\overline{\sigma}$ is the mean or average, σ_m is the median, and σ_{mode} is the mode of the random variable σ. The random variable σ is said to be lognormally distributed if the random variable $\ln \sigma$ is normally distributed. The relationship between the natural logarithm ln of a ratio, called natural units or nats, and 10 $\log_{10}$ of that ratio, called decibels or dB, is

$$\sigma(\text{nat}) = 0.23\ \sigma(\text{dB}) \tag{2.35}$$

The standard deviation s is the standard deviation of $\ln \sigma$, or s(nat) by definition.

By taking ln of both sides of equations (2.33) and (2.34) and applying equation (2.35), one obtains

$$\overline{\sigma}/\sigma_m(\text{in dB}) = 0.115\ [S(\text{dB})]^2 \tag{2.36}$$

and

$$\sigma_{\text{mode}}/\sigma_m\ (\text{in dB}) = -0.23\ [S(\text{dB})]^2 \tag{2.37}$$

An example showing the utility of equations (2.36) and (2.37) follows. If σ expressed in decibels is normally distributed and the standard deviation of the distribution of σ is 10 dB, the average or mean value of σ will exceed the median value by a factor of 11.5 dB and the median will exceed the mode (most probable value) by 23 dB.

The Appendix includes cumulative distributions for lognormal and Weibull statistics.

The Earth and Its Effects on Radar

2.9 Effects of the Earth's Curvature and Refraction

High frequency (HF) waves (3–30 MHz) literally reach around the earth because they propagate beyond the line-of-sight. This propagation is caused by diffraction along the curvature of the earth and by sky waves being refracted in the ionosphere. Although over-the-horizon (OTH) radars have been developed (Headrick and Skolnik 1974), most radars use microwaves. For microwaves the range limit is defined by the radar horizon discussed here and in sections 2.10 and 4.1.

Electromagnetic waves bend slightly because of atmospheric refraction. The refraction is caused by the fact that the index of refraction of the atmosphere is a function of height above the earth.

For a smooth earth, there is a demarcation in visibility because of the earth's curvature. The apparent demarcation (horizon) depends on the index of refraction. The index of refraction is essentially independent of radar (or radio) frequency. However, the refractive index for radar (or radio waves) is different from that for visible waves. Refraction causes the horizon to be at a greater distance than if the waves traveled in straight lines. The geometric horizon is the horizon that would exist if propagation were along perfectly straight lines.

For visible waves, the index of refraction depends only on temperature and pressure, but for radar the refractive index depends (in a complicated way) on temperature, pressure, and the water vapor content of the atmosphere. Let d_h, d_o, and d_g denote distances to the radar, optical, and geometric horizons. Then, for the case of a "standard" or average atmosphere

$$d_h = 1.07d_o = 1.15d_g$$

Therefore, under normal conditions when radar range is limited only by the horizon, microwave radar "sees" at greater distances than can be seen visually. For over-water paths, radar observations out to many times the d_h for a standard atmosphere have been reported (see sec. 2.12). This extended range is caused by ducting. Microwaves actually do propagate beyond the horizon because of diffraction from the earth (which serves as the radiating surface), but the signal strength due to diffraction alone is very weak.

Accurate computations for the tracking of targets can only be made with rather complicated calculations of the true angles of incidence and reflection at the earth, and the calculations must include effects of curvature in the propagation paths. Basic concepts for the analysis are given in section 4.1. The effects of the curved propagation paths are included by substituting an equivalent earth radius r_e (usually $r_e = 4r/3 = 5{,}280$ statute miles) for the true earth radius r; straight lines can then be used to calculate the various true angles and distances.

The reader is referred to figure 4.1 for a comparison of depression angle β_1 with grazing angle θ for propagation along path M_1. F.E. Nathanson (1969) gives the approximation

$$\beta_1 \approx \sin^{-1}\left(\frac{h}{R} + \frac{R}{2r_e}\right) \tag{2.38}$$

where r_e = the equivalent earth radius

h = the antenna height

R = the radar slant range

When using the symbols of figure 4-1, $R = M_1$ and $h = h_1$. Similarly, θ is given by the approximation

$$\theta \approx \sin^{-1}\left(\frac{h}{R} - \frac{R}{2r_e}\right) \tag{2.39}$$

Notice that the sines of β_1 and θ differ in the approximations of equations (2.38) and (2.39) because of the term $R/2r_e$. The reader is reminded that r_e is an equivalent radius that accounts for atmospheric refraction. As the atmosphere changes, r_e is changed, and the values of β_1 and θ calculated by equations (2.38) and (2.39) are changed.

For most geometries used in studies of RCS, β_1 and θ are nearly equal and can be approximated by using $\beta_1 \approx \theta \approx h/R$. For example, for an "average" or standard atmosphere ($r_e = 5{,}280$ statute miles) slant ranges R of 1 and 10 miles give values of $R/2r_e$ of approximately 10^{-4} and 10^{-3}, respectively. It is for this reason that the words grazing and depression are often used in this book as if β_1 and θ are interchangeable.

D.E. Kerr (1951) has provided a comprehensive treatment of propagation phenomena, and useful curves are included to facilitate calculations. H.R. Reed and C.M. Russell (1953) include excellent background material on the effects of the earth's curvature and atmospheric refraction. N.I. Durlach (1965) has presented an analysis of the various mathematical relationships between pertinent variables that are useful for computer programming.

2.10 The Effect of Interference on a Target

At short ranges, the earth can be assumed flat and the effects of refraction are negligible. Most of the ground-based experimental investigations on radar reflectivity of land and sea have been made under these conditions.

Even when the earth can be assumed to be flat, effects of reflections from the earth must be included. If spherical waves are reflected from a sufficiently large plane surface, the backscattered wave appears to come from an image of the transmitter antenna. In this case, received signal power varies with range R as $(2R)^{-2}$. Therefore, backscattered power from a large, flat reflecting surface (such as from the side of a ship) might be expected to vary as R^{-2} at short range. The transition range, that is, the range between that which yields the usual R^{-4} dependence and the R^{-2} dependence, is strongly influenced by surface shape. For a disc of diameter D and an antenna of diameter d, one would expect the transition range to be in the neighborhood of $2(D + d)^2/\lambda$. The reader will recognize this expression as that from antenna theory which is used to determine the transition between the Fresnel and Fraunhofer zones (see sec. 2.6).

For studies of targets over a reflecting surface, combinations of the two distinct paths must be considered: (1) the most direct path between antenna and target, and (2) the most direct path between antenna and target and for which the wave is reflected off the earth. Paths one and two will be referred to below as simply "direct" and "reflected," respectively. In figure 4-1, these paths are called M_1 and M_2.

At ranges for which backscattered power would vary as R^{-4} in free space, the effect of interference caused by a phase difference between the direct and reflected waves is sometimes observed. The difference ΔR in path length of these waves ia a major contributor to the phase difference. For a target (e.g., corner reflector) at constant height above water for which prominent scattering elements are confined to a small volume, nodes and antinodes are observed in received power as a function of range. For horizontal polarization, the reflection coefficient for a smooth water surface has a magnitude of essentially one and a phase of 180° for all angles of incidence. Therefore, the envelope of the antinodes would, in principle, vary as R^{-4}.

At even greater ranges (such that ΔR is less than $\lambda/2$), but for ranges not extending to the radar horizon, target power is expected to vary as R^{-8} because of the interference effect. An example of experimental data that depict the transition between the R^{-4} and R^{-8} zones is given in figure 4-16. Chapter 4 discusses forward scattering from the earth.

2.11 Nature of the Sea Surface and Wind Speed Statistics

Qualitatively, the basic characteristics of the sea surface are well known: large-scale, roughly periodic waves upon which ripples, foam, and spray are superimposed. The large waves are commonly called the macrostructure; ripples and the like are called the microstructure of the sea surface. The macrostructure of the sea is usually described by specifying the *sea* and the *swell*. *Sea* consists of relatively steep short-crested waves produced and driven by the winds in their locale and are called wind waves. *Swell* consists of waves of long wavelength, nearly sinusoidal in shape, and are produced by distant winds. The very irregular appearance of the sea surface is due to interference of the various wind and swell waves and to local atmospheric turbulence. Ocean currents have only a slight effect on the characteristics of the sea surface except along the coastlines. Near a coast, currents (usually tidal currents) may cause a considerable increase in wave heights due to their interference with wind and swell waves. Foam and spray are largely caused by interference, while ripples are usually caused by turbulent gusts of wind near the surface.

Since the crests of waves are steeper and narrower than the troughs, the mean level is lower than halfway between the crests and troughs. The vertical distance between trough and crest is called wave height. The horizontal distance between successive crests, measured in the direction of travel, is called wavelength. The time interval between passage of successive crests at a point that is fixed with respect to the earth is called wave period. The theoretical relationship between speed, wavelength, and period is shown in figure 5-23.

There have been two qualitative sea-state scales in general use. The one used exclusively before World War II (and a great deal of the time during and after) is the Beaufort scale. It would be better named "Beaufort wind scale" since it is based on wind force and not wave height. This scale is given in table 2-1. As noted in the table, wave height depends not only on local wind force, but also on the length of time and the length of sea over which the wind has been blowing.

The Douglas scale is more descriptive of the macrostructure than the Beaufort scale in that the sea is specified by wave height; in the scale's complete form, the sea and swell are given separately (Fairbridge 1966, p. 787). Table 2-1 includes the frequently used short form that gives the "sea" number only. For more information on sea state scales, see Nathaniel Bowditch (1966, app. R).

Since sea waves are in general irregular and unpredictable in detail, reported wave heights are often based on subjective and inexact observation. *Significant wave height* is defined as the average value of the peak-to-trough heights of the one-third highest waves in a given observation. Observers on ships use this definition of wave height to be representative of a quick guess. In other words, significant height $h_{1/3}$ is "supposed" to be the predominant height of the waves. The observer certainly neither records the heights of many waves nor picks out the highest one-third. Another definition of wave height used by oceanographers is the crest-to-trough wave height exceeded by 10 percent of the waves, $h_{1/10}$. The approximate relationships between significant wave height and other wave heights follow.

Wave Height	*Abbreviation*	*Relative Height*
Average	h_{av}	0.64
Significant	$h_{1/3}$	1.00
Height exceeded by 10 percent of the waves	$h_{1/10}$	1.29

Results of various analyses (Kinsman 1965; Watters 1953) of sea waves indicate that height distributions are approximately Gaussian. Blair Kinsman states that (1) for any phenomenon not affected by the small-scale components of wave motion, one need not hesitate to assume the Gaussian distribution, and (2) even if one is interested in fine details of the surface, the Gaussian assumption is often valid. The standard deviation Δh of surface height versus time (fixed position) provides another convenient parameter for estimating wave height. Since wave height distributions are approximately Gaussian, Δh for sea waves is expected to be approximately equal to the rms (root-mean-squared) wave height. It has been estimated

Table 2-1
Wind and Wave Scales[a]

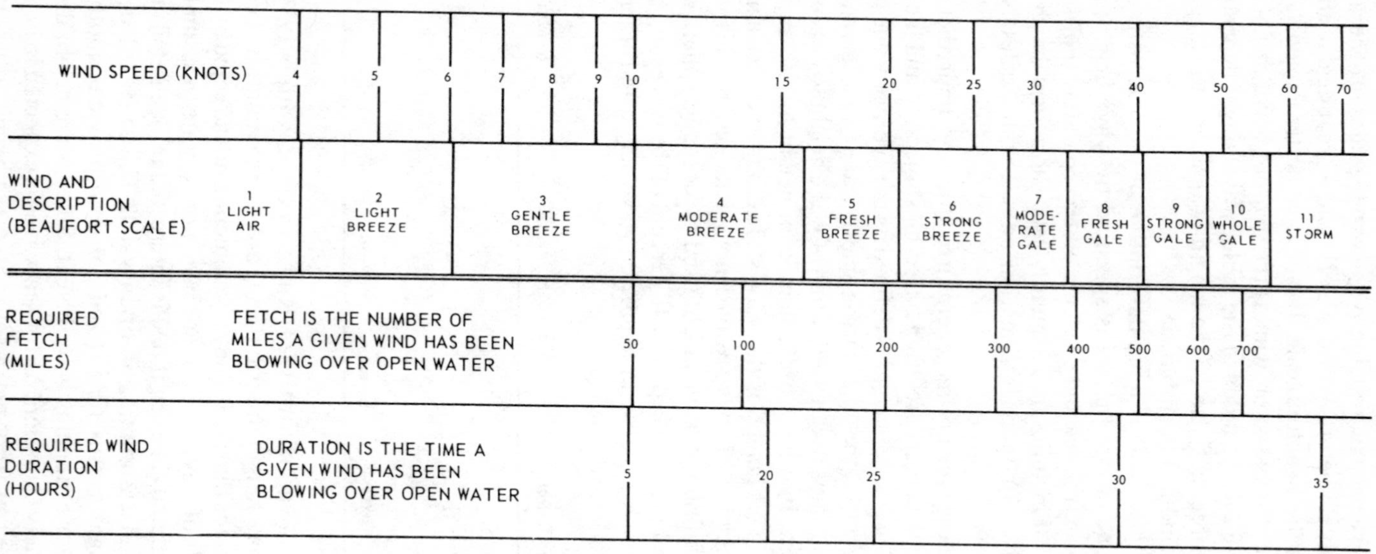

WIND SPEED (KNOTS)	4, 5, 6, 7, 8, 9, 10, 15, 20, 25, 30, 40, 50, 60, 70
WIND AND DESCRIPTION (BEAUFORT SCALE)	1 LIGHT AIR; 2 LIGHT BREEZE; 3 GENTLE BREEZE; 4 MODERATE BREEZE; 5 FRESH BREEZE; 6 STRONG BREEZE; 7 MODE-RATE GALE; 8 FRESH GALE; 9 STRONG GALE; 10 WHOLE GALE; 11 STORM
REQUIRED FETCH (MILES) — FETCH IS THE NUMBER OF MILES A GIVEN WIND HAS BEEN BLOWING OVER OPEN WATER	50, 100, 200, 300, 400, 500, 600, 700
REQUIRED WIND DURATION (HOURS) — DURATION IS THE TIME A GIVEN WIND HAS BEEN BLOWING OVER OPEN WATER	5, 20, 25, 30, 35

IF THE FETCH AND DURATION ARE AS GREAT AS THOSE INDICATED ABOVE, THE FOLLOWING WAVE CONDITIONS WILL EXIST. WAVE HEIGHTS MAY BE UP TO 10% GREATER IF FETCH AND DURATION ARE GREATER.

WAVE HEIGHT CREST TO TROUGH (FEET)	1	2	4 WHITE CAPS FORM	6	8	10	15	20	25	30	40	50	60

SEA STATE AND DESCRIPTION (DOUGLAS SEA NUMBER)	1 SMOOTH	2 SLIGHT	3 MODE-RATE	4 ROUGH	5 VERY ROUGH	6 HIGH	7 VERY HIGH	8 PRECIPITOUS

THE HEIGHT OF THE WAVES IS ARBITRARILY CHOSEN AS "THE HEIGHT OF THE HIGHEST 1/3 OF THE WAVES."

[a] Adapted from *Undersea Technology*, p. 37, May 1964.

(Pierson, Neumann, and James 1955) that $h_{1/3}$ and $h_{1/10}$ are $4\Delta h$ and $5\Delta h$, respectively. Therefore, wave height, as usually reported, ranges between four and five times the standard deviation of wave height or, alternatively, four and five times the rms wave height.

Another listing of wave height distributions, taken from a publication by H.B. Bigelow and W.T. Edmondson (1947), is shown in table 2-2. These distributions are based on more than 40,000 extracts from ships' log books.

The United States Navy Marine Climatic World Atlases (Naval Aerology Branch 1957) provide detailed surface wind data for the North Atlantic, North Pacific, and Indian Oceans for all months of the year. These data have been collected over a period of years from island, coastal, and ship weather stations. The surface wind data, presented in frequency distributions of Beaufort Force, were analyzed by averaging over all oceans and months of the year—the results are presented in figure 2-5. The surface wind probability density function is plotted in figure 2-5 (a); figures 2.5 (b) and 2.5 (c) were obtained from figure 2.5 (a).

2.12 Propagation Over the Horizon

For microwaves, extensions of range up to several times the horizon as calculated for the standard atmosphere have been observed over water. The guiding by refractive anomalies appears to be the only significant means by which coverage beyond the horizon can be obtained with microwaves.

An atmospheric duct is produced by a rapid decrease of index of refraction with altitude. This can be caused by an increase in temperature (temperature inversion) and/or a decrease in humidity; humidity gradients generally cause the most pronounced changes of refractive index. Dry air, introduced by a wind from a land mass, is a common cause for strong ducts.

Propagation can be strongly influenced by ducting if a radar antenna or target is near the water (antenna and/or target within duct). As a first-order approximation, over-water ducts can be considered as waveguides with a variable but maximum height of approximately 200 meters and a minimum height of about 50 meters. Elevated ducts have also been used for enhanced propagation over the horizon (Guinard et al. 1964).

For interpreting target and sea echo data for ranges out to the radar horizon, calculations necessary to predict effects of the atmosphere are complex and are not completely reliable. However, for short ranges (less than horizon for standard atmosphere), effects of trapping are probably negligible if antennas and targets are located within 100 feet of the water (within the duct). For an antenna not within a duct and a target near water (and conversely), trapping may cause maxima and minima in signal strength that depend on range (depression angle).

Table 2-2
Relative Frequency of Significant Wave Heights in Regions of the World[a]

Region	Height of Waves in Feet 0-3 (%)	3-4 (%)	4-7 (%)	7-12 (%)	12-20 (%)	>20 (%)
North Atlantic, between Newfoundland and England	20	20	20	15	10	15
Mid-equatorial Atlantic	20	30	25	15	5	5
South Atlantic, latitude of southern Argentina	10	20	20	20	15	10
North Pacific, latitude of Oregon and south of Alaskan Peninsula	25	20	20	15	10	10
East-equatorial Pacific	25	35	25	10	5	5
West wind belt of South Pacific, latitude of southern Chile	5	20	20	20	15	15
North Indian Ocean, Northeast monsoon season	55	25	10	5	0	0
North Indian Ocean, Southwest monsoon season	15	15	25	20	15	10
Southern Indian Ocean between Madagascar and northern Australia	35	25	20	15	5	5
West wind belt of southern Indian Ocean on route between Cape of Good Hope and southern Australia	10	20	20	20	15	15
Averages over all regions	22	23	20.5	15.5	9.5	9.0

[a]From M.W. Long et al. (1965).

In conclusion, propagation in a layered refractive medium has been studied both experimentally and theoretically by a large number of investigators. Many of the results obtained through World War II are given by D.E. Kerr (1951). In spite of the large amount of information available, experimental and theoretical methods required to obtain usefully accurate results for radar problems are unreliable and costly to undertake. The character of the propagation varies widely and models that can be applied by other than trained specialists have not been developed. References on this subject include Martin Katzin, R.W. Bauchman and William Binnian (1947), C.L. Pekeris (1947), F.A. Sabransky (1958), E.E. Gossard (1964), N.W. Guinard et al. (1964), François Ducastel (1966), I.M. Hunter and T.B.A. Senior (1966), and J.R. Wait (1970).

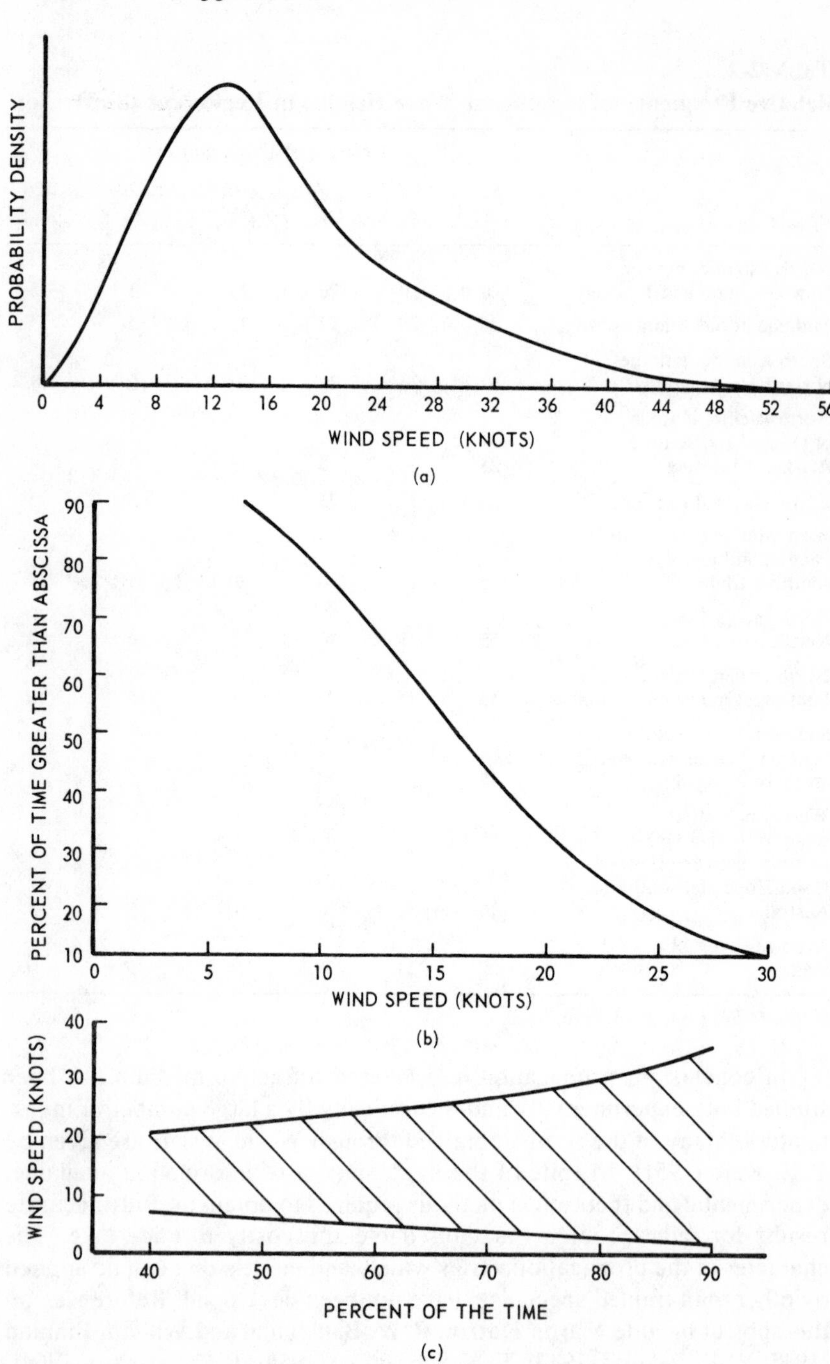

Source: Long et al. (1965).

Figure 2-5. Surface Wind Speed Statistics.

2.13 Attenuation and Scattering by the Atmosphere

Radar signals are attenuated in the air because of absorption by atmospheric gases, clouds, rain, and snow; the signals are also attenuated by reflections from the larger particles such as raindrops, hail, and snowflakes. Generally speaking, the effects of both absorption and scattering become more pronounced as the radar frequency is increased. In the case of absorption, resonance effects (fig. 2-6) are frequency sensitive and have peaks as determined by L.V. Blake (1962, 1968). This figure shows effects of the frequency sensitivity of oxygen and water vapor resonances, and it also illustrates a general trend toward increased attenuation with increased frequency. The physical dimensions of particulate matter in the atmosphere are generally small with respect to radar wavelengths, and under these conditions radar cross section is proportional to λ^{-4} (known as the Rayleigh scattering law). The reader is familiar with some effects of the Rayleigh law in the visible region. Since the scattered power varies inversely with wavelength, blue light is scattered more than red light. The blue color of the sky is thus explained as being due to the scattering of the light by the molecules of the atmosphere. At sunset the sun looks red because the light reaching the earth has more blue removed by scattering than red.

Kerr (1951) presents a comprehensive review of the knowledge available at the end of World War II on atmospheric absorption and scattering. It had then become apparent that the choice of the *K* band operating frequency (24 GHz) near the water vapor absorption peak (see fig. 2-6) prevented that band from being useful for radar surveillance under adverse weather conditions. Blake (1962, 1968) reviews in depth the effects of atmospheric attenuation on radar and presents useful curves that will facilitate calculations. For general references on the effects of weather, the reader is referred to L.J. Battan (1973) and P.L. Smith, Jr., K.R. Hardy, and K.M. Glover (1974).

Interest in the use of millimeter waves has led to a number of theoretical and experimental studies of propagation effects. The publications by A.B. Crawford and D.C. Hogg (1956), R.D. Hayes (1964), L.A. Hoffman, H.S. Winthroub, and W.A. Garber (1966), and W.I. Thompson, III (1971) will provide the reader with useful entries into the literature. R.K. Crane (1974) has measured attenuation and phase shift caused by rain simulated with a sprayer system that provides a stable drop size distribution. The measurements were made using vertical and horizontal polarizations at 7.9 GHz and vertical polarization at 33.9 GHz. The experimental results were in general agreement with the theoretical model used. For 7.9 GHz, the measured values of attenuation and phase shift are higher for horizontal polarization than for vertical.

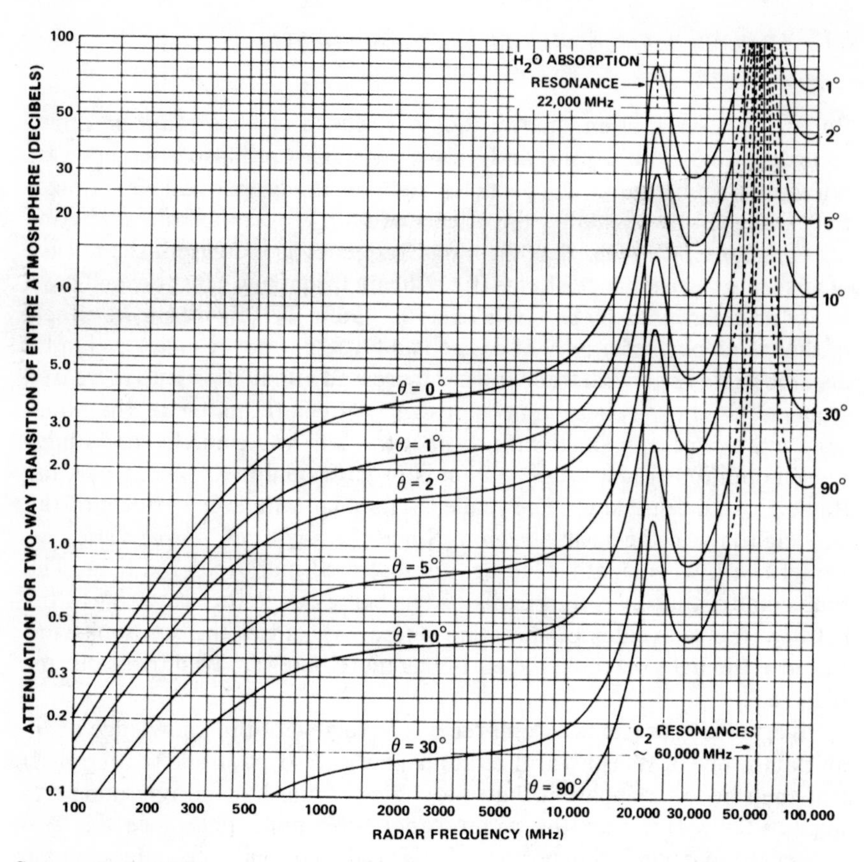

Source: Blake (1962).

Figure 2-6. Radar Attenuation for Propagation Through the Earth's Atmosphere. Ionospheric loss, which may be significant during the daytime below 500 MHz, is not included.

References

Aitchison, J. and J.A.C. Brown, *The Log-Normal Distribution,* Cambridge University Press, London, 1969.

Barrick, D.E., "A More Exact Theory of Backscattering from Statistically Rough Surfaces," Ohio State University Report 1388-18, August 1965.

Barton, D.K., *Radar System Analysis,* Prentice-Hall, Inc., Englewood Cliffs, New Jersey, 1964.

Battan, L.J., *Radar Observation of the Atmosphere,* University of Chicago Press, Chicago, Illinois, 1973.

Beckmann, Petr, *Elements of Applied Probability Theory,* Harcourt, Brace and World, Inc., New York, New York, 1967.

Bigelow, H.B. and W.T. Edmondson, *Wind Waves at Sea, Breakers and Surf,* U.S. Navy Hydrographic Office, H.O. Pub. No. 602, Washington, D.C., 1947.

Blake, L.V., "A Guide to Basic Pulse-Radar Maximum-Range Calculation Part 1—Equations, Definitions, and Aids to Calculation," Naval Research Laboratory Report 5868, December 28, 1962.

Blake, L.V., "Radio Ray (Radar) Range-Height-Angle Charts," *Microwave Journal*, vol. 11, p. 49, October 1968.

Bowditch, Nathaniel, "American Practical Navigator," U.S. Navy Hydrographic Office H.O. Publication No. 9, 1966.

Corriher, H.A., Jr., "Radar Reflectivity of Aircraft," Engineering Experiment Station, Georgia Institute of Technology, Final Report ONR Contract N00014-67-A-0159-0003, January 14, 1970.

Corriher, H.A., Jr., B.O. Pyron, R.D. Wetherington, and A.B. Abeling, *Radar Reflectivity of Sea Targets*, Engineering Experiment Station, Georgia Institute of Technology, Final Report, vol. 1, Contract Nonr-991(12), September 30, 1967.

Crane, R.K., "The Rain Range Experiment—Propagation Through a Simulated Rain Environment," *IEEE Transactions on Antennas and Propagation,* vol. AP-22, pp. 321-28, March 1974.

Crawford, A.B. and D.C. Hogg, "Measurement of Atmospheric Attenuation at Millimeter Wavelengths," Bell Telephone Monograph 2646, vol. 35, pp. 907-16, July 1956.

Daley, J.C. "An Empirical Sea Clutter Model," NRL Memorandum Report 2668, Naval Research Laboratory, Washington, D.C., October 1973.

Domville, A.R., "The Bistatic Reflection from Land and Sea of X-band Radio Waves," Part I, GEC (Electronics) Ltd. Memorandum SLM1802, July 1967.

Domville, A.R., "The Bistatic Reflection from Land and Sea of *X*-band Radio Waves," Part II, GEC (Electronics) Ltd. Memorandum SLM2116, July 1968.

Ducastel, François, *Tropospheric Radiowave Propagation Beyond the Horizon,* Pergamon Press, New York, New York, 1966.

Durlach, N.I., "Influence of the Earth's Surface on Radar," Lincoln Laboratory Technical Report 373, Massachusetts Institute of Technology, January 18, 1965.

Fairbridge, R.W. (Ed.), *The Encyclopedia of Oceanography,* Reinhold Publishing Corporation, New York, New York, 1966.

Goldstein, Herbert, "A Primer of Sea Echo," Report No. 157, U.S. Navy Electronics Laboratory, San Diego, California, 1950.

Gossard, E.E., "Radio Refraction by the Marine Layer and Its Effect on Microwave Propagation," Navy Electronics Laboratory Report 1240, September 21, 1964.

Guinard, N.W., J.T. Ransone, D.L. Randall, C.G. Purves, and P.L. Watkins, "Propagation Through An Elevated Duct: Tradewinds III," *IEEE Transactions on Antennas and Propagation,* vol. AP-12, pp. 479-90, July 1964.

Hayes, R.D., "Total Atmospheric Attenuation at Millimeter Wavelengths," Ph.D. Thesis, Georgia Institute of Technology, May 1964.

Hayes, R.D., J.R. Walsh, D.F. Eagle, H.A. Ecker, M.W. Long, J.G.B. Rivers, and C.W. Stuckey, "Study of Polarization Characteristics of Radar Targets," Engineering Experiment Station, Georgia Institute of Technology, Final Report, Contract DA-36-039-sc-64713, October 1958.

Headrick, J.M. and M.I. Skolnik, "Over-the-Horizon Radar in the HF Band," *Proceedings of the IEEE,* vol. 62, pp. 664-73, June 1974.

Hoffman, L.A., H.J. Winthroub, and W.A. Garber, "Propagation Observations at 3.2 Millimeters," *Proceedings of the IEEE,* vol. 54, pp. 449-54, April 1966.

Hunter, I.M. and T.B.A. Senior, "Experimental Studies of Sea-Surface Effects on Low-Angle Radars," *Proceedings of the IEE,* vol. 113, p. 1731, 1966.

Johnson, R.C., H.A. Ecker, and R.A. Moore, "Compact Range Techniques and Measurements," *IEEE Transactions on Antennas and Propagation,* vol. AP-17, pp. 568-76, September 1969.

Johnson, R.C., H.A. Ecker, and J.S. Hollis, "Determination of Far-Field Antenna Patterns from Near-Field Measurements," *Proceedings of the IEEE,* vol. 61, pp. 1668-94, December 1973.

Katzin, M., R.W. Bauchman, and William Binnian, "3- and 9-Centimeter Propagation in Low Ocean Ducts," *Proceedings of the IRE,* vol. 34, pp. 891-905, September 1947.

Kerr, D.E., *Propagation of Short Radio Waves,* Massachusetts Institute of Technology Radiation Laboratory Series, vol. 13, McGraw-Hill Book Company, Inc., New York, New York, 1951.

Kinsman, Blair, *Wind Waves,* Prentice-Hall, Englewood Cliffs, New Jersey, pp. 342-45, 1965.

Knott, E.F. and T.B.A. Senior, "How Far is Far?," *IEEE Transactions on Antennas and Propagation,* vol. AP-22, pp. 732-34, September 1974.

Kouyoumjian, R.G. and Leon Peters, Jr., "Range Requirements in Radar

Cross-Section Measurements," *Proceedings of the IEEE,* vol. 53, pp. 920-28, August 1965.

Lawson, J.L. and G.E. Uhlenbeck, *Threshold Signals,* Massachusetts Institute of Technology Radiation Laboratory Series, vol. 24, McGraw-Hill Book Company, Inc., New York, New York, 1950.

Long, M.W., R.D. Wetherington, J.L. Edwards, and A.B. Abeling, "Wavelength Dependence of Sea Echo," Engineering Experiment Station, Georgia Institute of Technology, Final Report, Contract N62269-3019, July 1965.

Nathanson, F.E., *Radar Design Principles,* McGraw-Hill Book Company, Inc., New York, New York, 1969.

Naval Aerology Branch of the Office of Chief of Naval Operations, *U.S. Navy Marine Climatic Atlas of the World,* vol. III, Indian Ocean, NAVAER 50-IC-530, 1957.

Norton, K.A., L.E. Vogler, W.V. Mansfield, and P.J. Short, "The Probability Distribution of the Amplitude of a Constant Vector Plus a Rayleigh-Distributed Vector," *Proceedings of the IRE,* vol. 43, pp. 1354-61, October 1955.

Pekeris, C.L., "Theoretical Interpretation of Propagation of 10-Centimeter and 3-Centimeter Waves in Low-Level Ocean Ducts," *Proceedings of the IRE,* vol. 35, pp. 453-62, May 1947.

Pierson, W.J., Jr., Gerhard Neumann, and R.W. James, *Observing and Forecasting Ocean Waves,* U.S. Navy Hydrographic Office Pub. No. 603, p. 11, Washington, D.C., 1955.

Reed, H.R. and C.M. Russell, *Ultra High Frequency Propagation,* Wiley, New York, New York, 1953.

Rice, S.O., "Mathematical Analysis of Random Noise," *Bell System Technical Journal,* vol. 23, p. 282, 1944.

Ridenour, L.N., *Radar System Engineering,* Massachusetts Institute of Technology Radiation Laboratory Series, vol. 1, McGraw-Hill Book Company, Inc., New York, New York, 1947.

Ruck, G.T., D.E. Barrick, W.D. Stuart, and C.K. Krichbaum, *Radar Cross Section Handbook,* Plenum Press, New York, New York, 1970.

Sabransky, F.A., "Investigation of Extended Over-Water Ranges of Low-Sited Radar," *Journal of Meteorology,* vol. 15, p. 303, June 1958.

Silver, Samuel, *Microwave Antenna Theory and Design,* Massachusetts Institute of Technology Radiation Laboratory Series, vol. 12, McGraw-Hill Book Company, Inc., New York, New York, 1949.

Skolnik, M.I., *Introduction to Radar Systems,* McGraw-Hill Book Company, Inc., New York, New York, 1962.

Smith, P.L., Jr., K.R. Hardy, and K.M. Glover, "Applications of Radar to

Meteorological Operations and Research," *Proceedings of the IEEE,* vol. 62, pp. 724-45, June 1974.

Thompson, W.I., III, "Atmospheric Transmission Handbook: A Survey of Electromagnetic Wave Transmission in the Earth's Atmosphere Over the Frequency Range 3 KHz-3,000 THz," Transportation Systems Center, Technical Report No. DOT-TSC-NASA-71-6, U.S. Department of Transportation, February 1971.

Undersea Technology, p. 37, May 1964.

Wait, J.R., *Electromagnetic Waves in Stratified Media,* Pergamon Press, New York, New York, 1970.

Watters, J.K.A., "Distribution in Height of Ocean Waves," *New Zealand Journal of Science and Technology,* vol. 34B, p. 408, March 1953.

3 Polarization, Depolarization and Theories of Scattering

Polarization and Depolarization

The polarization of an electromagnetic wave describes the orientation of the electric field strength vector at a given point in space during one period of oscillation. Depolarization refers to the change in polarization that an electromagnetic wave undergoes as a consequence of propagation, reflection, scattering, or diffraction. A radar wave is always polarized even though the electric vector may fluctuate rapidly and at random.

There are a number of good books that cover the subject of polarization,[a] but Petr Beckmann (1968) has authored the only book devoted exclusively to depolarization. D.B. Kanareykin, W.F. Pavlov, and V.A. Potekhin (1966) have written a book (in Russian) on the polarization of radar signals.

3.1 Polarization Scattering Matrix

George Sinclair (1948) introduced the polarization scattering matrix to the radar literature in 1948. Other references on this subject include E.M. Kennaugh (1952), C.D. Graves (1956), J.R. Copeland (1960), and M.W. Long (1966). The matrix provides a convenient method for displaying the multiple polarization properties of a radar target or a collection of targets. In other words, scattering is completely described for an electromagnetic target if all phases and amplitude terms of the matrix are specified.

When a target is illuminated by a conventional linear polarization radar, information obtained is limited to the scattering properties of the target for the polarization that is transmitted and received. The general situation includes different polarizations for transmitting and for receiving. To illustrate, suppose that the transmitted polarization is in direction y and the received polarization is in direction x. Assume that these directions are arbitrary except, of course, for the fact that they are perpendicular to the direction of propagation. We will later assume that x and y are also perpendicular to one another. Let E_x^t represent the electric field strength of the plane wave incident on a target and let E_y^r represent the reflected (re-

[a]See, for example, J.D. Kraus (1950).

radiated) electric field at the receiving antenna. Symbolically, these fields can be related to one another by a reflection coefficient[b] a_{yx} as follows:

$$E^r_{x_1} = a_{yx}\, E^t_y \tag{3.1}$$

Equation (3.1) can also be thought of as a relationship between transmitted and received electric fields, where range dependence either has been suppressed or is included in a_{yx}. The factor a_{yx} is complex (contains phase factors) because the process of reflection involves, in general, changes in phase as well as amplitude. The complete description of a given scattering process includes a complete exposition of the polarization properties. The following equations, in addition to (3.1), contain information about a target

$$E^r_{x_2} = a_{xx}\, E^t_x \tag{3.2}$$

$$E^r_{y_1} = a_{xy}\, E^t_x \tag{3.3}$$

$$E^r_{y_2} = a_{yy}\, E^t_y \tag{3.4}$$

The polarization of any plane wave can be described in terms of two orthogonal, linearly polarized components. If the polarization is linear, these components are in time phase; if the polarization is elliptical, the components will be out of time phase. Thus, for the general case of transmitting an arbitrary polarization, the reflected field can be described by combining equations (3.1) through (3.4) as follows:

$$\begin{aligned} E^r_x &= E^r_{x_1} + E^r_{x_2} = a_{xx}\, E^t_x + a_{yx}\, E^t_y \\ E^r_y &= E^r_{y_1} + E^r_{y_2} = a_{xy}\, E^t_x + a_{yy}\, E^t_y \end{aligned} \tag{3.5}$$

The matrix representation of equations (3.5) is

$$\begin{bmatrix} E^r_x \\ E^r_y \end{bmatrix} = \begin{bmatrix} a_{xx} & a_{yx} \\ a_{xy} & a_{yy} \end{bmatrix} \begin{bmatrix} E^t_x \\ E^t_y \end{bmatrix} \tag{3.6}$$

The two by two matrix is known as the polarization scattering matrix, or simply polarization matrix, and is represented by the symbol A. It may be noted that the order of subscripts is opposite to that customarily used with matrices so as to coincide with the practice in radar of first and second subscripts denoting transmit and receive polarizations, respectively.

Equation (3.6) contains neither more nor less information than is contained in equations (3.5). The important thing to notice is that for a static situation all polarization properties of a target are contained in the matrix. One must recognize that scattered energy depends on such parameters as transmitter frequency, target orientation, and proximity of target with

[b]The first subscript denotes transmitted component and the second subscript denotes received component.

other objects. Therefore, the magnitudes and relative phases of the a's are often variables, even for a given target.

In general, a_{xy} and a_{yx} are not equal. However, R.S. Berkowitz (1965) shows that for conventional monostatic radar (backscatter) the matrix A is symmetric; that is, $a_{xy} = a_{yx}$. The plausibility of this type of reciprocity can be visualized by mentally interchanging the transmitting and receiving roles with two co-located antennas.

The polarization of an electromagnetic wave can be completely characterized by two orthogonal components which are, in general, elliptical. Linear and circular polarizations are two special cases of elliptical polarization and are the most widely used for reference systems. In the case of a circular reference system, the matrix equation may be expressed as

$$\begin{bmatrix} E_1^r \\ E_2^r \end{bmatrix} = \begin{bmatrix} c_{11} & c_{21} \\ c_{12} & c_{22} \end{bmatrix} \begin{bmatrix} E_1^t \\ E_2^t \end{bmatrix} \tag{3.7}$$

E_1 and E_2 are orthogonal circular components of the electric field in the plane perpendicular to the direction of propagation. As for linear polarization, the matrix is symmetric ($c_{12} = c_{21}$) for the case of backscatter.

The matrix coefficients c_{ij} can be expressed in terms of the a_{ij} if the transmitted and received circularly polarized waves are expressed in terms of the respective linearly polarized waves (Long 1966). From equation (3.7), for the conditions E_1^t and E_2^t separately equaling zero, the c_{ij} can be obtained in terms of E_1^t, E_2^t, E_1^r, and E_2^r. For the equations that follow, E_1^t is defined as the circularly polarized electric field that exists if $E_y^t = jE_x^t$ (in this case E_2^t is zero), and E_2^t is defined as the circularly polarized electric field that exists if $E_y^t = -jE_x^t$ (in this case E_1^t is zero); $|E_1^t|$ and $|E_2^t|$ are equal to $\sqrt{2}|E_x|$. From reciprocity, a bilateral antenna receives the same polarization that it radiates. Therefore, upon reception, the antenna outputs are proportional to

$$E_1^r = \frac{E_x^r + jE_y^r}{\sqrt{2}} \quad \text{and} \quad E_2^r = \frac{E_x^r - jE_y^r}{\sqrt{2}}$$

E_x^r and E_y^r (for E_1^t and E_2^t each equaling zero) can be obtained from equation (3.6). One finally sees that

$$|c_{11}| = \left| \frac{a_{xx} - a_{yy}}{2} + ja_{xy} \right|$$

$$|c_{12}| = \left| \frac{a_{xx} + a_{yy}}{2} \right| \tag{3.8}$$

and

$$|c_{22}| = \left| \frac{a_{xx} - a_{yy}}{2} - ja_{xy} \right|$$

For circular polarization, rotation sense is defined as that seen when looking in the direction of propagation. Echoes are often described simply as "same" or "opposite"; this means that the terminus of the electric vector is rotating in either the same or opposite direction as that of the transmitted wave. To discuss data for horizontal and vertical polarizations (denoted later by H and V), assume that the x and y directions for equations (3.8) point to the right and upward, respectively. Then, the definitions of E_1^t and E_2^t result in directions 1 and 2 being left circular and right circular respectively.

3.2 Relationships between Linear and Circular Polarizations

As previously stated, the a's and c's are complex quantities that are proportional to electric fields. To equate the various radar cross sections (proportional to electric field squared), care must be taken to account for phase appropriately in equations (3.8). It should be apparent to the reader that $\sigma_{ij} = K|a_{ij}|^2$ and $\sigma_{ij}/\sigma_{kl} = |a_{ij}|^2/|a_{kl}|^2$. Therefore,

$$\sigma_{11} = K \left| \frac{a_{xx} - a_{yy}}{2} + ja_{xy} \right|^2$$

$$\sigma_{12} = K \left| \frac{a_{xx} + a_{yy}}{2} \right|^2 \qquad (3.9)$$

and

$$\sigma_{22} = K \left| \frac{a_{xx} - a_{yy}}{2} - ja_{xy} \right|^2$$

For a perfectly conducting surface having radii of curvature that are large compared to a wavelength, one would expect that a_{xx} is equal to a_{yy}. This means that at all times the phase and amplitude change upon reflection is the same for both polarizations. Then, equations (3.9) would indicate that

$$\sigma_{11} = \sigma_{22} = \sigma_{xy}$$

and (3.10)

$$\sigma_{12} = \sigma_{xx} = \sigma_{yy}$$

One expects that σ_{xy} would be small for a large, smooth scatterer. Edges might produce some depolarized echo—surfaces with radii small compared to a wavelength, such as sharp edges, are expected to be the major contributors to σ_{xy}.

Radar cross section σ for any combination of transmitted and received polarizations can be expressed in terms of the a's. Horizontal and vertical polarizations will be denoted by H and V, respectively; left and right circular polarizations will be denoted by L and R, respectively. In terms of the a's, radar cross section for transmitting circular and receiving horizontal and vertical polarizations can be expressed (Long 1967) as

$$\begin{aligned} \sigma_{LH} &= K\frac{|a_{HH} + ja_{HV}|^2}{2} \\ \sigma_{LV} &= K\frac{|a_{HV} + ja_{VV}|^2}{2} \\ \sigma_{RH} &= K\frac{|a_{HH} - ja_{HV}|^2}{2} \\ \sigma_{RV} &= K\frac{|a_{HV} - ja_{VV}|^2}{2} \end{aligned} \tag{3.11}$$

In general, none of the σ's are equal on an instantaneous basis because the phase and amplitude of each of the a's are fluctuating functions of time.

For computing average values, certain simplifications can be made for targets for which the a's are statistically independent of one another. Let a bar denote a time average. If a_{HH}, a_{VV}, and a_{HV} are statistically independent (see sec. 7.2) of one another, then (from eqs. (3.8)) average cross section can be expressed as

$$\overline{\sigma}_{11} = \overline{\sigma}_{22} = \frac{\overline{\sigma}_{HH}}{4} + \frac{\overline{\sigma}_{VV}}{4} + \overline{\sigma}_{HV}$$

and (3.12)

$$\overline{\sigma}_{12} = \overline{\sigma}_{21} = \frac{\overline{\sigma}_{HH}}{4} + \frac{\overline{\sigma}_{VV}}{4}$$

For scatterers of this type, it is obvious that σ_{11} would exceed σ_{12}.

Simplifications also can be made for equations (3.11) if the a's are statistically independent. In this case, average cross sections may be expressed as

$$\overline{\sigma}_{LH} = \overline{\sigma}_{RH} = \frac{\overline{\sigma}_{HH} + \overline{\sigma}_{HV}}{2}$$

and (3.13)

$$\overline{\sigma}_{LV} = \overline{\sigma}_{RV} = \frac{\overline{\sigma}_{VV} + \overline{\sigma}_{HV}}{2}$$

I.M. Hunter and T.B.A. Senior (1966) give results obtained at X band for transmitting right circular and receiving H and V polarizations. They reported that for a limited series of measurements on sea echo the ratio $\overline{\sigma}_{RV}/\overline{\sigma}_{RH}$ was essentially the same as $\overline{\sigma}_{VV}/\overline{\sigma}_{HH}$. They then made an extensive series of simultaneous measurements of $\overline{\sigma}_{RV}$ and $\overline{\sigma}_{RH}$ in order to obtain relative magnitudes for $\overline{\sigma}_{VV}$ and $\overline{\sigma}_{HH}$. As may be seen from equations (3.13),

$$\overline{\sigma}_{VV} = \overline{\sigma}_{HH} \quad \text{if} \quad \overline{\sigma}_{RV} = \overline{\sigma}_{RH}$$

$$\overline{\sigma}_{VV} > \overline{\sigma}_{HH} \quad \text{if} \quad \overline{\sigma}_{RV} > \overline{\sigma}_{RH} \tag{3.14}$$

and

$$\overline{\sigma}_{VV} < \overline{\sigma}_{HH} \quad \text{if} \quad \overline{\sigma}_{RV} < \overline{\sigma}_{RH}$$

Thus it is clear that $\overline{\sigma}_{RV}/\overline{\sigma}_{RH}$ is not, in principle, equal to $\overline{\sigma}_{VV}/\overline{\sigma}_{HH}$. However, as discussed elsewhere (Long 1967), the technique of measuring the ratio $\overline{\sigma}_{RV}/\overline{\sigma}_{RH}$ is a valid means for determining whether $\overline{\sigma}_{VV}$ is less than, equal to, or greater than $\overline{\sigma}_{HH}$—providing that the two terms in each pair ((a_{HH}, a_{HV}) and (a_{VV} a_{VH})) are statistically independent.

3.3 A Randomly Oriented Dipole

Assume that the thin, infinitesimal wire illustrated in figure 3-1 is constrained to lie in the X-Y plane. It will be shown below that the radar cross section for the case of a plane wave incident from the negative Z direction is a function of β for linear polarization, but independent of β for circular polarization.

For such a dipole, the current induced is along the length of the wire and is proportional to the projection of the electric field along the wire. Therefore, the reradiated field is proportional to the current and polarized along the wire. From above, it may then be seen that

$$a_{xx} = \cos^2\beta \qquad a_{yy} = \sin^2\beta$$

$$a_{xy} = \sin\beta\,\cos\beta \qquad a_{yx} = \sin\beta\,\cos\beta$$

$$A = \begin{bmatrix} a_{xx} & a_{yx} \\ a_{xy} & a_{yy} \end{bmatrix} = \begin{bmatrix} \cos^2\beta & \sin\beta\,\cos\beta \\ \sin\beta\,\cos\beta & \sin^2\beta \end{bmatrix}$$

Notice that if the dipole is either horizontal or vertical ($\beta = 0°$ or $\beta =$

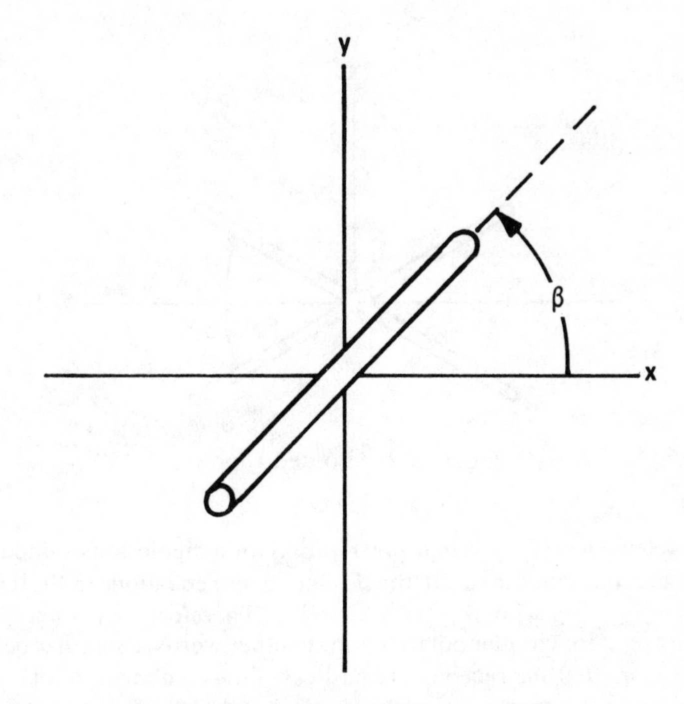

Figure 3-1. A Dipole in the *X-Y* Plane.

90°), the diagonal terms are zero. Under these conditions there is not a cross-polarized component in the reflected field, if the incident field is either vertically or horizontally polarized. In general terms, if a target has symmetry, a set of orthogonal polarization axes can be established to reduce the matrix to diagonal form; that is

$$A = \begin{bmatrix} a_{xx} & 0 \\ 0 & a_{yy} \end{bmatrix}$$

To illustrate, consider the two crossed, thin conductors in figure 3-2. Assume that horizontal polarization is transmitted. Because the crossed wires are symmetrical about the horizontal axis and the line-of-sight to the target, there must be an upward component of vertical reflection equal to a downward component. This would produce a zero resultant in the vertical component; thus, $a_{HV} = a_{VH} = 0$.

Since radar cross section is proportional to the square of received electric field (square of the a's),

$$\sigma_{xx} = \sigma_{\max} \cos^4 \beta \qquad \sigma_{yy} = \sigma_{\max} \sin^4 \beta$$

$$\sigma_{xy} = \sigma_{yx} = \sigma_{\max} \sin^2 \beta \cos^2 \beta$$

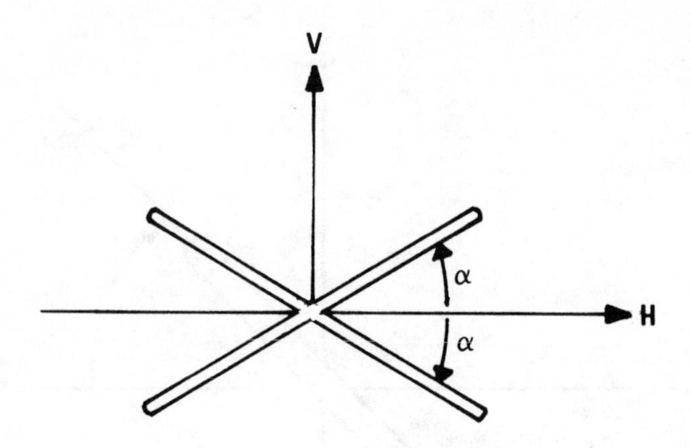

Figure 3-2. Crossed Dipoles.

To determine σ for circular polarization for a dipole constrained to the $X-Y$ plane, one can substitute the a's above into equations (3.8). It is then seen that $|c_{11}| = |c_{22}| = |c_{12}| = |c_{21}| = 1/2$. Therefore, $\sigma_{11} = \sigma_{22} = \sigma_{12} = \sigma_{21} = 1/4\,\sigma_{max}$ for circular polarization. In other words, if circular polarization is transmitted and received, regardless of the combinations of circularity used, the radar cross section for an infinitesimal dipole is one-fourth the maximum value for linear polarization.

The reader can readily visualize why the cross section for circular polarization is independent of β by recalling that a circularly polarized wave can be resolved into two orthogonal, linearly polarized waves. Since the wire "sees" only the linearly polarized component that is polarized along the wire, only half the incident power is scattered by a wire. Further, the reradiated wave is linearly polarized and one circularly polarized component (of the two comprising the reradiated linearly polarized wave), or half the scattered power, is received. Therefore, for transmitting and receiving circular polarizations, received power is one-fourth that which would be received for maximized conditions with linear polarizations.

3.4 A Diplane Reflector

A dihedral corner or diplane reflector consists of two plane surfaces at right angles. The change of polarization caused by a reversal in the direction of E_y, caused by reflection from a perfect conductor, is illustrated in figure 3-3. Since the same path length exists for both polarizations, there is a 180° change in relative phase between the vertically polarized component (y direction) and the horizontally polarized component (x direction).

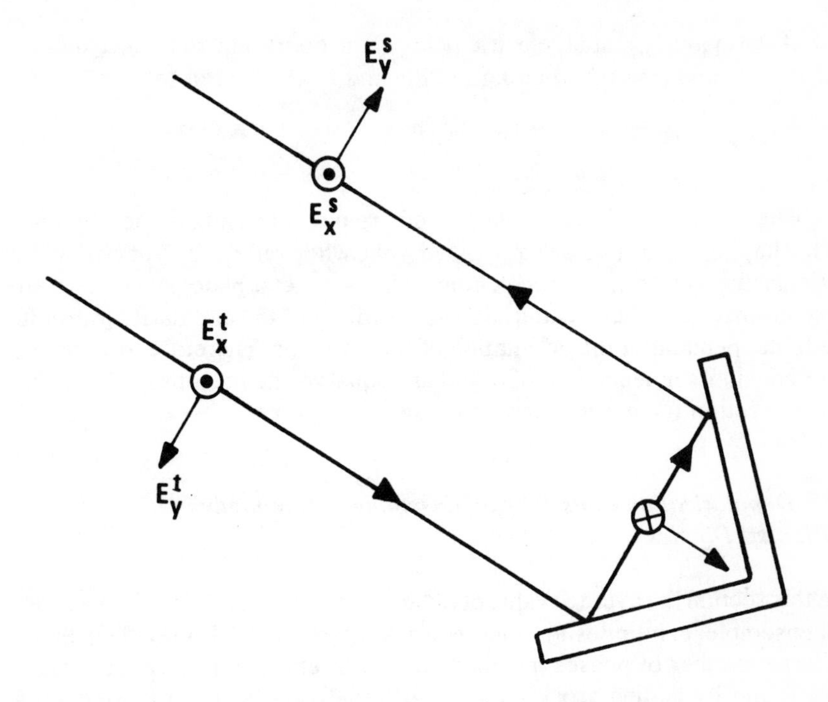

Figure 3-3. Polarization Change Caused by a Diplane.

By resolving the E^s into horizontal and vertical components, it can be shown that the polarization matrix for linear polarization is

$$A = \begin{bmatrix} a_{xx} & a_{yx} \\ a_{xy} & a_{yy} \end{bmatrix} = \begin{bmatrix} \cos 2\alpha & \sin 2\alpha \\ \sin 2\alpha & -\cos 2\alpha \end{bmatrix} \tag{3.15}$$

Notice that if the seam (symmetry axis) is either horizontal or vertical ($\alpha = 0°$ or $\alpha = 90°$), the diagonal terms are zero and there is not a cross-polarized component in the reflected field if the incident field is either vertically or horizontally polarized.

The a's from equation (3.15) can be substituted into equations (3.8) to determine radar cross section for circular polarization. This yields

$$|c_{11}| = \left| \frac{a_{xx} - a_{yy}}{2} + ja_{xy} \right| = |\cos 2\alpha + j \sin 2\alpha| = 1$$

$$|c_{12}| = \left| \frac{a_{xx} + a_{yy}}{2} \right| = 0$$

$$|c_{22}| = \left| \frac{a_{xx} - a_{yy}}{2} - ja_{xy} \right| = |\cos 2\alpha - j \sin 2\alpha| = 1$$

As previously stated, electric fields are proportional to the magnitudes of the a's and c's. By squaring the a's and c's, it is seen that

$$\sigma_{HH} = \sigma_{VV} = \sigma_{\max}\cos^2 2\alpha \qquad \sigma_{11} = \sigma_{22} = \sigma_{\max}$$

$$\sigma_{HV} = \sigma_{VH} = \sigma_{\max}\sin^2 2\alpha \qquad \sigma_{12} = \sigma_{21} = 0$$

The results for circular polarization are not surprising if one examines what happens to a circularly polarized wave when reflected. Notice that the circularity is reversed for each bounce from a metal plate; since there are two bounces the back-scattered field is entirely of the circularity transmitted, independent of the orientation of the reflector. Therefore, σ_{11} and σ_{22} are constants independent of α and are equal to the maximum obtainable cross section for linear polarization and σ_{12} is zero.

3.5 Depolarization Caused by an Ensemble of Randomly Oriented Dipoles

In this section the average value of radar cross section for a large collection, or ensemble, of infinitesimal dipoles is taken from M.W. Long (1965). Since a large number of phases (positions of scatterers) are involved, averages are found by adding cross sections. Alternatively, since any orientation angle is equally as likely as any other, average radar cross section can be obtained by multiplying the number of dipoles by average cross section for one dipole.

Consider an infinitesimal dipole oriented along OA of figure 3-4. Assume that a plane wave propagating in the negative z direction is incident on the dipole and that a wave scattered in the positive z direction is received.

Let $\sigma_{xx}(\alpha,\beta)$ represent the radar cross section of the dipole OA for a transmitted wave linearly polarized along the x axis and a received wave also linearly polarized along the x axis. A dipole reradiates that component of incident field parallel to its axis and a receiver of linear polarization accepts only that component of field parallel to its direction of polarization. Recall that cross section is proportional to received power and that power is proportional to the square of the amplitude of the field. Then,

$$\sigma_{xx}(\alpha,\beta) = \sigma_{\max}\sin^4\alpha\cos^4\beta \tag{3.16}$$

where $\sigma_{\max} = \sigma_{xx}(\pi/2, 0)$ is the radar cross section of the dipole when oriented along the x axis.

Similarly, let σ_{xy} represent the radar cross section of the dipole OA for a transmitted wave linearly polarized along the y axis. Then,

$$\sigma_{xy}(\alpha,\beta) = \sigma_{\max}\sin^4\alpha\sin^2\beta\cos^2\beta \tag{3.17}$$

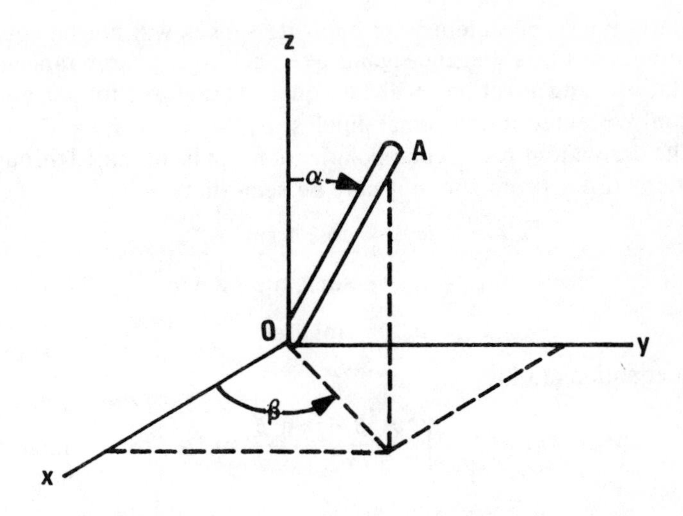

Figure 3-4. Scattering Geometry.

To calculate the ratio of these cross sections for a large collection of randomly oriented dipoles, consider the average of each of these cross sections over all possible orientations (α, β). Let a bar denote those averages, then if any orientation is as likely as any other

$$\frac{\overline{\sigma_{xx}}}{\overline{\sigma_{xy}}} = \frac{\int_{\beta=0}^{2\pi} \int_{\alpha=0}^{\pi} \sigma_{xx}(\alpha, \beta) \sin\alpha \, d\alpha \, d\beta}{\int_{\beta=0}^{2\pi} \int_{\alpha=0}^{\pi} \sigma_{xy}(\alpha, \beta) \sin\alpha \, d\alpha \, d\beta}$$

$$= \frac{\int_{\beta=0}^{2\pi} \cos^4 \beta \, d\beta \int_{\alpha=0}^{\pi} \sin^5 \alpha \, d\alpha}{\int_{\beta=0}^{2\pi} \sin^2 \beta \cos^2 \beta \, d\beta \int_{\alpha=0}^{\pi} \sin^5 \alpha \, d\alpha} = 3$$

Therefore, neglecting possible effects of multiple scattering, the ratio $\overline{\sigma}_{xx}/\overline{\sigma}_{xy}$ for an ensemble of independent, randomly oriented infinitesimal dipoles is expected to be 3 or 4.8 dB.

The ratio of averages is unity for receiving both circular polarizations $(\overline{\sigma}_{11}, \overline{\sigma}_{12})$ if a circular is transmitted. Recall that for a single dipole the back-scattered field is linearly polarized along the dipole, independent of β (see figure 3-1). For an ensemble of dipoles with random and variable positions, the resultant of the sum of linearly polarized waves will produce an elliptically polarized back-scattered field. The signals received by two circularly polarized antennas of opposite senses will not be equal on an instantaneous basis, but the signals averaged over a large range of dipole orientations and positions will be equal. Therefore, for an esemble of randomly oriented infinitesimal dipoles, $\overline{\sigma}_{11} = \overline{\sigma}_{22} = \overline{\sigma}_{12} = \overline{\sigma}_{21}$.

The derivations for circular polarization can be obtained through use of equations (3.8). From above it may be seen that

$$a_{xx} = \cos^2\beta \sin^2\alpha$$

$$a_{yx} = a_{xy} = \sin\beta \cos\beta \sin^2\alpha$$

$$a_{yy} = \sin^2\beta \sin^2\alpha$$

From equations (3.8)

$$|c_{12}| = |c_{21}| = \left| \frac{\cos^2\beta + \sin^2\beta}{2} \sin^2\alpha \right| = \frac{1}{2} \sin^2\alpha$$

and

$$|c_{11}| = |c_{22}| = \left| \frac{\cos^2\beta - \sin^2\beta}{2} + j \sin\beta \cos\beta \right| (\sin^2\alpha) = \frac{1}{2} \sin^2\alpha$$

Therefore, $\sigma_{11} = \sigma_{22} = \sigma_{12} = \sigma_{22}$.

By using the relationships developed for linear polarization and recognizing that

$$\frac{1}{2\pi} \int_0^{2\pi} \cos^4\alpha \, d\alpha = \frac{3}{8}$$

and

$$\frac{1}{2\pi} \int_0^{2\pi} \sin^2\alpha \cos^2\alpha \, d\alpha = \frac{1}{8}$$

it may be seen that

$$\overline{\sigma}_{11} = \overline{\sigma}_{22} = \overline{\sigma}_{12} = \overline{\sigma}_{21} = (2/3)(\overline{\sigma}_{xx}) = (2/3)(\overline{\sigma}_{yy}) = 2\overline{\sigma}_{xy} = 2\overline{\sigma}_{yx}$$

for an ensemble of independent, randomly oriented infinitesimal dipoles.

Theories for Radar Cross Section of Rough Surfaces

Several theoretical concepts that are useful for discussing echo from land and sea are reviewed in sections 3.6 through 3.15. The reader is referred to Petr Beckmann and André Spizzichino (1963) and G.T. Ruck et al. (1970) for comprehensive treatments of theories for electromagnetic scattering.

3.6 Simplified Models

Early radar theories assumed that for a uniformly rough area, $\sigma^\circ(\theta) = \sigma^\circ(0) \sin^2 \theta$. According to elementary texts in optics, the power scattered at an angle θ from a diffuse surface, such as a blotter or freshly fallen snow, is proportional to $\sin \theta$. This is known as Lambert's law. Since power density incident on a planar surface is also proportional to $\sin \theta$, Lambert's law for backscatter is $\sigma^\circ = \sigma^\circ(0) \sin^2 \theta$. Based on theoretical considerations, W.S. Ament (1956) determined that $\sigma^\circ = 4 \sin^2 \theta$ for a diffuse Lambert surface.

R.E. Clapp (1946) was probably the first person to report on the Lambert functional relationship for radar. He rejected it on the basis that it did not fit his data. A dependence of the form $\sigma^\circ(\theta) = \sigma^\circ(0) \sin \theta$ is usually found to be a reasonably good fit over a wide range (10° to 80°) of θ.

The second model considered by Clapp was composed of a single layer of spheres (N per unit area) each with a scattering cross section A_k, giving

$$\sigma^\circ = NA_k \qquad \text{or} \qquad \gamma = NA_k/\sin\theta \tag{3.18}$$

where

$$\gamma = \sigma^\circ/\sin\theta$$

The terminology γ was used by R.L. Cosgriff, W.H. Peake, and R.C. Taylor (1960). Clapp also rejected this model because it was not consistent with his measurement results. However, as pointed out by Cosgriff, Peake, and Taylor (1960), the model could apply to certain types of vegetation, such as a field of oats in head where the heads of grain might represent a single layer of rather large scatterers (see figures 6-12 and 6-13 for grass in head).

In his third model, Clapp represented the ground by many layers of spheres, each of which absorbed a fraction $(1 - \delta)$ of the energy; the remainder was reradiated isotropically. Then

$$\sigma^\circ = \delta \sin\theta \qquad \gamma = \delta$$

where δ is a constant. This model agreed with Clapp's measurements and with many subsequent measurements of the return from vegetation-covered terrain. The data in figures 3-5 and 3-6 fit this model closely.

A.H. Schooley (1956) presented results of an analysis showing some limiting values of σ° versus θ for perfectly smooth and perfectly rough surfaces. He states that the words *perfectly rough* are ambiguous for a surface because roughness is a matter of definition. Schooley considered the rough surface to be one which the incident energy is reradiated uniformly over 2π steradians.

Let the radius of a pencil beam be a. If the rough surface were replaced

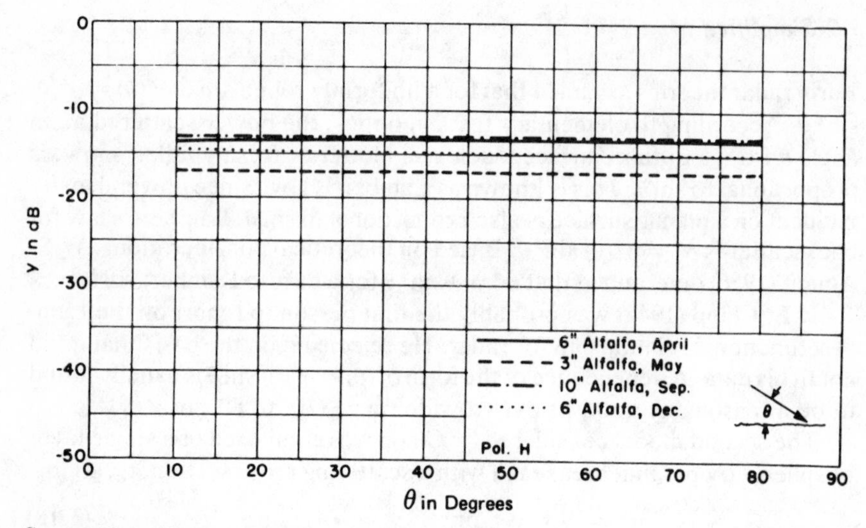

Source: Cosgriff, Peake, and Taylor (1960).

Figure 3-5. Seasonal Changes of Alfalfa at K_a Band, *HH* Polarization.

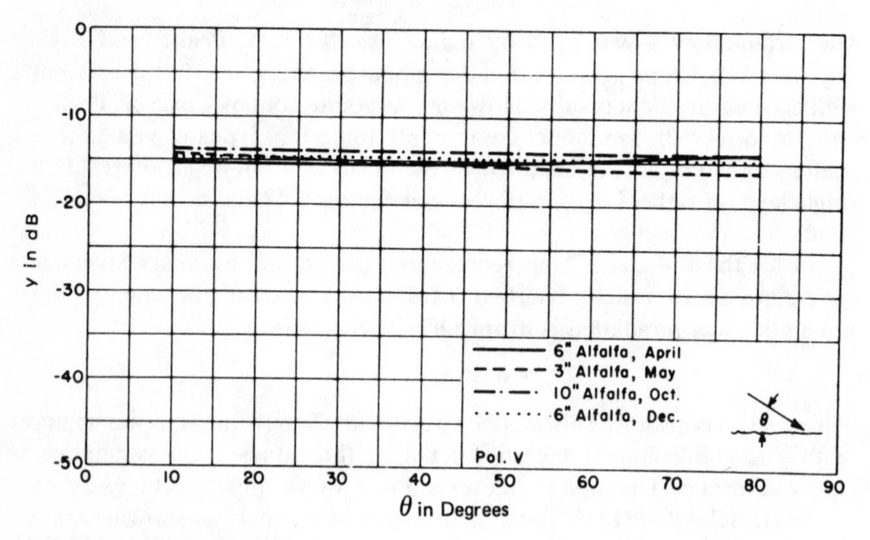

Source: Cosgriff, Peake, and Taylor (1960).

Figure 3-6. Seasonal Changes of Alfalfa at K_a Band, *VV* Polarization.

by a sphere of radius a ($\sigma = \pi a^2$), the energy would be scattered uniformly in all directions (4π steradians). Therefore, the cross section of the rough surface is twice that of the sphere or $\sigma_s = 2\pi a^2$. According to Schooley, the surface area illuminated by the beam is an ellipse of area A $= \pi ab$, where $b = a/\sin\theta$. Therefore, $\sigma°$ for Schooley's rough surface is

$$\sigma° = \sigma_s / A = 2 \sin\theta$$

It is often found (figures 3-5 and 3-6 are examples) that the relationship

$$\sigma°(\theta) = \sigma°(0) \sin\theta \tag{3.19}$$

fits data over a wide range of θ, but generally the measured values of $\sigma°(0)$ are considerably less than 2.

3.7 Classical Interference Theory

Sometimes the behavior of sea echo at small grazing angles is explained on the basis of destructive interference between the direct and reflected electromagnetic fields that illuminate the scatterers (secs. 4.16 and 6.6). The so-called "critical angle" separates the near-grazing and plateau regions (sec. 6.15), and it is an angle (within a range of angles) for which destructive interference ceases to be a dominant factor.

The critical angle depends on several factors. Since the effect of destructive interference is usually more pronounced for *HH* than for *VV* polarization, the critical angle is usually different for *HH* and *VV* polarizations. The critical angle tends to decrease as the roughness of the sea increases; also, the critical angle is larger for longer wavelengths.

Herbert Goldstein (Kerr 1951, pp. 522-26) reported that the observed polarization dependence of $\sigma°$ can be explained by an interference phenomenon. He considered the scatterers causing the echo to be spray droplets thrown upwards from the waves. However, Goldstein knew that droplets could not be the principal scatterers because their cross section is too small (Goldstein 1946).

Martin Katzin (1957) also used the multipath interference effect to explain sea echo, but for his model the scatterers are facets that overlie the large-scale wave pattern or swell. In Katzin's model the sea is considered to be the superposition of sets of smooth surfaces of various sizes, with orientations distributed about the mean-sea contour (swell).

For this *classical* interference effect to exist, the reflecting surface must be smooth enough to produce a significant amount of forwardly scattered power (see eq. 4.29). Therefore, this effect cannot account for polarization dependencies at depression angles for which the sea is rough in terms of the Rayleigh criterion. However, the "local" interference effect (sec. 4.18) has been used to account for polarization dependencies under rough sea conditions (see sec. 6.17).

3.8 The Tangent Plane Approximation

The tangent plane approximation is a standard method for relating incident and scattered fields at a surface. With the tangent plane approximation, it is assumed that the current flowing at each point on a surface is the same as would flow (at each point) if the surface were flat and tangent to the actual surface. Then, in order to represent scattering from a surface, scattered fields are constructed by adding the complex (amplitudes and phases) fields with phase distributions set by the distances of the individual surface elements from the radar. The tangent plane approximation is restricted to surfaces having radii of curvature large compared to a wavelength and, therefore, models based on this approximation are not representative of most natural surfaces. Because of the large radii of curvature, scattering from models based on the approximation does not have a depolarized component, but actual scattering can be depolarized even for a surface with irregularities that are small compared to wavelength.

A surface is called "slightly rough" if the rms height of roughness Δh is such that $\Delta h \sin \theta << \lambda$. At the opposite extreme, the surface is called "very rough" if $\Delta h \sin \theta >> \lambda$. There have been a number of efforts to explain the reradiation from slightly rough surfaces by using the tangent plans approximation (Beckmann and Spizzichino 1963), and it seems that the first such published analysis was by H. Davies (1954). A requirement of the tangent plane approximation (that surfaces have radii of curvature large compared to a wavelength) plus the definition for a slightly rough surface describes a smooth undulating surface with hills much less than a wavelength in height and separated by hundreds of wavelengths. Clearly, such a surface is not typical of most terrain.

3.9 Very Rough Surfaces

Rough surfaces are usually synthesized by adding many individual surfaces that satisfy the tangent plane approximation. For most cases of practical interest, the rms roughness height will be much larger than a wavelength. In those cases, the scattered power from a synthesized surface is incoherent; that is, the phase angle of the scattered field is uniformly distributed between 0 and 2π. The average cross section of the surface is therefore equal to the sum of the cross sections of the individual surfaces.

D.E. Barrick and W.H. Peake (1967) grouped the various theoretical approaches on rough surfaces into three classes (physical optics, ray optics, and geometrical optics or stationary phase). They state that the results obtainable from the three approaches are similar. In these approaches, the backscattering comes only from surfaces that are oriented normal to the incident beam; in other words, the average RCS is proportional to the probability of all surface tangents being normal to the incident beam. Therefore, the models yield backscattering predictions that show no depolarization and yield the same cross section for *HH* and *VV* polarizations.

The restrictions on the various methods for rough surfaces are such that multiple scattering and shadowing are neglected. Models for very rough surfaces give estimates for σ that are useful for angles near vertical incidence (Barrick 1974; Beckmann 1973; Beckmann and Spizzichino 1963, chap. 18; Ruck et al. 1970, chap. 9).

3.10 The Facet Model

In the facet model the earth is represented by many small segments (usually flat surfaces) called facets. The commonly observed glitter of sunlight reflected from rippled water is caused by facets. Martin Katzin (1957) used the dependence of reradiation patterns on hypothesized facet size to develop a model for the sea. Isadore Katz and L.M. Spetner (1958, 1960) used wavelength dependence of RCS as a parameter for establishing the number of facets required to simulate terrain. For facets that are large with respect to wavelength, most of the return occurs at normal incidence; whereas for small facets, the reradiation is almost insensitive to orientation. For example, the glitter of sunlight from water is very directional because the lateral dimensions of the facets are large with respect to an optical wavelength.

The meaning of facet size can be ambiguous because natural surfaces are not perfectly flat. A.H. Schooley (1955) reported measured curvature for sea wave facets in order to quantify the amount of departure from flatness. The RCS of a curved surface can be less sensitive to changes in wavelength than a flat one. For example, D.E. Kerr (1951, pp. 463-68) shows that the RCS for quartic surfaces (with departures of flatness exceeding several wavelengths) is independent of wavelength.

3.11 The Slightly Rough Planar Surface

A number of developments have evolved from the rigorous mathematical treatment of scattering formulated by S.O. Rice (1951). W.H. Peake (1959) extended Rice's formulation so as to compute radar cross section. Rice's formulation, which does not use the tangent plane approximation, is a boundary perturbation approach based on an expansion of both the surface and the scattered fields into Fourier series. The results of Peake are valid at arbitrary scattering angles. However, a specular component is not explicitly contained in the solution; therefore, near the specular direction a strong, reflected field must be added to the solution.

For back scattering, according to Peake, the average scattering cross section per unit area can be expressed as

$$\sigma^{\circ}_{ij} = 4\pi(2\pi/\lambda)^4 \sin^4\theta \, |T_{ij}|^2 \, W \tag{3.20}$$

where i and j refer to the incident and scattered polarizations. The terms T_{ij} and W depend on incidence angle; but if θ is sufficiently small, the $\sin^4\theta$ dependence dominates as is the case for the interference theory (sec. 3.7). The T_{ij} also depend on the electrical (dielectric and magnetic) properties of the surface and are, therefore, sensitive to polarization and incidence angle. W depends on surface roughness, wavelength, and incidence angle; it is said to be the "roughness height spectral density" of the surface. A number of investigators, other than Peake, have studied W for both land and sea (Barrick and Peake, 1968; Bass et al., 1968). Some of the results of Wright are outlined in section 3.12.

Within the approximations of a slightly rough surface, the mathematical solution of Peake for σ is precise in that it does not neglect effects of shadowing and multiple scattering as does the tangent plane approximation. This means that the solution of Peake should be valid for small grazing angles for which effects of shadowing and multiple scattering can be appreciable, S.O. Rice (1951) and W.H. Peake (1959) considered only nonmagnetic surfaces ($\mu_r = 1$), but D.E. Barrick and W.H. Peake (1967) extended the derivations for any value of relative permeability μ_r. For horizontal and vertical polarizations, the T_{ij} given by Barrick and Peake are

$$\begin{aligned} T_{HH} &= \frac{(\mu_r - 1)[(\mu_r - 1)\sin^2\theta + \varepsilon_r\mu_r] - \mu_r^2(\varepsilon_r - 1)}{[\mu_r \cos\theta + \sqrt{\varepsilon_r\mu_r - \sin^2\theta}]^2} \\ T_{VV} &= \frac{(\varepsilon_r - 1)[(\varepsilon_r - 1)\sin^2\theta + \varepsilon_r\mu_r] - \varepsilon_r^2(\mu_r - 1)}{[\varepsilon_r \cos\theta + \sqrt{\varepsilon_r\mu_r - \sin^2\theta}]^2} \end{aligned} \tag{3.21}$$

with $T_{HV} = T_{VH} = 0$. The symbol ε_r represents the relative complex dielectric constant ($\varepsilon_r = 1$ for vacuum). It should be noted that the factors T_{HH} and T_{VV} play a role similar to that of the Fresnel coefficients in reflection from a plane (sec. 4.3) and contain the polarization dependence of the echo.

For an isotropic surface (Barrick and Peake 1967), the function W is

$$W = \frac{2\overline{z^2}}{\pi} \int_0^\infty r\rho(r) J_0 \, (2kr\cos\theta) r \, dr$$

where θ is the grazing angle and k is the propagation constant in air ($k = 2\pi/\lambda$), and r is a measure of position on the planar (isotropic) surface. The factor $\overline{z^2}$ is the mean-square surface roughness defined by

$$\overline{z^2} = \lim_{L\to\infty} (1/L^2) \int_{-L/2}^{L/2} \int_{-L/2}^{L/2} f^2(x, y)\, dx\, dy \tag{3.22}$$

where $f(x, y)$ is the elevation of the surface above the average surface (i.e., $f = 0$). The function $\rho(r)$ is the autocorrelation function for the surface

$$\rho(r) = \rho(\sqrt{x^2 + y^2})$$

$$= \lim_{L\to\infty} (1/L^2) \int_{-L/2}^{L/2} \int_{-L/2}^{L/2} \frac{f(\xi,\eta) f(\xi + x\eta + y)}{\overline{z^2}}\, dx\, dy \tag{3.23}$$

and it has been assumed that the surface is isotropic. $J_o(2kr\cos\theta)$ is the Bessel function of zero order and is written as

$$J_o(2kr\cos\theta) = 1 - \frac{(2kr\cos\theta)^2}{2^2} + \frac{(2kr\cos\theta)^4}{2^2 \cdot 4^2} - \frac{(2kr\cos\theta)^6}{2^2 \cdot 4^2 \cdot 6^2} + \ldots$$

Solutions using equation (3.20) are valid only to within what is called the first order in the perturbation parameters. This means that scattering for any given angle of incidence is caused exclusively by those scatterers that satisfy the first-order Bragg condition,[c] a well-known principle in optics. The first-order mathematical approximation does not admit cross-polarized scattering, but cross-polarized scattering has been measured even for slightly rough surfaces. The second-order approximation to the problem, which does contain cross-polarized terms, has been derived by G.R. Valenzuela (1967); that solution is complicated and is not readily interpretable.

Comparisons of calculations for σ° using equation (3.20 with experimental values obtained by R.C. Taylor (1959) have been reported by R.L. Cosgriff, W.H. Peake, and R.C. Taylor (1960) for concrete and asphalt roads. Cosgriff, Peake, and Taylor (1960) report values of mean-square surface roughness $\overline{z^2}$ and the correlation function of distance $\rho(r)$ obtained from an analysis of plaster casts made from sample areas of roads. Figures 3-7 and 3-8 compare measured and calculated results for an asphalt parking lot. From these figures it is seen that the angular dependence, the frequency dependence, and the polarization dependence of the measured values are well represented by the theory.

3.12 Ripples on Water

J.W. Wright (1966) extended the theoretical approaches of Peake and Rice

[c]The Bragg equation is $n\lambda = 2d \cos\theta$, where n is an integer. This equation specifies the directions of reradiation from point scatterers. The first order condition described here states that back-scattering is caused by those scatterers separated by a distance d such that $\lambda = 2d \cos\theta$.

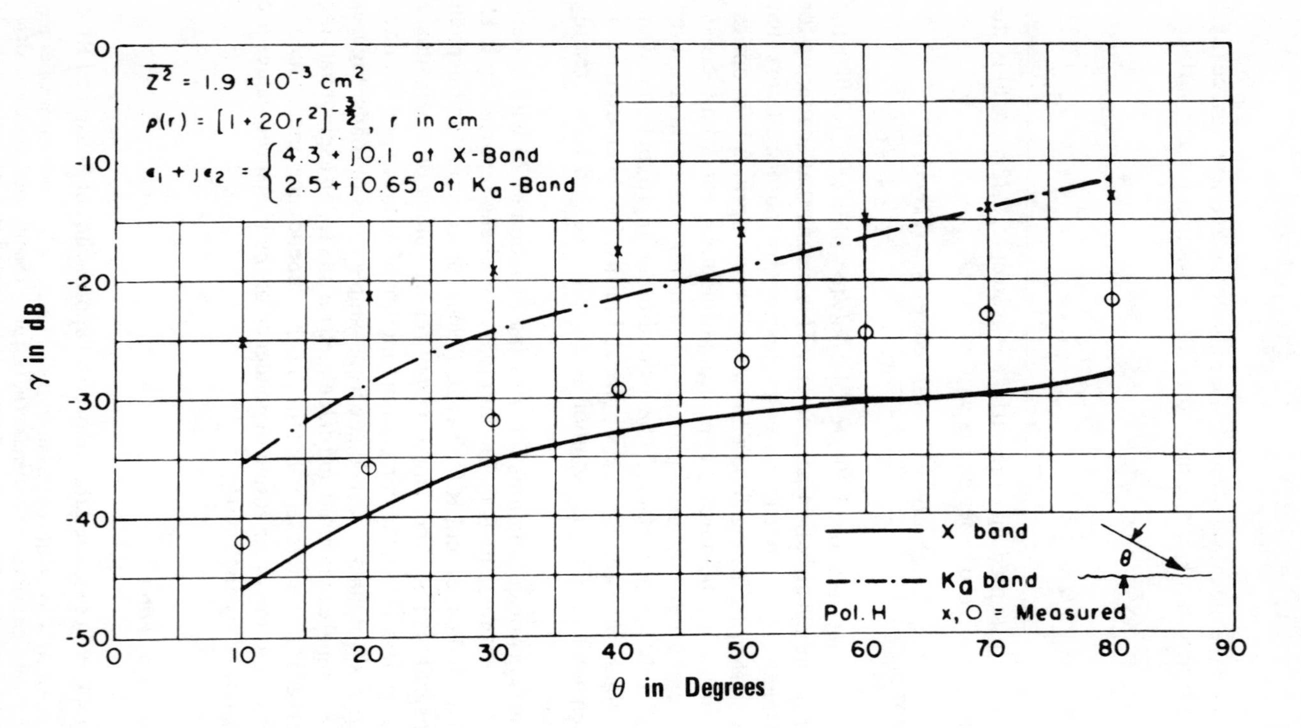

Source: Cosgriff, Peake, and Taylor (1960).

Figure 3-7. Calculated and Measured Return from an Asphalt Parking Lot, *HH* Polarization.

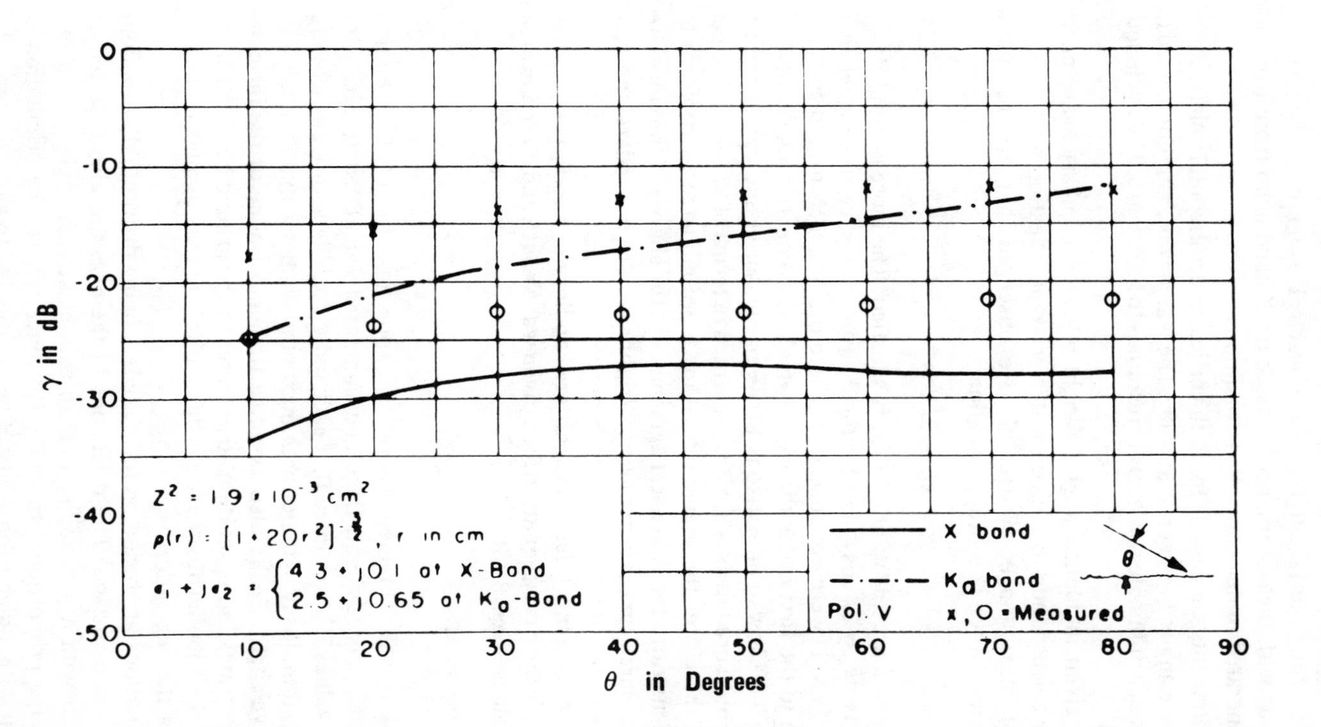

Source: Cosgriff, Peake, and Taylor (1960).

Figure 3-8. Calculated and Measured Return from an Asphalt Parking Lot, *VV* Polarization.

for a slightly rough surface to the case of water. He compared the resulting calculations with measurements he made on X-band echo from capillary waves generated under controlled conditions.

Measurements were made for both vertical and horizontal polarizations for incidence angles between 4° and 80°. The water-wave amplitude used (a few tenths of a millimeter or less) corresponds to less than 0.01 wavelength at X band.

For a given incidence angle, Wright adjusted capillary wavelength λ_w, which corresponds to a given water-wave propagation number $k_w = 2\pi/\lambda_w$, for maximum echo strength. He observed that the maximum strength exists when the Bragg equation

$$k_w = 2k\cos\theta \tag{3.24}$$

where k_w and k are $2\pi/\lambda_w$ and $2\pi/\lambda$, is satisfied. The more general Bragg condition is $nk_w = 2k\cos\theta$; but for the measurements, results higher than first order ($n = 1$) were not observed. Therefore, the measured results are applicable to the first-order theory to which he compared his results.

Figure 3-9 shows comparisons between measured and calculated results. The measured results are not presented in terms of cross section per unit area; therefore, the magnitudes shown are adjusted to conform to calculated magnitudes. A general agreement in the shapes of the calculated and the measured curves is apparent. Wright extended the understanding of sea echo by showing that the scattering elements of primary importance for depression angles much less than 90° are capillary or short-gravity waves which satisfy the Bragg equation for a given wavelength and direction of the incident field (see sec. 3.15).

3.13 Vegetation Model

W.H. Peake (1959) obtained theoretical expressions for $\sigma°$ of vegetation such as grass, weeds, and flags and compared calculated results with measured values. He assumed that each individual blade or stem scatters like a long, thin, lossy cylinder with maximum diameter that is much less than a wavelength. Peake also assumed tha almost none of the incident energy penetrates the vegetation so as to reach the ground below. Therefore, in Peake's model absorption is due entirely to the complex dielectric constant of the vegetation.

The equations are based on the cylinders being distributed at random over the terrain surface. The directions of the axes are also oriented at random, but with a probability that an axis makes an angle χ with the vertical being proportional to $\cos^2\chi$. In other words, the orientation is random but with a high probability of being oriented toward the vertical.

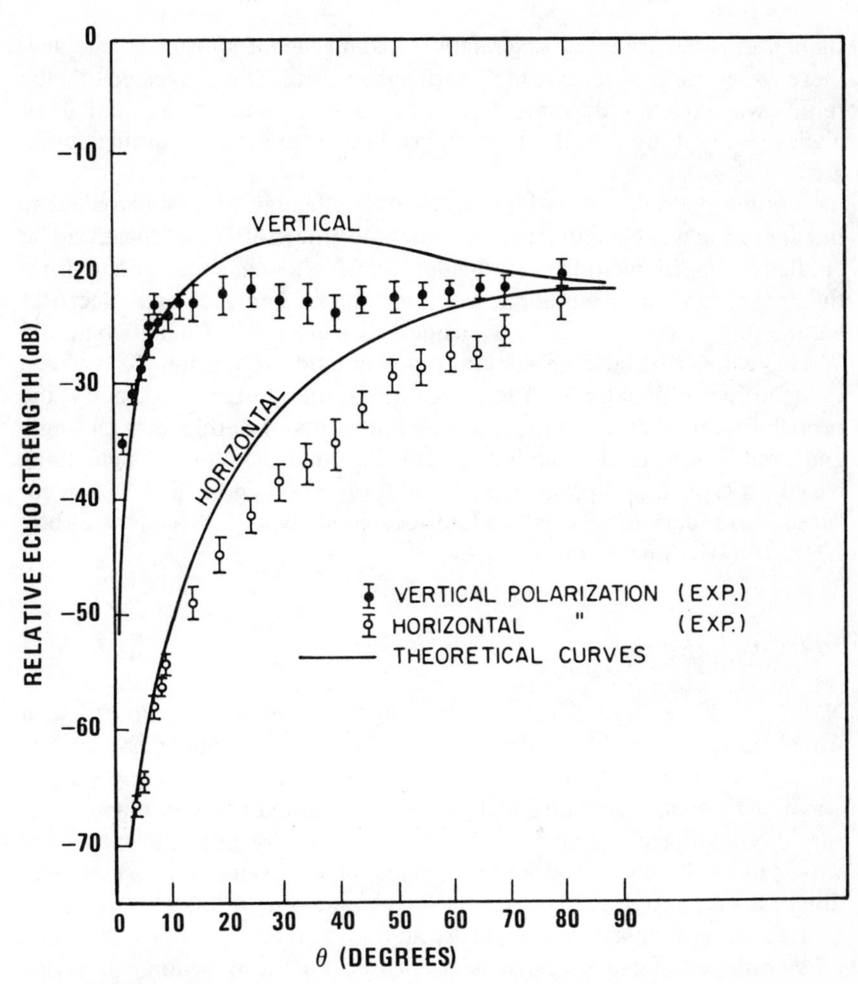

Source: Wright (1966).

Figure 3-9. Radar Cross Section Versus Depression Angle for Capillary Waves on Tap Water.

Cosgriff, Peake and Taylor (1960) reported experimental observations that support Peake's assumptions. For example, they reported that $\sigma°$ for vegetation changes very little when the vegetation is cut; that is, the echo is essentially independent of the depth of the vegetation. They also observed seasonal variations in $\sigma°$ for bare ground and compared these variations with seasonal variations in $\sigma°$ for short grass. It was clear that at K_a band the echo was primarily from vegetation. (The relative contributions of the vegetation and the underlying earth were not as predominant at X band.) Seasonal variations also caused significant changes; for example, even though grass was kept at essentially constant height by regular mowing, there was a marked increase in σ° during the spring. This suggested that the return was strongly dependent on water content and on the number of blades per unit area, both of which are known to increase during spring growth.

In the upper microwave frequency range, leaves are often larger than the operating wavelength. Hence, sometimes the scattering foliage can be studied by approximating the individual leaf as a planar sheet with uniform thickness, random orientation, arbitrary shape, and the same electrical parameters as the leaf itself. By using this leaf model, Cheng Donn and W.H. Peake (1974) use an integral representation to account for various polarizations of incident and scattered fields, the scattering geometry, the probability distribution of the leaf orientation, and the effect of both singly and doubly scattered radiation. According to the authors, calculations based on typical leaf parameters show fairly good agreement with measured cross sections of green and desiccated soybeans at frequencies of 2 GHz, 10 GHz, and 35 GHz.

3.14 Composite Surfaces

Natural surfaces with roughness in height that is large compared to a wavelength almost always possess a smaller scale of roughness (i.e. the surface is not locally smooth). To be specific consider sea waves. Sea waves are typically several feet high, the surfaces of the waves might under rare circumstances be smooth, but typically are covered by small wind waves or capillaries. Therefore, a surface might be viewed as a combination of a very rough surface (that is locally smooth) with a slightly rough surface superimposed on the locally smooth wave.

A number of papers have been written on the development of the composite theory: (Barrick and Peake 1968; Bass et al. 1968; Burrows 1973; Fung and Chan 1969; Guinard and Daley 1970; Peake et al. 1970; Valenzuela 1967, 1968; Wright 1966, 1968). One of the difficulties in developing the theory is to obtain adequate statistical descriptions of surfaces. Much progress has been made in describing the sea by J.W. Wright (1968) who used techniques outlined by O.M. Phillips (1966). In this work, a slightly rough surface is superimposed on the sea swell structure; the effect of the swell structure is to tilt the scattering (slightly rough) surface. Effects of the swell are otherwise neglected.

N.R Guinard and J.C. Daley (1970) extended the technique to estimate maximum echo strength expected under rough conditions, and some of their results of comparing theory with experiment for horizontal and vertical polarizations are given in figures 6.32 through 6.35. A closer correspondence between theory and experiment is obtained for VV than for HH polarization. The model is very significant because it provides numbers for absolute cross section that are close to those actually observed, but the model does not account for well-known observations that average σ_{HH} sometimes exceeds average σ_{VV}. Averages (and medians) of σ_{HH} have been reported to exceed averages (and medians) of σ_{VV} for microwave and millimeter waves, for moderate and rough seas, and for small depression angles (sec. 6.17).

The composite theory described above includes the facets (smooth areas on sea waves) only in the sense of varying the slope of the waves on which the ripples are located. Therefore, the need exists for a sea echo theory that is applicable to small grazing angle and includes the effects of the facets. A two-scatterer concept fo sea echo (Long 1965, 1974) has been introduced to explain experimental results. The two principal scatterers are: (1) a wind-dependent fine structure of the sea (ripples) interspersed with (2) smooth surfaces (facets) that are contained within the wave structure. Differences in the magnitudes and polarization properties of fluctuations of the two types of echoes have been identified (sec. 5.12). Averages for σ_{HH} and σ_{VV} of the facets, per se, are discussed in sections 6.17 and 6.21.

3.15 Doppler Spectra of Sea Echo

An entirely new concept of sea scatter was triggered as a result of observations by D.D. Crombie (1955) at a very long radar wavelength ($\lambda = 20$ meters). Crombie and, somewhat later, R.P. Ingalls and M.L. Stone (1957) reported that doppler shifts from the sea indicate that the velocity of the radar scatterers is that of the gravity waves that have a spacing (wavelength) L equal to one-half the radar wavelength. For the extremely low radar frequencies used in the studies (13 to 25 MHz), the doppler shift is about 0.5 Hz and the observed line widths were narrow (0.01 to 0.03 Hz).

The early experimenters were able to predict the doppler shift to a high degree of precision by assuming that the wave trains (sea) contributing most strongly are those traveling toward or away from the radar and having a spacing L equal to $\lambda/2$. The wave velocity used for the calculations was obtained from the classical hydrodynamic formula for gravity wave velocity,

$$v = \sqrt{gL/2\pi} \tag{3.25}$$

where $g = 9.81\ \text{m/s}^2$ is the acceleration caused by gravity. According to this model, the doppler beat frequency between the transmitted and received signals is

$$f_D = 2v/\lambda = \sqrt{g/\pi\lambda} = \sqrt{gf/\pi c} \tag{3.26}$$

Notice that the doppler shift (eq. (3.26)) varies as the square root of radar frequency rather than directly with frequency as is usual for doppler shifted signals. This is because the scatterer velocity is a function of radar frequency; that is, the sea surface consists of a continuum of wavelengths formed by waves traveling toward and away from the radar with velocity v given by equation (3.25). Therefore, the scattering for a given radar wavelength is not principally from the predominant sea-wave pattern. Instead, the scattering is a resonant phenomenon caused by wave components that are spaced at a distance $\lambda/2$.

The grazing angles used in the early measurements were close to zero, but from subsequent studies it has been found that the resonance condition to be satisfied is $L = \lambda/2 \cos\theta$ where θ is the incidence angle measured from the horizontal. This relationship is the same as the first-order Bragg condition given in equation (3.24). Recently, the second- and higher-order Bragg reflections have been measured and analyzed for HF and VHF frequencies (Barrick, 1971; Barrick et al. 1974).

J.W. Wright (1966) made measurements (sec. 3.12) on X-band echo from capillary waves for incidence angles between 4° and 80°. He also observed that the maximum strength occurred when the first-order Bragg equation was satisfied. According to Wright, effects of surface tension on wave velocity must be included in calculations of doppler shift for microwaves. To include these effects, equation (3.26) is replaced by

$$f_D = \sqrt{\frac{g\cos\theta}{\lambda\pi} + \frac{16\pi s\cos^3\theta}{\lambda^3}} \tag{3.27}$$

where s is the ratio of surface tension to water density.

F.G. Bass et al. (1968) reported the results of measurements over a wide range of frequencies (table 3.1) made from land at several grazing angles between 1° and 20°. It was reported that the spectrum width and the shift of the central frequency of the spectrum, for a given wavelength,depend on sea roughness and on an angle between the radar-beam direction and the sea-wave direction. Narrow spectra were observed for calm seas (wave height less than 0.2 to 0.3 meter). Bass et al. also reported that central spectral frequencies for calm seas had the least values among those in all the set of experiments, and that they practically do not depend on the azimuth and other radiation conditions. According to Bass et al., high winds create additional increase in the doppler shift that is said to be caused by wind drift of the surface layer of water.

Some values of f_D reported for calm seas are given below. Bass et al. have shown that the measured shifts agreed closely with values that are

Table 3-1
Doppler Shifts Measured for Calm Seas

Radar Frequency (MHz)	*Measured Shift (Hz)*	*Reference Source*
13.56	0.38	Crombie (1955)
18.39	0.425	Ingalls & Stone (1957)
24.70	0.50	Ingalls & Stone (1957)
75.0	~0.85	Bass et al. (1968)
200.0	~1.4	Bass et al. (1968)
600.0	~2.5	Bass et al. (1968)
3,000	~6.5	Bass et al. (1968)
9,375	18	Bass et al. (1968)

predicted by equation (3.27). The measured data for 3,000 MHz and lower frequencies agree closely with the values obtained from equation (3.26).

Wright's tank measurements on wind waves have been continued at X and K bands for wind speeds between 4 and 15 knots. From these measurements J.W. Wright and W.C. Keller (1971) report that doppler bandwidth is proportional to the wind speed and to the Bragg wave number $2k \cos \theta$. The fetch used was 2.75 meters. These measurements have been extended to 70 GHz and to longer fetches and higher winds at X and K bands (Duncan, Keller, and Wright 1974). Doppler spectra and radar cross section data are given as a function of polarization, windspeed, fetch, and depression angle in the latter paper, and the doppler bandwidth is approximately accounted for by the rms particle velocity of the dominant waves.

Investigations on first-order and higher-order Bragg scattering has progressed to the point that it is clear that doppler records from large, ground-based radar can be used to determine the magnitudes and directions of waves and surface winds. Because of this, there is now promise for monitoring large areas of the sea with radar operating in the 1 to 30 MHz frequency region that permits OTH (over-the-horizon) propagation by the ionosphere (see sec. 5.13). Wave direction can be obtained with first-order Bragg data. The second-order Bragg data seem to be useful for estimating wave height and wind speed, and it seems that surface currents and current (depth) gradients can be inferred from the doppler sea-echo records.

References

Ament, W.S., "Forward and Back-Scattering by Certain Rough Surfaces," *IRE Transactions on Antennas and Propagation*, vol. AP-4, pp. 369-73, July 1956.

Barrick, D.E., "Second-Order Bragg Scatter from the Ocean at HF/VHF" Spring Meeting, International Union of Radio Science, April 1971.

Barrick, D.E., "Wind Dependence of Quasi-Specular Microwave Sea Scatter," *IEEE Transactions on Antennas and Propagation,* vol. AP-22, pp. 135-36, January 1974.

Barrick, D.E. and W.H. Peake, "Scattering from Surfaces with Different Roughness Scales: Analysis and Interpretation," Battelle Memorial Institute Research Report, November 1, 1967.

Barrick, D.E. and W.H. Peake, "A Review of Scattering from Surfaces with Different Roughness Scales," *Radio Science*, vol. 3 (new series), pp. 865-68, August 1968.

Barrick, D.E., J.M. Headrick, R.W. Bogle, and D.D. Crombie, "Sea Backscatter at HF: Interpretation and Utilization of the Echo," *Proceedings of the IEEE*, vol. 62, pp. 673-80, June 1974.

Bass, F.G., I.M. Fuks, A.I. Kalmykov, I.E. Ostrovsky, and A.D. Rosenberg, "Very High Frequency Radiowave Scattering by a Disturbed Sea Surface, Part II: Scattering from an Actual Sea Surface," *IEEE Transactions on Antennas and Propagation*, vol. AP-16, pp. 560-68, September 1968.

Beckmann, P., *The Depolarization of Electromagnetic Waves*, The Golem Press, Boulder, Colorado, 1968.

Beckmann, P., "Scattering by Non-Gaussian Surfaces," *IEEE Transactions on Antennas and Propagation*, vol. AP-21, pp. 169-75, March 1973.

Beckmann, Petr and André Spizzichino, *The Scattering of Electromagnetic Waves from Rough Surfaces*, The Macmillan Company, New York, New York, 1963.

Berkowitz, R.S., *Modern Radar,* Wiley, p. 560, 1965.

Burrows, M.L., "On the Composite Model for Rough-Surface Scattering," *IEEE Transactions on Antennas and Propagation*, vol. AP-21, pp. 241-43. March 1973.

Clapp, R.E., "A Theoretical and Experimental Study of Radar Ground Return," MIT Radiation Laboratory Report No. 1024, April 1946.

Copeland, J.R., "Radar Target Classification by Polarization Properties," *Proceedings of the IRE*, vol. 48, pp. 1290-96, July 1960.

Cosgriff, R.L., W.H. Peake, and R.C. Taylor, *Terrain Scattering Properties for Sensor System Design*, (Terrain Handbook II), Engineering Experiment Station Bulletin, vol. 29, No. 3, The Ohio State University, 1960.

Crombie, D.D., "Doppler Spectrum of Sea Echo at 13.56 Mc/s," *Nature*, vol. 175, p. 681, April 1955.

Davies, H., "The Reflection of Electromagnetic Waves from a Rough Surface," *Proceedings of the IEE*, part 4, vol. 101, pp. 209-14, 1954.

Donn, Cheng and W.H. Peake, "The Generalized Lommell-Seeliger Cross Section of Foliage Environment," Meeting of International Union of Radio Science, Georgia Institute of Technology, June 1974.

Duncan, J.R., W.C. Keller, and J.W. Wright, "Fetch and Wind Speed Dependence of Doppler Spectra," *Radio Science*, vol. 9, pp. 809-19, October 1974.

Fung, A.K., and H.L. Chan, "Backscattering of Waves by Composite Rough Surfaces," *IEEE Transactions on Antennas and Propagation*, vol. AP-17, pp. 590-97, September 1969.

Goldstein, Herbert, "Frequency Dependence of the Properties of Sea Echo," *Physics Review*, vol. 70, p. 938, 1946.

Graves, C.D., "Radar Polarization Power Scattering Matrix," *Proceedings of the IRE*, vol. 44, pp. 248-52, February 1956.

Guinard, N.R. and J.C. Daley, "An Experimental Study of a Sea Clutter Model," *Proceedings of the IEEE*, vol. 58, pp. 543-51, April 1970.

Hunter, I.M., and T.B.A. Senior, "Experimental Studies of Sea-Surface Effects on Low-Angle Radars," *Proceedings of the IEE*, vol. 113, p. 1731, November 1966.

Ingalls, R.P. and M.L. Stone, "Characteristics of Sea Clutter at HF," *IRE Transactions on Antennas and Propagation*, vol. AP-5, pp. 164-65, January 1957.

Kanareykin, D.B., N.F. Pavlov, and V.A. Potekhin, *The Polarization of Radar Signals* (in Russian), Sovyetskoye Radio, Moscow, 1966.

Katz, Isadore and L.M. Spetner, "A Functional Relationship Between Radar Cross Section of Terrain and Depression Angle," *IRE Transactions on Antennas and Propagation,* vol. AP-6, p. 310, July 1958.

Katz, Isadore and L.M. Spetner, "Polarization and Depression-Angle Dependence of Radar Terrain Return," *National Bureau of Standards Journal of Research—D. Radio Propagation*, vol. 64D, p. 483, September-October 1960.

Katzin, Martin, "On the Mechanisms of Radar Sea Clutter," *Proceedings of the IRE*, vol. 45, p. 44, January 1957.

Kennaugh, E.M., "Polarization Properties of Radar Reflections," Ohio State University Antenna Laboratory Report 389-12, March 1952.

Kerr, D.E. (Ed.), *Propagation of Short Radio Waves*, Massachusetts Institute of Technology, Radiation Laboratory Series, vol. 13, McGraw-Hill Book Company, Inc., New York, New York, 1951.

Kraus, J.D., *Antennas,* McGraw-Hill Book Company, Inc., New York, New York, 1950.

Long, M.W., "On the Polarization and the Wavelength Dependence of Sea Echo," *IEEE Transactions on Antennas and Propagation,* vol. AP-13, pp. 749-54, September 1965.

Long, M.W., "Backscattering for Circular Polarization," *Electronics Letters*, vol. 2, p. 351, September 1966.

Long, M.W., "Polarization and Sea Echo," *Electronics Letters*, vol. 3, p. 51, February 1967.

Long, M.W., "On a Two-Scatterer Theory of Sea Echo," *IEEE Transactions on Antennas and Propagation*, vol. AP-22, pp. 667-72, September 1974.

Peake, W.H., "Theory of Radar Return from Terrain," *1959 IRE Convention Record*, vol. 7, pp. 27-41.

Peake, W.H., D.E. Barrick, A.K. Fung, and H.L. Chan, "Comments on 'Backscattering of Waves by Composite Rough Surfaces,'" *IEEE Transactions on Antennas and Propagation,* vol. AP-18, pp. 716-26, September 1970.

Phillips, O.M., *The Dynamics of the Upper Ocean*, Cambridge University Press, London, 1966.

Rice, S.O., "Reflection of Electromagnetic Waves by Slightly Rough Surfaces," *Communications on Pure and Applied Math,* vol. 4, pp. 351-78, 1951.

Ruck, G.T., D.E. Barrick, W.D. Stuart, and C.K. Krichbaum, *Radar Cross Section Handbook*, Plenum Press, New York, New York, 1970.

Schooley, A.H., "Evaluation of Radar Sea Clutter," *Tele-Tech and Electronic Industries,* vol. 14, p. 70, March 1955.

Schooley, A.H., "Some Limiting Cases of Radar Sea Clutter Noise," *Proceedings of the IRE,* vol. 44, pp. 1043-47, August 1956.

Sinclair, George, "Modification of the Radar Range Equation for Arbitrary Targets and Arbitrary Polarization," Project Report No. 302-19, Ohio State University Research Foundation, September 25, 1948.

Taylor, R.C., "Terrain Return Measurements at X, K_u and K_a Band," *1959 IRE National Convention Record*, vol. 7, pp. 19-26.

Valenzuela, G.R., "Depolarization of EM Waves by Slightly Rough Surfaces," *IEEE Transactions on Antennas and Propagation*, vol. AP-15, pp. 552-57, July 1967.

Valenzuela, G.R., "Scattering of Electromagnetic Waves from a Tilted Slightly Rough Surface," *Radio Science,* vol. 3 (new series), pp. 1057-66, November 1968.

Wright, J.W., "Backscattering from Capillary Waves with Application to Sea Clutter," *IEEE Transactions on Antennas and Propagation,* vol. AP-14, pp. 749-54, November 1966.

Wright, J.W., "A New Model for Sea Clutter," *IEEE Transactions on Antennas and Propagation,* vol. AP-16, pp. 217-23, March 1968.

Wright, J.W. and W.C. Keller, "Doppler Spectra in Microwave Scattering from Wind Waves," *The Physics of Fluids*, vol. 14, pp. 466-74, March 1971.

4 Effects of the Earth's Surface

Fundamental Concepts

The presence of terrain affects the cross section of scatterers. The resulting effects cause changes in the range and depression angle dependencies of echo strength. Range and depression angle dependencies are functions of height of scatterers above the terrain, vertical extension of the scatterers, and whether or not the scatterers fill the antenna beam.

The reflection of microwaves from the earth's surface is an important factor in a number of phenomena associated with communications and radar systems. Comparisons will be given in this chapter of received signal strength for the cases of one- and two-way propagation as a function of polarization. Surface roughness causes marked departures from the results predicted theoretically for a smooth surface and the extent of the departures is related to incidence angle, polarization, and the size of the surface roughness elements relative to wavelength.

4.1 Effects of the Earth's Curvature and Refraction

Radar reflectivity is affected by the fact that the earth is not flat, because the earth's curvature causes the path length difference between the direct and reflected waves to be decreased, and it also decreases the amplitude of the reflected waves. The change from the plane to the spherical geometry is equivalent to a reduction in the height of radar and target as illustrated in figure 4-1 ($h_1' < h_1$, $h_2' < h_2$). Another effect of the earth's curvature is to decrease the amplitude of the reflected wave, because the reflected waves are spread out or diverged into a larger range of angles by the convex surface of the earth as illustrated in figure 4-2.

The above discussion neglects effects of the earth's atmosphere. Electromagnetic waves propagating within the earth's atmosphere do not travel in straight lines but bend slightly because of refraction. One effect of refraction is to extend the distance to the horizon, thus increasing radar coverage; another effect is the introduction of errors in the measurement of elevation angle. The classical method of accounting for atmospheric refraction in computations is to replace the actual earth of radius a by an equivalent earth of radius ka and by replacing the actual atmosphere by a

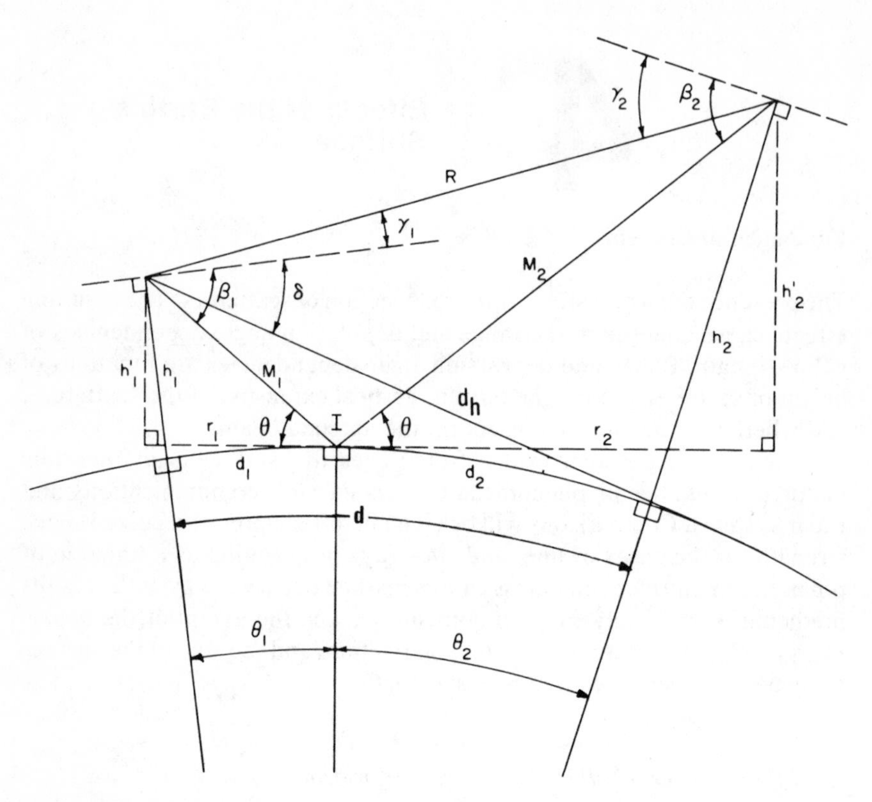

Source: Durlach (1965).

Figure 4-1. Spherical Earth Parameters.

homogeneous atmosphere in which the waves propagate in straight lines (see sec. 2.9). It is customary to use a value of $k = 4/3$ as simply a convenient means for approximating the effects of refraction within the United States continent. Larger and smaller values of k have been observed (U.S. Department of Commerce 1966). For propagation over water, values of nearly 10 (Hunter and Senior 1966) have been reported (see sec. 2.12).

For distances encountered with microwave radar for airborne, shipborne, or ground installations, the reduced heights h_1' and h_2' can be calculated from

$$h_1' = h_1 - \Delta h_1 = h_1 - d_1^2/2ka$$

$$h_2' = h_2 - \Delta h_2 = h_2 - d_2^2/2ka$$

where d_1 and d_2 are shown in figure 4-1.

As is already noted, the effect of average or standard atmospheric refraction is allowed for by increasing the earth's radius by a factor of

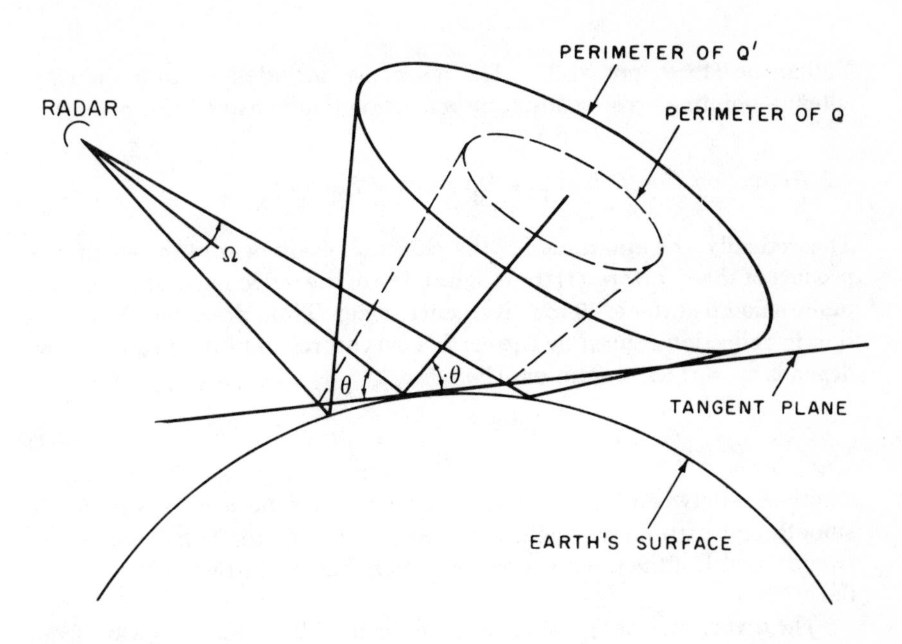

Source: Durlach (1965).

Figure 4-2. Divergence Factor.

$k = 4/3$. It so happens that with this factor, and expressing height in feet and distances in statute miles (5,280 feet), the height reductions due to refraction and curvature take the simple form

$$\Delta h_{1,2} = d_{1,2}^2/2$$

The distance d_1 represents the radar horizon if h_1' is zero. Therefore, the radar horizon is at a distance d_h such that

$$d_h = \sqrt{2(ka)h}$$

where h is antenna height above an idealized smooth earth. Therefore, for the "standard" conditions for which $k = 4/3$

$$d_h \text{ (statute miles)} = \sqrt{2h \text{ (feet)}}$$

For example, suppose a radar antenna were 50 feet and then 5,000 feet, respectively, above the sea. Then the radar horizon would be at a distance of 10 and then 100 statute miles, respectively.

Curves for the solution of d_1 or d_2 in terms of h_1, h_2, and d are available elsewhere (Kerr 1951), and equations for computing the various variables of figure 4-1 are given by N.I. Durlach (1965). Curves showing grazing angle in terms of antenna height and slant range are given by F.E.

Nathanson (1969, pp. 31-32). The book also includes a figure showing antenna pointing errors caused by refraction (Nathanson 1969, p. 218)

4.2 Reflection Coefficient and Divergence Factor

Theoretically, the magnitude of the reflection coefficient of the earth is a product of three factors: (1) the magnitude ρ of the reflection coefficient of a plane smooth surface; (2) the divergence factor D that describes the reduction in reflection caused by the earth's curvature; and (3) a factor $\mathcal{R}$ that depends on surface roughness. Therefore,

$$\left| \begin{matrix} \text{Reflection} \\ \text{Coefficient} \end{matrix} \right| = \rho D \mathcal{R} \tag{4.1}$$

where $\mathcal{R}$ is between 1 and 0 depending on whether the surface is perfectly smooth and extremely rough, respectively. The factor D has values between 0 and 1; if the radar geometry is such that the earth can be assumed flat, D is 1.

The divergence or spreading of waves caused by a spherical wave front being incident upon a spherical earth has been analyzed theoretically with much mathematical detail (Bremmer 1949). The divergence D is equivalent to a purely geometric factor that describes additional spreading of a beam of rays due to reflection from a spherical surface as illustrated in figure 4-2. More specifically,

$$D = \lim_{\Omega \to 0} \left(\frac{Q}{Q'} \right)^{1/2}$$

where Q' and Q are the corresponding curved and flat earth cross sections of the reflected beam, and Ω is the solid angle subtended by area Q.

It is noted that (1) D is always between zero and unity, (2) D approaches unity as k becomes very large or as h_1 or h_2 approach zero, and (3) D approaches zero as θ (fig. 4-2) approaches zero.

D.E. Kerr (1951) and Petr Beckmann and André Spizzichino (1963) give curves for determining D. An approximate formula usually valid (Kerr, 1951, p. 406; Posejsil et al., 1961; Blake, 1980, pp. 270-279) out almost to the horizon is

$$D = \left(1 + \frac{2 d_1 d_2}{k a d \sin \theta} \right)^{-1/2} \tag{4.2}$$

4.3 Reflection Coefficient for a Flat, Smooth Earth

An understanding of the basic phenomena of reflection from the earth can be obtained by considering the reflection of plane waves from a smooth dielectric surface. The reflections from a smooth earth can be determined from Fresnel's equations from which it is seen that the magnitude and phase of the reflection depends on frequency, polarization, and the angle of incidence of the wave, and on the electrical properties of the earth (dielectric constant and conductivity).[a] These equations are given for the two cases of electric vector perpendicular to or in the plane of incidence, corresponding to "horizontal" and "vertical" polarization in figure 4-3. The terminology of horizontal and vertical polarization is not precise but it corresponds to that used in radio and radar engineering. For calculations involving any other polarization, the wave is resolved into its linearly polarized components parallel and perpendicular to the surface (for earth, horizontally and vertically polarized components) and the problem may become more complicated than might at first be suspected.

Assume that the electromagnetic wave incident on the dielectric is propagating in free space and the dielectric surface is non-magnetic. Then, following figure 4-3, the reflection coefficients for horizontal and vertical polarizations are given by Fresnel's equations as follows:

$$\Gamma_H = \frac{\sin\alpha - (\varepsilon - \cos^2\alpha)^{1/2}}{\sin\alpha + (\varepsilon - \cos^2\alpha)^{1/2}} = \rho_H e^{-j\phi_H} \tag{4.3}$$

$$\Gamma_V = \frac{\varepsilon \sin\alpha - (\varepsilon - \cos^2\alpha)^{1/2}}{\varepsilon \sin\alpha + (\varepsilon - \cos^2\alpha)^{1/2}} = \rho_V e^{-j\phi_V} \tag{4.4}$$

where α is measured perpendicular to the surface normal n, and ε is the complex dielectric constant of the surface. The complex dielectric constant is given in terms of the permittivity K and conductivity σ by

$$\varepsilon = K/\varepsilon_0 - j\sigma/\omega\varepsilon_0 = \varepsilon' - j\varepsilon'' \tag{4.5}$$

where ε_0 is the dielectric constant of freespace, f is radar frequency, and ω equals $2\pi f$. Values of the various terms for "typical" samples of the earth are given in table 4-1.

A dependence of equations (4.3) and (4.4) on frequency enters through the dependence of ε'' on frequency. In addition, the parameters K and σ are also functions of frequency, by virtue of the frequency dependence (dispersion) of water. This dispersion is strong in the microwave region; therefore, reflection coefficients depend on water content and are frequency dependent.

For normal incidence ($\alpha = \pi/2$), the two polarizations become identical and equations (4.3) and (4.4) become

$$\Gamma_H = \frac{1 - \varepsilon^{1/2}}{1 + \varepsilon^{1/2}} \tag{4.6}$$

[a]See any standard book on electromagnetic theory, for example, J.A. Stratton (1941).

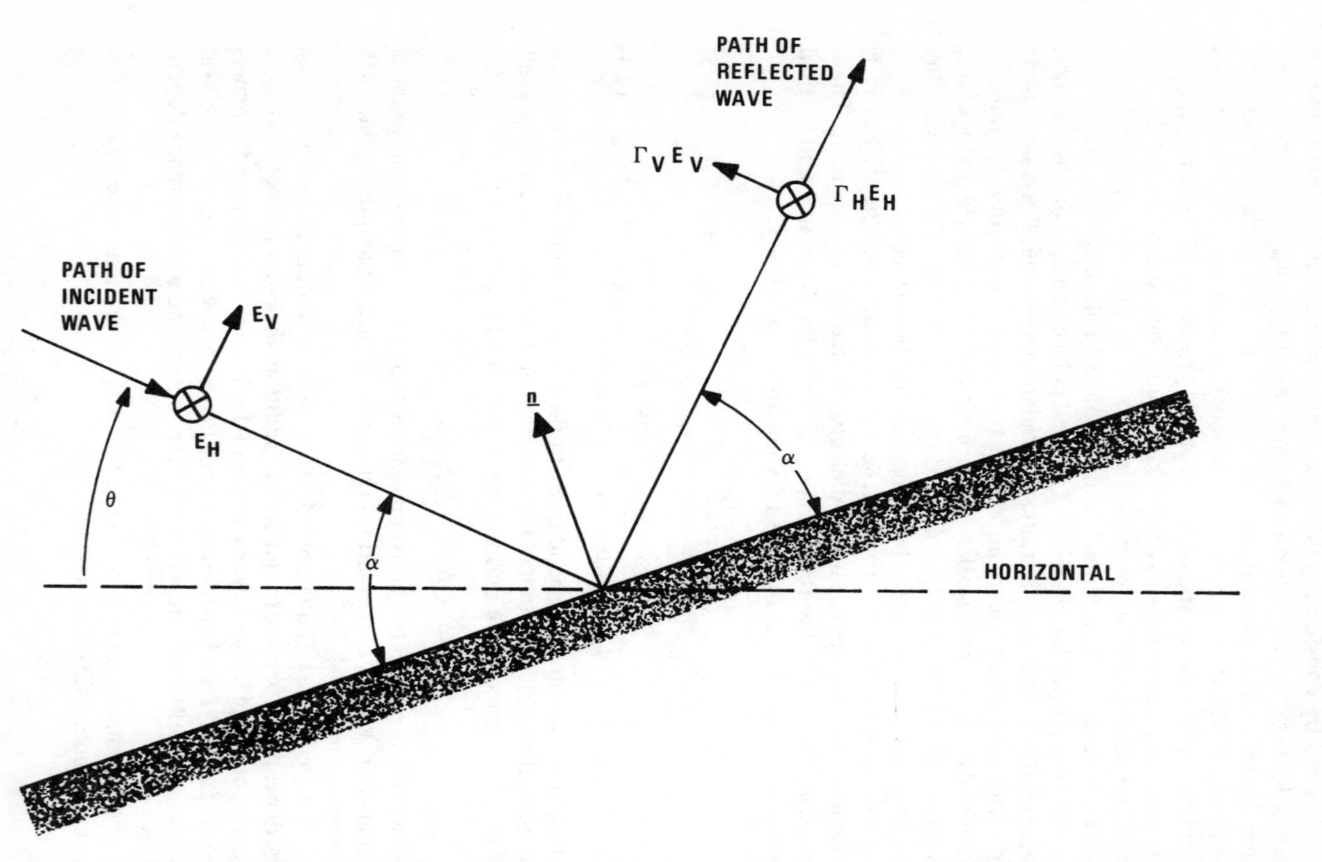

Figure 4-3. Reflection by a Smooth Plane Dielectric Surface. E_H and E_V are the electric vectors for horizontal and vertical polarizations, respectively.

Table 4-1
Approximate Electromagnetic Properties of Soil and Water[a]

Medium	λ	σ mho/m	ε'	ε''
Sea water	3 m-20 cm	4.3	80	774 52
20°-25°C	10 cm	6.5	69	39
28°C	3.2 cm	16	65	30.7
Distilled water, 23°C	3.2 cm	12	67	23
Fresh-water lakes	1 m	10^{-3}-10^{-2}	80	0.06 0.60
Very dry sandy loam	9 cm	0.03	2	1.62
Very wet sandy loam	9 cm	0.6	24	32.4
Very dry ground	1 m	10^{-4}	4	0.006
Moist ground	1 m	10^{-2}	30	0.6
Arizona soil	3.2 cm	0.10	3.2	0.19
Austin, Tex., soil, very dry	3.2 cm	0.0074	2.8	0.014

[a]From D.E. Kerr (1951, p. 398).

$$\Gamma_V = \frac{\varepsilon - \varepsilon^{1/2}}{\varepsilon + \varepsilon^{1/2}} = -\Gamma_H \tag{4.7}$$

The reason for the sign difference between Γ_V and Γ_H is apparent by referring to the directions assigned to the fields in figure 4-3. A minus sign is an indication that the true field is reversed compared to the assumed field; Γ_H and Γ_V must therefore differ in sign if the reflected fields have the same phase.

Figures 4-4 through 4-9 show graphs of ρ_H, ρ_V, ϕ_H, and ϕ_V versus wavelength and angle α for typical smooth land and water surfaces. These graphs depict major differences between the reflecting properties for horizontal and vertical polarizations. For horizontal polarization there is only a slight variation in magnitude and phase with depression angle. For vertical polarization, however, there is a variation caused by increased transmission into the surface that occurs near the Brewster angle. That angle, α_B, is where ρ_V reaches a minimum and there ϕ_V is $\pi/2$—α_B itself also depends on frequency. Although the dielectric properties of water and land depend on incidence angle and wavelength, for many purposes ρ_H and ϕ_H can be approximated by unity and π, respectively. For very small depression angles, ρ_V and ϕ_V are nearly unity and π, respectively; for large depression angles ϕ_V approaches zero. The angles ϕ_H and ϕ_V differ negligibly from 180° for microwaves and depression angles of 1° or less. These characteristic behaviors for smooth surfaces play significant roles in the reflecting prop-

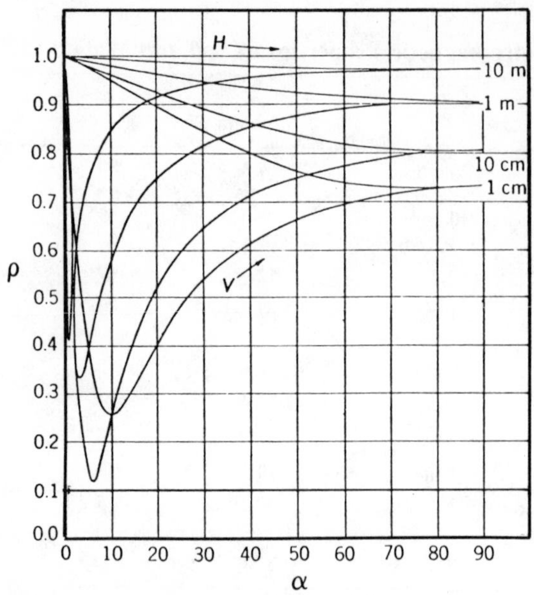

Source: Povejsil, Raven, and Waterman (1961, p. 184).

Figure 4-4. Magnitude of Reflection Coefficient for Sea Water (Temperature = 10°C) as a Function of Incidence Angle.

erties of radar targets for cases of small depression angles, even under conditions for which the land or water surface is physically rough.

4.4 Effect of Terrain on Target Echo

Various effects can be generated by the interference of the direct and reflected waves. Referring to figure 4-10, if both the direct and indirect paths are illuminated equally by the transmitting antenna, the resultant electric field at P is

$$E = E_d \left[1 + \Gamma e^{-j(2\pi/\lambda)\,\Delta R} \right] = E_d \left[1 + \rho e^{-j[(2\pi/\lambda)\,\Delta R + \phi]} \right] \tag{4.8}$$

where E_d is the field due to the direct wave, ΔR is the difference in path lengths for the direct and indirect paths, and ϕ is the phase delay caused by the reflection at the terrain.

By referring to figure 4-10 and by using the flat earth approximation, ΔR can be expressed as

$$\Delta R = 2h \sin\theta \tag{4.9}$$

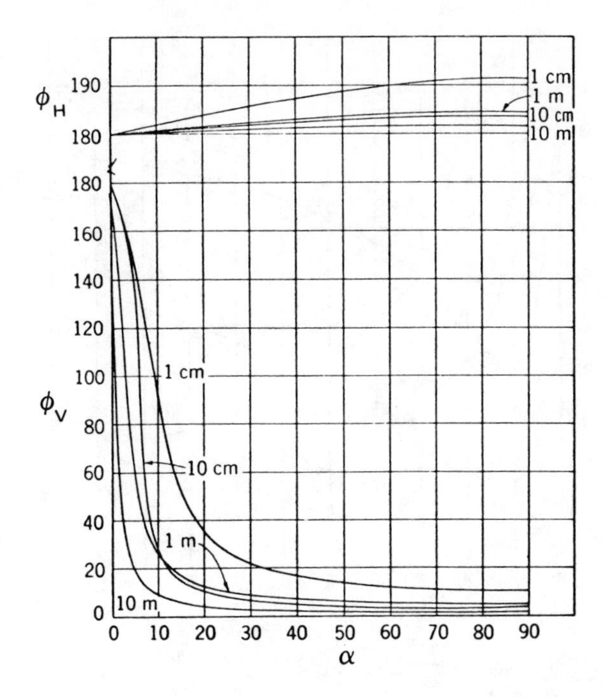

Source: Povejsil, Raven, and Waterman (1961, p. 184).

Figure 4-5. Phase of Reflection Coefficient for Sea Water (Temperature = 10°C) as a Function of Incidence Angle.

where h is target height and θ is measured between pointing direction of the antenna and the reflecting surface.

The ratio E/E_d represents the modification of the electric field strength at the target caused by the presence of the reflecting surface. That ratio is called the propagation factor F. Since the return paths (target back to antenna) are the same as the incident paths, echo power from a point target at point P is proportional to $|F|^4$. Thus, the radar equation can be written as

$$P_r = \frac{P_t G^2 \lambda^2 \sigma'}{(4\pi)^3 R^4} |F|^4 \qquad (4.10)$$

where σ' is emphasized as being the free-space cross section. The product $\sigma' |F|^4$ may be thought of as effective cross section σ.

A target may extend over a large enough area that F, which is complex (a phasor), may be a function of position on the target. Therefore, care must be exercised to acquire the proper equivalent value for $|F|^4$ for equation (4.10). If the target can be described as a number of discrete targets, then equation (4.17) can be used to calculate σ.

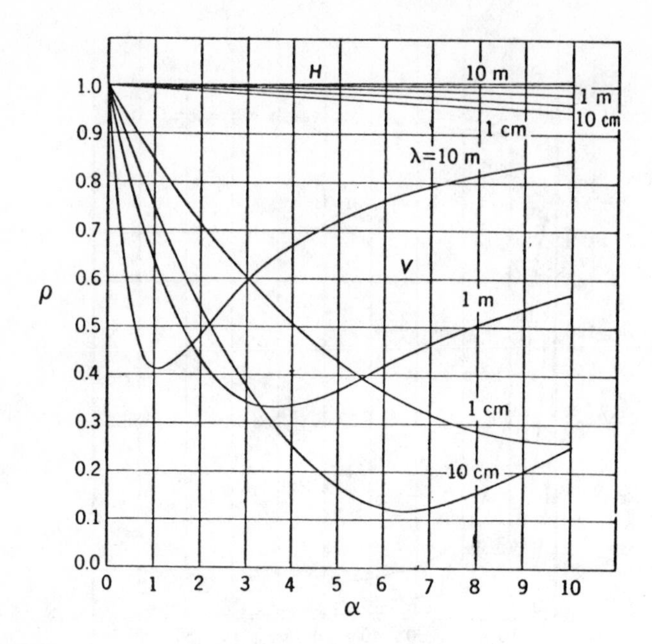

Source: Povejsil, Raven, and Waterman (1961, p. 185).

Figure 4-6. Expanded Plot of Figure 4-4 for Incidence Angles Between 0° and 10°.

The phenomenon of interference between the direct and indirect waves produces a polarization sensitive lobe structure above the reflecting earth. The manner in which the set of lobes is formed can be noted from equation (4.8). Let θ be fixed and let h be continuously increased. The resultant field (or magnitude of F) will pass through alternate maxima and minima and the height interval between an adjacent maximum and minimum is

$$\delta h = \lambda / 4 \sin \theta \tag{4.11}$$

In general, the positions of the lobes do not coincide for horizontal and vertical polarizations because of differences in ϕ_H and ϕ_V; differences in ρ_H and ρ_V also cause the magnitudes of $|F|$ to be different. Examples are given in figures 4-11 and 4-12.

For all but short radar ranges, effects of the earth's curvature and atmospheric refraction must also be included when determining the propagation factor. As already mentioned, the reflection coefficient of a surface is a product of three factors: (1) the reflection coefficient of a plane, smooth surface that may be calculated with Fresnel's equations; (2) the divergence factor that describes a reduction in reflection caused by the earth's curvature; and (3) a factor that depends on surface roughness.

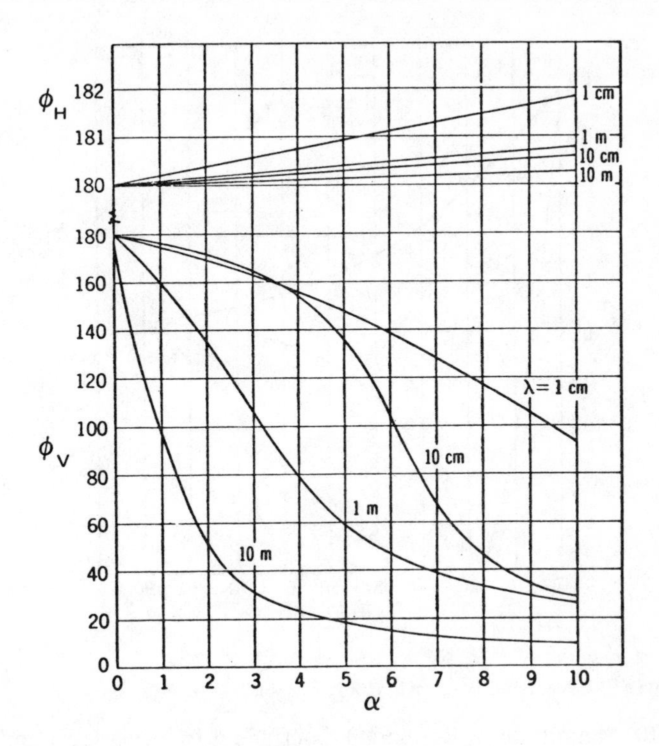

Source: Povejsil, Raven, and Waterman (1961, p. 185).

Figure 4-7. Expanded Plot of Figure 4-5 for Incidence Angles Between 0° and 10°.

Lord Rayleigh suggested a way of distinguishing between the extreme of a "smooth" or "rough" surface. Rayleigh's criterion for a surface to be considered smooth is sometimes given as

$$\Delta h \sin\theta < \lambda/8$$

where Δh is the rms height of surface irregularities, and θ is the angle between incident rays and a plane surface representing the average of the irregularities. For applications involving search radar, $\sin\theta$ might be as small as 10^{-2} or 10^{-3}. Therefore, according to the Rayleigh criterion, a very rough sea[b] can appear as a smooth surface for microwaves. Calculations of $|F|^4$ for various polarizations and smooth surfaces are made in the section that follows. The calculations are followed by a discussion of expected effects of surface roughness and some experimental results.

Strictly speaking, reflection is not from a point even for a perfectly flat

[b]Relationships between wave height and rms surface roughness are given in section 2.11.

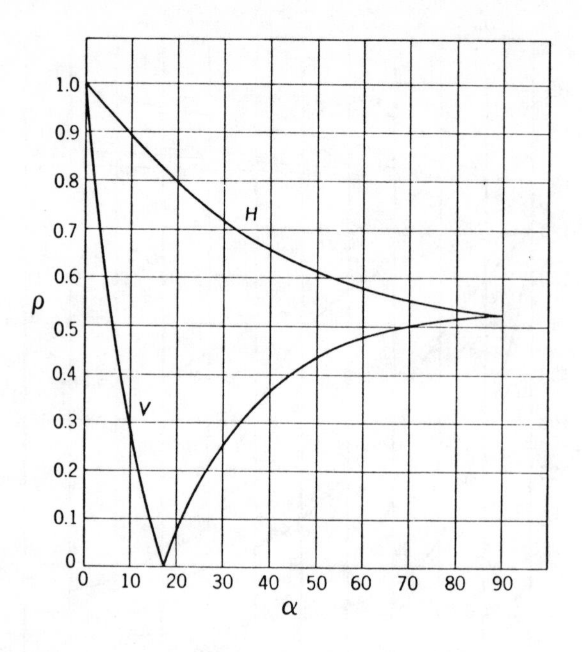

Source: Povejsil, Raven, and Waterman (1961, p. 186).

Figure 4-8. Magnitude of Reflection Coefficient for Average Land ($\varepsilon' = 10$, $\sigma = 1.6 \times 10^{-3}$ mho/m) as a Function of Incidence Angle.

surface. The extent of the region that contributes significantly to forward scatter depends in a complex way on geometry, and as a first approximation it is usually assumed that reflections are entirely from the first Fresnel zone (Barton 1974; Beckma[illegible]51). That zone is an ellipse with its center [illegible]e 4-1. Various detailed analyses of the effe[illegible]and roughness have been made. Although [illegible]ections from a point often give valid result[illegible]hat the surface characteristics in an extenc[illegible]I contribute to the propagation along the [illegible]

Echo from Targets That are Above a Flat, Smooth Earth

Effects of a flat, smooth earth on echo from targets are considered in sections 4.5 through 4.8. Effects of surface roughness on forward scattering are reviewed in sections 4.9 through 4.11, and the review is followed by a

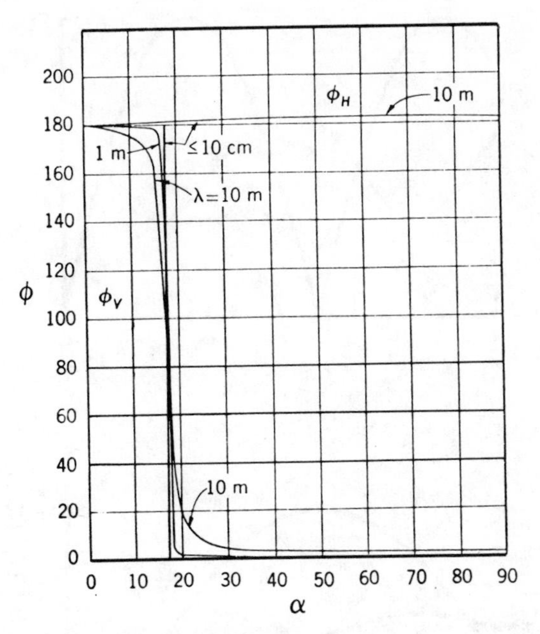

Source: Povejsil, Raven, and Waterman (1961, p. 186).

Figure 4-9. Phase of Reflection Coefficient for Average Land ($\varepsilon' = 10$, $\sigma = 1.6 \times 10^{-3}$ mho/m) as a Function of Incidence Angle.

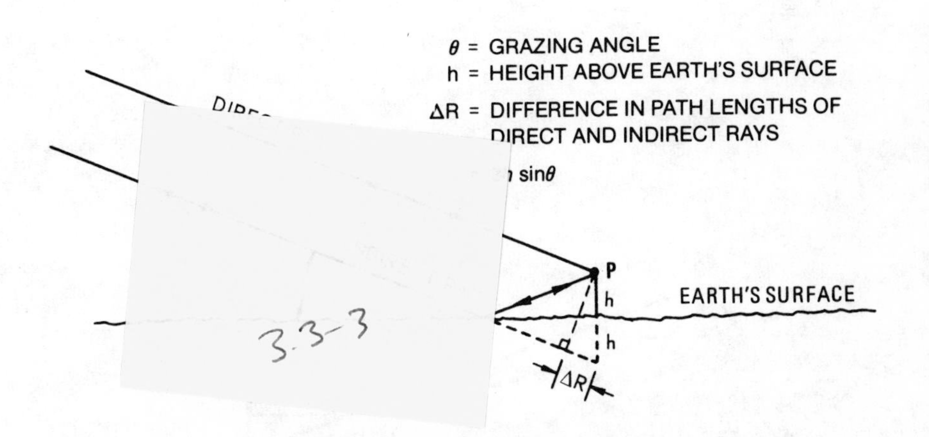

Figure 4-10. Interference Geometry.

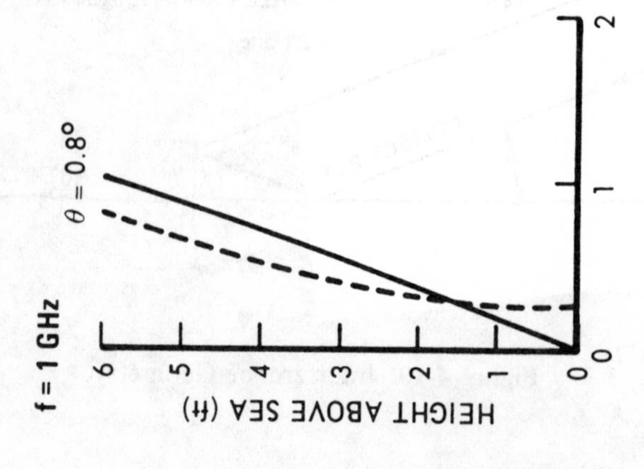
f = 1 GHz
θ = 0.8°
HEIGHT ABOVE SEA (ft)
0
1
2
3
4
5
6
0
1
2

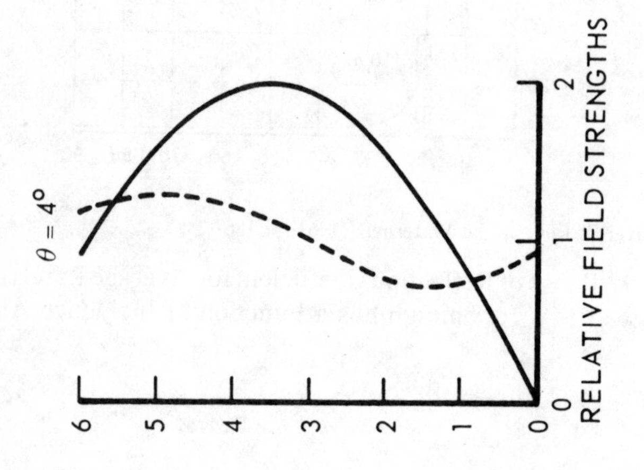
θ = 4°
0
1
2
3
4
5
6
0
1
2
RELATIVE FIELD STRENGTHS

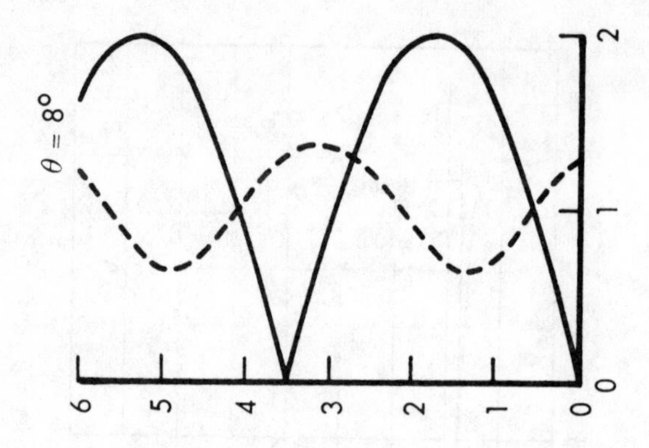
θ = 8°
0
1
2
3
4
5
6
0
1
2

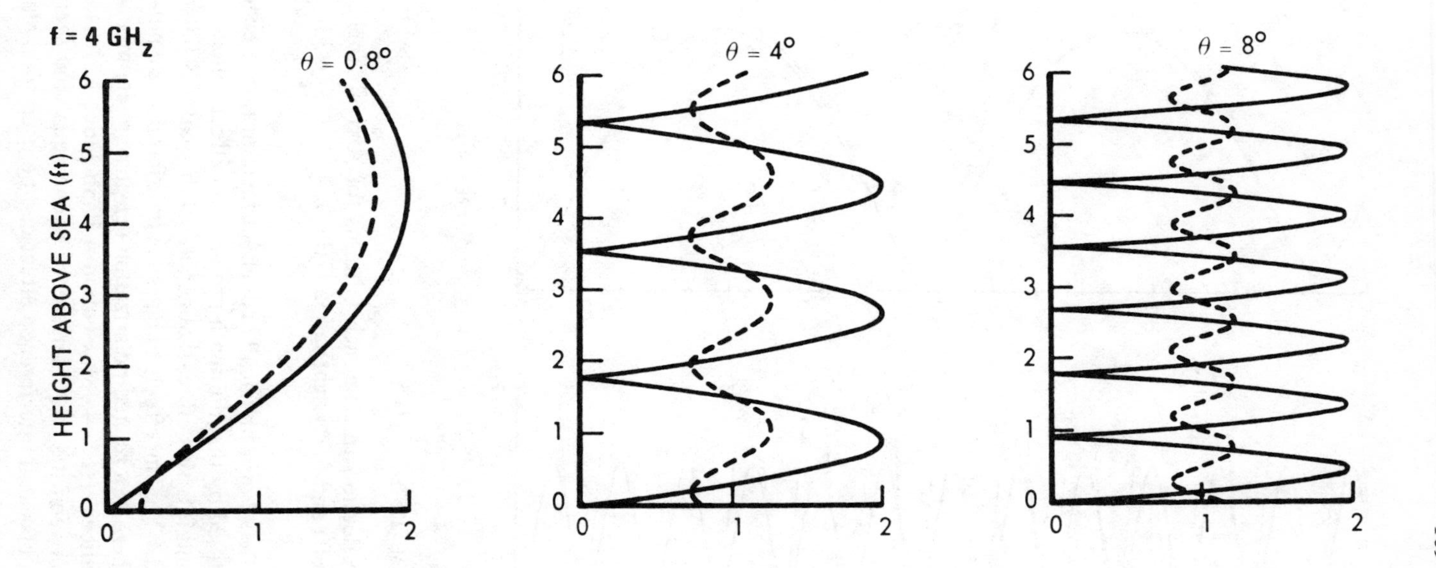

Source: Long et al. (1965).

Figure 4-11. Electric Field Amplitude Patterns above a Smooth Sea Relative to the Field Amplitude in Free Space, at 1 and 4 Gigahertz for Grazing Angles 0.8°, 4°, and 8°. Solid lines indicate horizontal polarization; broken lines indicate vertical polarization.

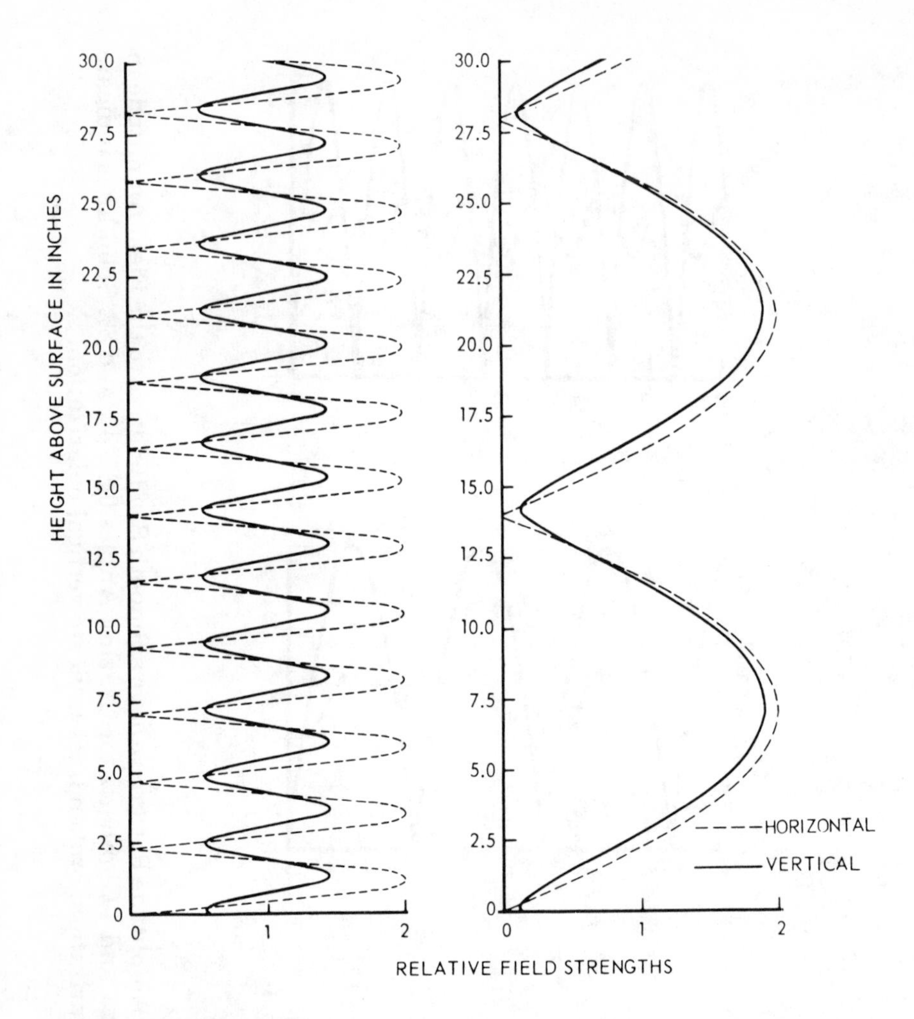

Source: Boring et al. (1957).

Figure 4-12. Reflection Interference Patterns for Arrival Angles of 4.0° (left) and 0.7° (right) at 35 GHz. $\varepsilon = 22.4 - j33.0$.

discussion of effects of a physically rough surface on target echo. Echo strength is modified by the multipath propagation effect called "classical interference" and it is this effect that generates the patterns illustrated in figures 4-11 and 4-12. The effects of classical interference are polarization sensitive, but there is a related polarization sensitive phenomenon called "local interference" (sec. 4.18). Section 4.19 includes a discussion of comparisons between the two interference phenomena, and section 4.17 shows how the classical interference effect can be used to explain rapid changes in σ° versus range.

4.5 Range and Depression Angle Dependence for a Small Object above a Smooth Earth

From equations (4.8) and (4.10) it may be seen that

$$F = 1 + \Gamma e^{-j(2\pi/\lambda)\,\Delta R} = 1 + \rho e^{-j[(2\pi/\lambda)\,\Delta R + \phi]} \tag{4.12}$$

where $\Delta R = 2h \sin\theta$ and h is target height (see fig. 4-10). For the flat earth approximation $\sin\theta$ is equal to H/R, where H is antenna height and R is distance between antenna and target.

From the law of cosines,

$$|F|^2 = (1 + \rho^2 + 2\rho\cos\psi) \tag{4.13}$$

where $\psi = (2\pi/\lambda)\,\Delta R + \phi$. Therefore, the cross section for a point target with free space value σ' is

$$\sigma = \sigma' \, |F^4| = \sigma' \, (1 + \rho^2 + 2\rho\cos\psi)^2 \tag{4.14}$$

Although the dielectric properties of land and sea depend on incidence angle and wavelength, for some purposes ρ and ϕ can be approximated by unity and π, respectively. This approximation is usually valid for horizontal polarization and water; using this, one gets

$$F = 1 - e^{-j(2\pi/\lambda)\Delta R}$$

and (4.15)

$$|F| = 2\,|\sin(\pi/\lambda)\Delta R|$$

Nodes and antinodes of received power versus range are frequently observed; these are given by equation (4.15). Therefore, the envelope of antinodes (maxima) of received power for one-way transmission or for radar vary in principle as R^{-2} or R^{-4}, respectively. In general, the minima in echo power versus range are not sharp nulls because ρ is usually less than one. In that case equation (4.14) is used to calculate echo power. Examples of range dependence of $|F|$ for a curved earth and a standard atmosphere are given in figure 4-13.

For ranges beyond the most distant antinode equation (4.15) can be approximated by

$$|F|^4 \approx (2\pi\Delta R/\lambda)^4 \approx [4\pi\,(h/\lambda)\sin\theta]^4 \approx [4\pi\,(h/\lambda)\,H]^4 \, R^{-4}$$

Thus, for these more distant ranges, received power varies as R^{-8} if $\rho = 1$ and $\phi = \pi$. (See fig. 4-16.)

From figures 4-6 through 4-9, it may be seen that ρ_H and ϕ_H are close to unity and π for very small grazing angles for land and sea. At angles smaller than the Brewster angle, ρ_V and ϕ_V can be approximated by

$$\rho_V \approx 1 - a\theta$$

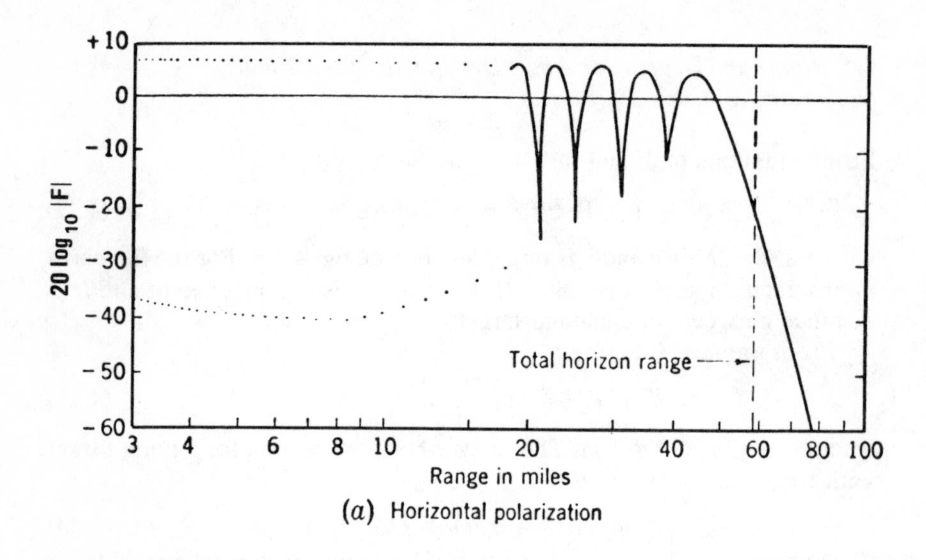

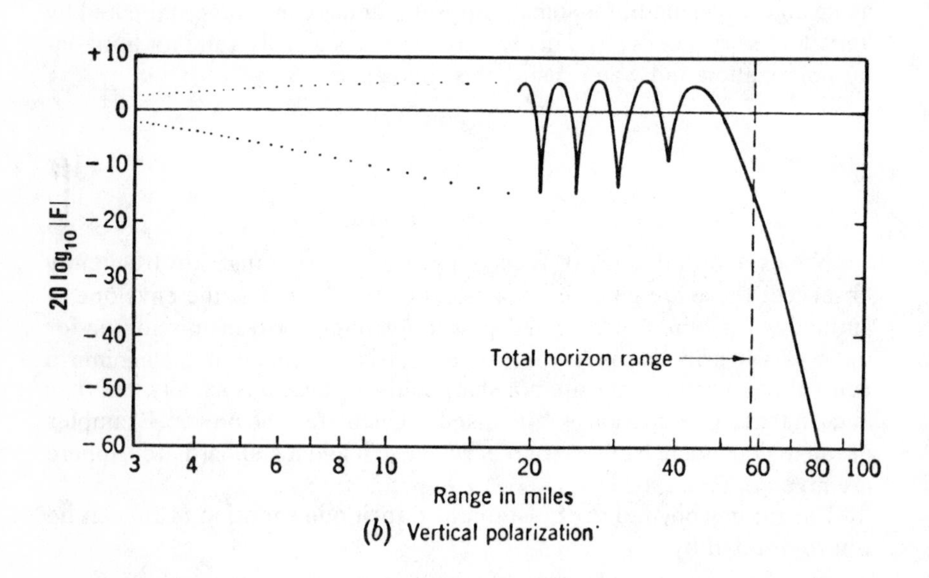

Source: D.E. Kerr, *Propogation of Short Radio Waves*. Copyright 1951, McGraw-Hill Book Company, Inc. Used with permission of McGraw-Hill Book Company.

Figure 4-13. Variation of Field Strength Relative to Free-space Field ($20 \log_{10} |F|$) from an Isotropic Antenna for a Wavelength of 10 cm and Terminal Heights of 90 and 1,000 Feet Over a Smooth Sea. The "total horizon range" represents the limit of line-of-sight propagation for a standard atmosphere.

and

$$\phi_V \approx \pi - b\theta$$

Then from equation (4.13)

$$|F|^2 = 1 + (1 - a\theta)^2 + 2(1 - a\theta)\cos[(4\pi h/\lambda)\sin\theta + \pi - b\theta]$$

In the limit of a very small value of θ, the cosine term equals approximately minus one. Thus,

$$\lim_{\theta \to 0} |F|^2 = 1 + (1 - a\theta)^2 - 2(1 - a\theta) = (1 - \rho_V)^2 = a^2\theta^2$$

and for θ small and less than θ_B,

$$|F|^4 = a^4\theta^4 = a^4H^4R^{-4} \tag{4.16}$$

It is therefore seen that for a point target above a smooth earth surface, echo power for geometries involving very small grazing angles is usually expected to vary as R^{-8}.

4.6 Vertically Extensive Objects above a Smooth Earth

Since F is a complex function of position, care must be exercised in integrating if the proper equivalent value of $|F|^4$ is to be obtained for an extended target. For example, if the target were composed of several scatterers each with free-space cross section σ_i', one must add the reradiated fields from all scatterers while appropriately accounting for differences in relative phase:

$$\sigma = \left| \sum_i F_i^2 [\sqrt{\sigma_i'}\, e^{j\psi_i}] \right|^2 \tag{4.17}$$

The phase factor $e^{j\psi_i}$ includes phase change on reflection and phase delay because of distance from radar. Notice that the term in brackets is independent of range if the target is in the far zone (sec. 2.6). The F_i are equal to

$$F_i = 1 + \rho e^{-j[(4\pi/\lambda)h_i \sin\theta + \phi]}$$

and are therefore dependent on a number of parameters including target height. Since the h_i are extended in height, the nulls in range dependence (that exist for a point target) will tend to be filled. The effect of this null filling depends on vertical extent of the target relative to $\lambda/4\sin\theta$, the height of an interference lobe.

For an extended target, F is not constant over the target because the individual scatterers are, in general, distributed in height on an interference pattern of the kind illustrated in figure 4-11. If target range is considerably beyond the point where effective target height is less than $\lambda/4 \sin\theta$, echo power is expected to vary as R^{-8}. This subject is discussed further in section 4.13.

In general, a target is composed of many scatterers. Since F is a complex function of position, appropriate care must be exercised in integrating equation (4.17) if a good estimate of $|F|^4$ is to be obtained for an extended target. Certain simplifying assumptions are possible if only an average is needed for a large number of randomly positioned scatterers. For example, if all F_i were unity (no reflecting plane), the time average or average for many different positions of the scatterers would be

$$\langle \sigma \rangle = \sum_i \sigma_i' = \langle \sigma' \rangle$$

In general, the average cross section for an ensemble of independent scatterers may be expressed as

$$\langle \sigma \rangle = \sum_i |F_i|^4 \sigma_i' = \langle |F|^4 \rangle \langle \sigma' \rangle = \langle \mathscr{F} \rangle \langle \sigma' \rangle \tag{4.18}$$

where for brevity $\mathscr{F}$ denotes $|F|^4$ and $\langle \sigma' \rangle$ denotes average cross section for the scatterers in absence of a reflecting plane. Equation (4.18) illustrates effects of a reflecting plane $\langle \mathscr{F} \rangle$ on average cross section $\langle \sigma \rangle$ of a collection of scatterers in terms of average free-space cross section $\langle \sigma' \rangle$.

Now consider area extensive scatterers such as rain or trees near a smooth earth. Such targets can usually be considered as a collection of random scatterers that are uniformly distributed with heights extending over at least several interference lobes. Under these conditions, the average value of $|F|^4$ is expected to be the same as that averaged over one or more complete lobes. Therefore, for transmitting and receiving either vertical or horizontal polarization, the average of $|F|^4$ is

$$\frac{1}{2\pi}\int_0^{2\pi} |F|^4 d\psi = \frac{1}{2\pi}\int_0^{2\pi} (1 + \rho^2 + 2\rho\cos\psi)^2 d\psi$$

$$= 1 + 4\rho^2 + \rho^4$$

where

$$\psi = (2\pi/\lambda)\Delta R + \phi$$

and the dependence of ρ on ψ has been neglected. Thus, it is clear that

$$\sigma_{VV} = \sigma_{VV}' \left[1 + 4\rho_V^2 + \rho_V^4\right]$$

and (4.19)

$$\sigma_{HH} = \sigma_{HH}' \left[1 + 4\rho_H^2 + \rho_H^4\right]$$

Note that σ_{VV}/σ_{VV}' and σ_{HH}/σ_{HH}' can be as large as 6 (7.8 dB); therefore, rain or tree return may be as much as 7.8 dB larger in the presence of, than in the absence of, a smooth earth. Results of using equations (4.19) for obtaining effects of a smooth sea on a large vertically extensive collection of scatterers are shown in figure 4-14.

The primary cause discussed for changes of $|F|$ with range has been changes in the locations of scatterers within the interference pattern. A secondary effect discussed is changes in the reflection coefficient caused by changes in depression angle.

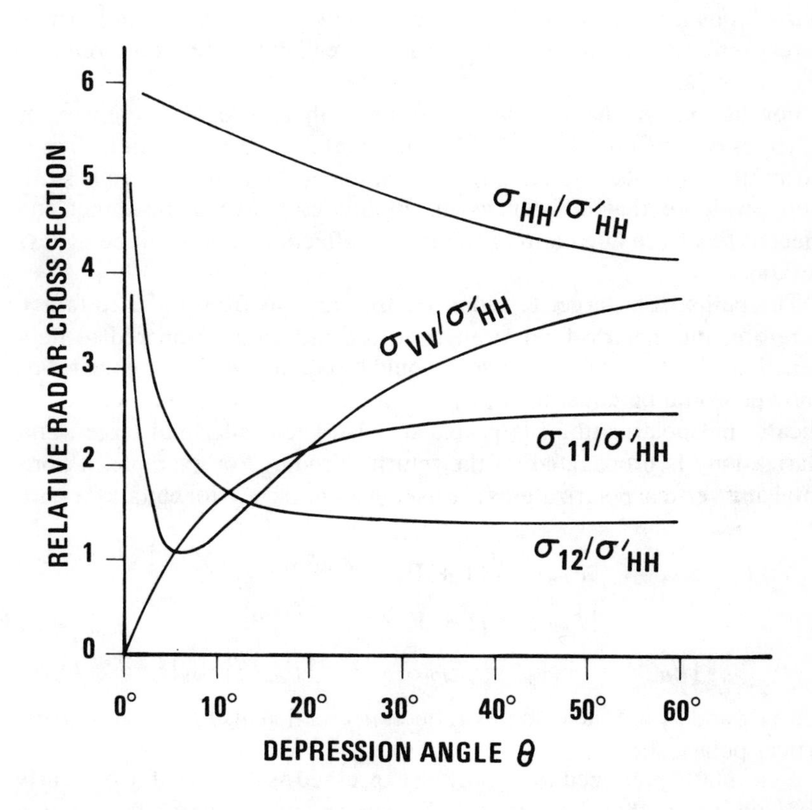

Source: Long and Zehner (1970).

Figure 4-14. Relative Radar Cross Sections for a Large Collection of Nondepolarizing Scatterers. Calculations were made for a transmitted wavelength of 10 cm and a perfectly smooth surface.

4.7 Propagation Factors for Circular, Horizontal, and Vertical Polarizations above a Smooth Earth

If either horizontal or vertical polarization is transmitted, there is no depolarization ($\sigma_{VH} = \sigma_{HV} = 0$) for a target above a smooth sea if the target itself does not depolarize. However, even though there is no depolarization in the sense of linear polarization, the echo for transmitting circular polarization will in general be mixed; that is, there will be left and right circularly polarized components in the echo. This depolarization in the sense of circular polarization is caused by the fact that in general the complex values (phase factors) for the propagation factors for horizontal and vertical polarizations are not equal. It is shown below that for sufficiently small depression angles a smooth sea will have negligible effect on circularly polarized echo.

For the analysis that follows, it is assumed that no depolarization exists due to reflections off the sea if either horizontal or vertical polarization is transmitted. It is also assumed that the antenna elevation pattern is sufficiently wide so that differences in amplitudes between the direct and reflected fields are due entirely to the sea reflection coefficient being less than unity.

The paths from target to radar are the same as from radar to target. Therefore, the electric field strength at the radar (as compared to field strength if target were in free space) would be equal to $F_{ii}F_{jj}$, and the ratios of power would be equal to $|F_{ii}F_{jj}|^2$. The indices i and j are included to indicate that polarization i is propagated between radar and target, and polarization j is propagated on the return to radar. For example, if horizontal and vertical polarizations are used, values of $|F|^4$ for equation (4.10) are

$$|F_{HH}|^4 = |1 + \Gamma_H e^{-j(2\pi/\lambda)\,\Delta R}|^4$$

$$|F_{VV}|^4 = |1 + \Gamma_V e^{-j(2\pi/\lambda)\,\Delta R}|^4$$

$$|F_{HV}|^4 = |F_{VH}|^4 = |F_{HH}F_{VV}|^2 = |F_{HH}|^2\,|F_{VV}|^2 \tag{4.20}$$

where Γ_H and Γ_V are the complex reflection coefficients for horizontal and vertical polarizations.

A circularly polarized wave may be expressed as the sum of two linearly polarized waves that are orthogonal in space and for which the relative time-phase angle is 90°. Therefore, when a circularly polarized wave is reflected from a dielectric surface, it becomes elliptically polarized because reflection coefficients for horizontal and vertical polarizations are in general different. Elliptical waves can be described as the sum of two perpendicular linearly polarized waves, or alternatively, as the sum of two circularly polarized waves of opposite rotational sense. For some purposes circularly polarized waves are described simply as "same" or "opposite"; this means that the rotation direction for the received wave is either the same as, or opposite to, that of the transmitted wave.

Isadore Katz (1963) gives equations for the two circularly polarized waves that exist as a result of direct and indirect paths for one-way propagation, and he also reviews effects of surface roughness. For one-way propagation, the ratios (with respect to free-space fields) of same and opposite sense circular fields can be expressed as

$$F_{11} = 1 + \frac{(\Gamma_H + \Gamma_V)}{2} e^{-jk\Delta R} = \frac{1}{2}(F_{HH} + F_{VV}) \tag{4.21}$$

and

$$F_{12} = \frac{(\Gamma_V + \Gamma_H)}{2} e^{-jk\Delta R} = \frac{1}{2}(F_{VV} - F_{HH})$$

For circular polarizations, $|F|^4$ for equation (4.10) is *not* $|F_{11}|^4$ or $|F_{12}|^4$. As indicated by equations (4.21) the circular wave is transformed into two oppositely rotating circular waves on propagating to the target, and on propagating from target to radar each of the two reradiated circulars are also transformed into circulars and mixed. The problem of determining $|F|^4$ is difficult to visualize because (a) a target will, in general, cause a circularly polarized incident wave to be reradiated as an elliptical wave (sum of two circulars), and (b) two circulars are generated by reflecting off a dielectric surface.

The equations developed in chapter 3 to relate target scattering properties for linear polarizations to those for circular are useful for determining effects of reflections off the earth's surface. Let a prime denote "target in absence of reflecting plane (free space)." Then, if the first subscript represents "transmitted polarization" and the second "received polarization,"

$$a_{ij} = a'_{ij} F_{ii} F_{jj}$$

By using equations (3.8) received field strengths for circular polarization in terms of those for linear polarization can be expressed as

$$|c_{11}| = \left| \frac{a'_{HH}F^2_{HH} - a'_{VV}F^2_{VV}}{2} + j\, a'_{HV}F_{HH}F_{VV} \right|$$

$$|c_{12}| = |c_{21}| = \left| \frac{a'_{HH}F^2_{HH} + a'_{VV}F^2_{VV}}{2} \right| \tag{4.22}$$

$$|c_{22}| = \left| \frac{a'_{HH}F^2_{HH} - a'_{VV}F^2_{VV}}{2} - j\, a'_{HV}F_{HH}F_{VV} \right|$$

Recall that the F's and the a's contain phase factors; therefore, the equations above are, in general, difficult to interpret.

To obtain insight into the effects of the indirect path on the circularly polarized echoes, assume that ΔR is essentially constant over the target and that $a_{VV} = a_{HH}$ and $a_{VH} = a_{HV} = 0$. For such a target in free space, $\sigma'_{12} = \sigma'_{21} = \sigma'_{VV} = \sigma'_{HH}$ and $\sigma'_{11} = \sigma'_{22} = \sigma'_{VH} = \sigma'_{HV} = 0$. A smooth metallic surface with curvature large compared with wavelength or a conventional three-sided corner reflector used as a navigational buoy will generally satisfy these conditions. Use of equations (4.22) for this class of targets gives (Long 1968)

$$
\begin{aligned}
\sigma_{11} = \sigma_{22} &= \frac{\sigma'_{HH}}{4} \left| F^2_{HH} - F^2_{VV} \right|^2 = \frac{\sigma'_{12}}{4} \left| F^2_{HH} - F^2_{VV} \right|^2 \\
\sigma_{12} = \sigma_{21} &= \frac{\sigma'_{HH}}{4} \left| F^2_{HH} + F^2_{VV} \right|^2 = \frac{\sigma'_{12}}{4} \left| F^2_{HH} + F^2_{VV} \right|^2
\end{aligned}
\tag{4.23}
$$

Recall that rotation sense is defined as viewed along the direction of propagation; therefore, reflection off a sphere reverses circularity. Thus, the cross section for a sphere (in free space) is zero if "same" polarization is received. A target for which σ for same circular is zero is sometimes called a single (or odd) bounce target; one for which opposite circular is zero is called a double (or even) bounce target. Equations (4.23) are for an odd bounce target and, by comparing these with equation (4.22), values of $|F|^4$ can be obtained. Table 4-2 contains values of $|F|^4$ for odd bounce as well as even bounce targets; most real targets are a mixture of the two.

Equations (4.23) are useful for considering the effects of the earth on a point target. Since ϕ_H and ϕ_V are in general different, nulls in F_V and F_H do not occur at the same range. Depths of the nulls in F_H and F_V depend on ρ_H and ρ_V which are, in general, also different.

For microwaves and depression angles of 1° or less, ϕ_H and ϕ_V differ negligibly from 180°. Under these conditions, the peaks and nulls in F_{HH} and F_{VV} occur for ranges such that values of ΔR are $N\lambda/2$ where N is odd and even, respectively; for these conditions, peaks and nulls in F for opposite circular occur at the same ranges as do those for F_{HH} and F_{VV}. Table 4-3 contains values of extrema in $|F|^4$ if ϕ_H and ϕ_V are nearly equal. For a smooth sea, ρ_H will be nearly unity for depression angles such that ϕ_V is nearly 180°; however, sea roughness will decrease both ρ_H and ρ_V.

To illustrate the effect of reflections from a smooth water surface for these small depression angles, assume that ρ_H and ρ_V are 1.0 and 0.8, respectively. Then for a range such that ΔR is $\lambda/2$ (a maximum in $|F|^4$), the ratio of $|F|^4$ for opposite to same sense circular would be approximately 20 dB. This means that for a sphere or a corner reflector the ratio σ_{12}/σ_{11} would be approximately 20 dB, and σ_{12} would be only a fraction of a decibel less than if the water had absolutely no effect on circularity. For these conditions, σ_{HH}, σ_{VV}, and σ_{12} would be nearly equal.

Table 4-2
Propagation Factors for Circular Polarization and Simple Point Targets

Received Polarization	*Target*	$\lvert F \rvert^4$
Same circular	Odd bounce	$1/4 \lvert F_{HH}^2 - F_{VV}^2 \rvert^2$
Opposite circular	Odd bounce	$1/4 \lvert F_{HH}^2 + F_{VV}^2 \rvert^2$
Same circular	Even bounce	$1/4 \lvert F_{HH}^2 + F_{VV}^2 \rvert^2$
Opposite circular	Even bounce	$1/4 \lvert F_{HH}^2 - F_{VV}^2 \rvert^2$

Table 4-3
Extrema in F^4 for Circular Polarization and Depression Angles for which ϕ_V and ϕ_H are Nearly Equal to 180°

Received Polarization	*Target*	F^4_{max}	F^4_{min}
Same circular	Odd bounce	$1/4\,[(1+\rho_H)^2-(1+\rho_V)^2]^2$	$1/4\,[(1-\rho_H)^2-(1-\rho_V)^2]^2$
Opposite circular	Odd bounce	$1/4\,[(1+\rho_H)^2+(1+\rho_V)^2]^2$	$1/4\,[(1-\rho_H)^2+(1-\rho_V)^2]^2$
Same circular	Even bounce	$1/4\,[(1+\rho_H)^2+(1+\rho_V)^2]^2$	$1/4\,[(1-\rho_H)^2+(1-\rho_V)^2]^2$
Opposite circular	Even bounce	$1/4\,[(1+\rho_H)^2-(1+\rho_V)^2]^2$	$1/4\,[(1-\rho_H)^2-(1-\rho_V)^2]^2$

4.8 Propagation Factors for a Cloud of Scatterers above a Smooth Earth

Equations (4.19) give the effects of a smooth earth on σ_{HH} and σ_{VV} for a cloud of scatterers. By a cloud is meant a collection of randomly located scatterers that extend to heights at least several interference lobes above the earth's surface. In this section, equations for circular polarizations are derived that are applicable to such a cloud of scatterers providing the individual scatterers do not of themselves depolarize the echo.

Raindrops cause only negligible depolarization and therefore this section is of importance for considering the advantage of circular polarization for improving detectability of objects in rain (also see sec. 4.13). Equations (4.23) can be used for such scatterers. If it is assumed that the scatterers are randomly located in a uniform sense and overlap several interference lobes, then from equations (4.23) cross section for circular polarization will be

$$\sigma_{11} = \frac{\sigma'_{HH}}{4}\left\{\frac{1}{2\pi}\int_0^{2\pi} |F^2_{HH} - F^2_{VV}|^2 d\psi\right\}$$

and (4.24)

$$\sigma_{12} = \frac{\sigma'_{HH}}{4}\left\{\frac{1}{2\pi}\int_0^{2\pi} |F^2_{HH} + F^2_{VV}|^2 d\psi\right\}$$

where

$$\sigma'_{HH} = \sigma'_{VV} = \sigma'_{12} \quad \text{and} \quad \psi = (2\pi/\lambda)\,\Delta R + \phi \tag{4.25}$$

The integrands above can be readily expressed as the sum of trigonometric functions that are easy to integrate:

$$|F^2_{HH} - F^2_{VV}|^2 = [2\rho_H \cos\psi_H + \rho_H^2 \cos 2\psi_H - 2\rho_V \cos\psi_V - \rho_V^2 \cos 2\psi_V]^2 + [2\rho_H \sin\psi_H + \rho_H^2 \sin 2\psi_H - 2\rho_V \sin\psi_V - \rho_V^2 \sin 2\psi_V]^2$$

and (4.26)

$$|F^2_{HH} + F^2_{VV}|^2 = [2 + 2\rho_H \cos\psi_H + \rho_H^2 \cos 2\psi_H + 2\rho_V^2 \cos\psi_V + \rho_V^2 \cos 2\psi_V]^2 + [2\rho_H \sin\psi_H + \rho_H^2 \sin 2\psi_H + 2\rho_V \sin\psi_V + \rho_V^2 \sin 2\psi_V]^2$$

Using equations (4.25) and (4.26), one finds (Long and Zehner 1970)

$$\sigma_{11} = \sigma'_{HH}\left[\rho_H^2 + \rho_V^2 - 2\rho_H\rho_V \cos(\phi_H - \phi_V) + \frac{\rho_H^4 + \rho_V^4 - 2\rho_H^2\rho_V^2 \cos 2(\phi_H - \phi_V)}{4}\right]$$

(4.27)

$$\sigma_{12} = \sigma'_{HH}\left[1 + \rho_H^2 + \rho_V^2 + 2\rho_H\,\rho_V \cos(\phi_H - \phi_V) + \frac{\rho_H^4 + \rho_V^4 + 2\rho_H^2\rho_V^2 \cos 2(\phi_H - \phi_V)}{4}\right]$$

Results of using equations (4.19) and equations (4.27) for calculating effects of a smooth sea on the linear and circular polarization cross sections of a large vertically extensive collection of scatterers are shown in figure 4-14. The phase and amplitude characteristics of reflection coefficient versus depression angle θ which were used are those given for sea water and a 10 cm wavelength (see figs. 4-4 and 4-5). The slow fall-off of σ_{HH}/σ'_{HH} with θ corresponds to the gradual decrease of ρ_H from its maximum value of unity at $\theta = 0°$. The minimum in σ_{VV}/σ'_{HH} is due to the pseudo-Brewster-angle effect for vertical polarization. Curves for circular polarization show the combined effects of the linear quantities as determined by equations (4.27). The results are complex, but it is seen that echo from rain and other scatterers that are essentially nondepolarizing above the earth is strongly dependent on transmitted and received polarizations and depression angle.

Equations (4.27) are not, in principle, valid unless the sea is perfectly smooth and flat. For example, a horizontally or vertically polarized wave reflected from a flat surface is not depolarized, but the field scattered from an actual surface would be partly depolarized. Further, the field reflected from the sea is expected to fluctuate to some extent even though the surface is "smooth" in terms of the Rayleigh roughness criterion. This subject is covered in section 4.15, and figures 4-17 and 4-18 help illustrate the added complexities caused by the sensitivity of echo strength to the earth's surface roughness expressed with respect to transmitter wavelength.

Effects of Surface Roughness on Forward Scattered Fields

4.9 Reflection Coefficient for Rough Surfaces

According to the Rayleigh roughness criterion (sec. 4.4), effective roughness is proportional to $\Delta h \sin\theta/\lambda$. Therefore, one would expect the apparent reflection coefficient to be a function of surface roughness, depression angle, and transmitter wavelength. Available experimental data indicate that for small values of the parameter $\Delta h \sin\theta/\lambda$, reflection coefficient increases with increases in that parameter. However, for large values ($\Delta h \sin\theta/\lambda >> 1/8$) it seems that reflection coefficient is independent of $\Delta h \sin\theta/\lambda$ for a given surface.

Various theoretical and experimental investigations have indicated that the field scattered by a rough surface can be considered to be the sum of two components: a specular component and a so-called diffuse component.[c] The reflection coefficients are designated as R_s and R_d, respectively. The existence of the two components of the reflected field is real—it is not simply the result of the theoretical analyses, and the two components have been observed by several different experimental methods.

A specularly reflected field is not a strong function of position. Therefore, when added to a "direct" field it will produce a periodically varying field if sampled in the vertical dimension, as illustrated in figure 4-11. The amplitude of an idealized diffusely scattered field is Rayleigh distributed and the phase is distributed uniformly. Therefore, a diffuse field added to a "direct" field will produce a randomly varying field if sampled in the vertical dimension.

The diffusely scattered field is expected from a large number of small scatterers that will be caused, for example, by tree tops of a thick forest. It would be expected that the reflections would be almost totally diffuse for microwaves and independent of $\Delta\Phi$ for certain ranges in values of θ and λ.

---

[c]Alternatively, these fields are called coherent and incoherent, respectively (see sec. 2.5, chapter 2).

However, diffuse scattering is not always independent of $\Delta\Phi$, because in the limit of a perfectly smooth surface, diffuse scattering will not exist. A specular field may also be caused by many scatterers since it can be produced by the sum of many fields that are of themselves specular. Actual fields are, of course, the sum of both a specular and a diffuse component.

The modification in the reflection coefficient caused by a Gaussian distributed surface has been calculated by a number of authors, including W.S. Ament (1953) and by Petr Beckmann and André Spizzichino (1963). For these calculation effects of sharp edges and shadowing are neglected. The results yield an expression for magnitude of reflection coefficient ρ_s of a surface that is the product of two factors:

$$\rho_s = \mathcal{R}_s \rho_0 \tag{4.28}$$

where ρ_0 is the reflection coefficient of a plane, smooth surface and $\mathcal{R}_s$ is the specular "scattering coefficient." The mean square value of $\mathcal{R}_s$ for a Gaussian height distribution is given by

$$\langle \mathcal{R}_s^2 \rangle = e^{-(\Delta\Phi)^2} \tag{4.29}$$

with

$$\Delta\Phi = 4\pi \, \Delta h \sin\theta/\lambda$$

where θ is the incidence angle and Δh represents the standard deviation, or rms value, of the Guassian distribution of heights. As will be described below, results of experiments indicate that equations (4.28) and (4.29) are applicable to "specularly" reflected fields.

Figure 4-15 shows values of $\langle \mathcal{R}_s^2 \rangle$ given by Beckmann and Spizzichino (1963, p. 318) that were calculated from equation (4.29); the crosses represent experimental data given by various other authors. Spizzichino used calculated values of ρ_0 and D (see eq. (4.1)) with reported values of ρ_s to obtain $\mathcal{R}_s$. Spizzichino states that the correlation of $\mathcal{R}_s$ with $\Delta\Phi$ is good considering that (a) the reflection data are often given with little precision, (b) the evaluation of ρ is simply an estimate, and (c) the values for Δh he had to use were at best only guesses. Also, Spizzichino postulates that the points with $\Delta\Phi$ exceeding 0.4π probably correspond either to diffuse scattering or to a combination of specular and diffuse because authors sometimes have given values of reflection coefficient without stating to which of the two types of reflection the data correspond. The points in figure 4-15 were obtained from data measured over land and over sea and for plane and hilly conditions. Moreover, the data were obtained by various measurement methods.

For purposes of definition of terms, assume that the reflection coefficient ρ_d of the diffusely scattered component is given by

$$\rho_d = \mathcal{R}_d \rho_0 \tag{4.30}$$

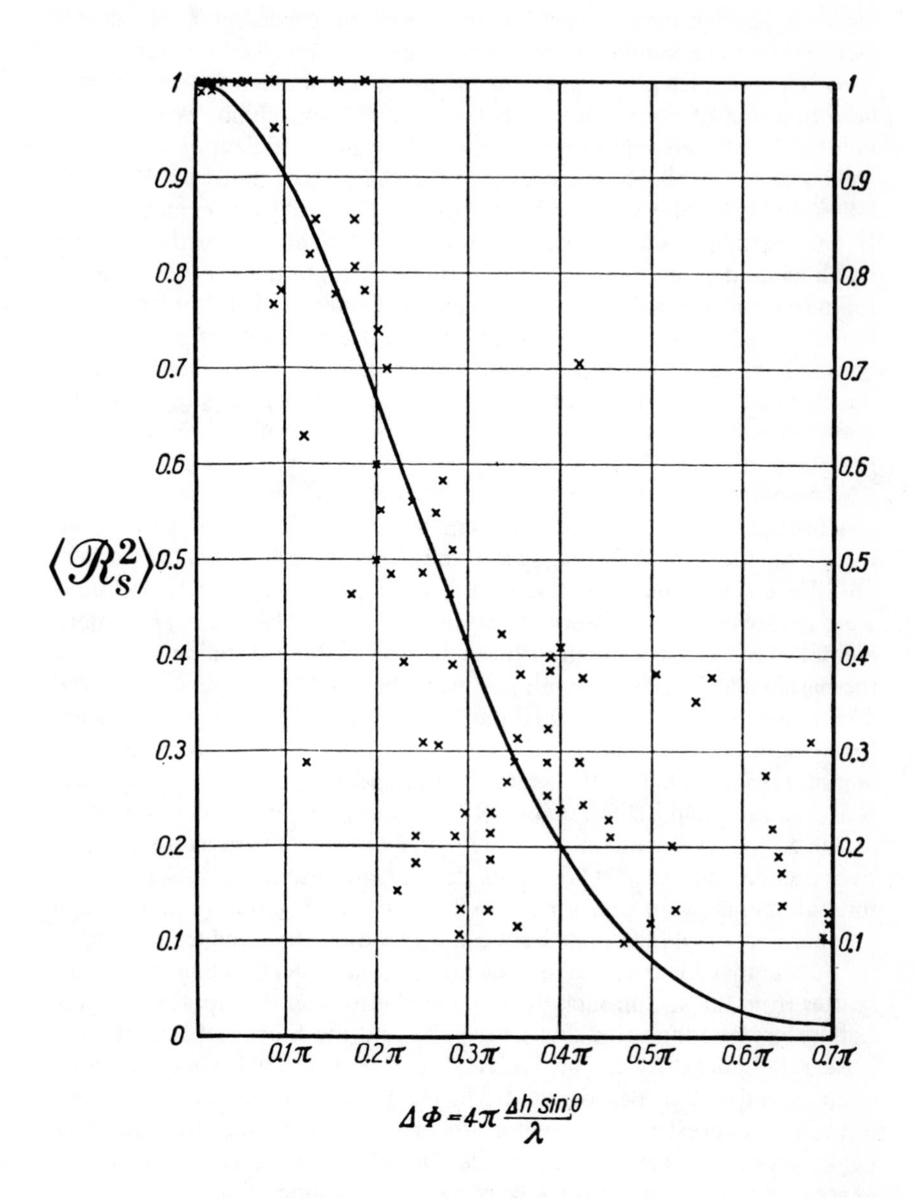

Source: Beckmann and Spizzichino (1963, p. 318).

Figure 4-15. Dependence of the Scattering Coefficient on the Phase Variance. The full curve plots equation (4.29).

where ρ_0 is the reflection coefficient of a smooth earth and $\mathcal{R}_d$ is a coefficient that depends only on the irregularities of the reflecting surface.

The percentage power contained in the specular and diffuse fields may be a function of polarization (Beckmann and Spizzichino 1963, p. 339). Polarization dependent effects on the reflection coefficient of samples of soil covered by vegetation have been reported by L.H. Ford and R. Oliver (1946) and E.M. Sherwood and E.L. Ginzton (1955) at wavelengths of 9 and 10 cm. Vertical blades of grass reflect in a polarization sensitive manner much as do dipoles: if the grass is sufficiently high, the wave is diffusely scattered and the reflection coefficient ρ_d is small, of the order of 0.1. However, a horizontally polarized wave is reflected as if the ground were bare and the reflection is predominantly specular or diffuse depending on the irregularities of the ground, per se. If the vegetation is more dense and completely covers the ground, ρ_d is of the order of 0.1 regardless of polarization.

The book by Beckmann and Spizzichino (1963) contains several chapters of theoretical and experimental material on reflection of waves scattered from the ground. In that book three series of measurements by others on diffuse scattering made under very different experimental conditions were reviewed; it was observed that in spite of vast differences in experimental conditions, for rough surfaces the results showed similar values for the parameter $\mathcal{R}_d$. For example, $\mathcal{R}_d$ was observed to be widely scattered about a mean value between 0.30 and 0.35 and most of the values of $\mathcal{R}_d$ were between 0.2 and 0.4. Results considered included those by Kenneth Bullington (1954), C.I. Beard, Isadore Katz, and L.M. Spetner (1956), and R.E. McGavin and L.J. Maloney (1959). The experimental conditions were at 4,000 MHz over cultivated and inhabited terrain, at various frequencies over the sea, and at 1,000 MHz over desert land, respectively. Some of the measurements were performed on links between ground stations and another part employed links between a ground station and an airplane.

Substantial advances have been made in the understanding of forward scatter from the sea through the combined efforts of the Applied Physics Laboratory at Johns Hopkins University and the Electrical Engineering Research Laboratory at the University of Texas. This work has been summarized by C.I. Beard (1961). The electric field scattered in a forward direction is expressed as the vector sum of a coherent field and a fluctuating field (incoherent) having zero mean. The phase of the coherent field is expressed as if the sea surface were smooth; the amplitude of this field is reduced by a multiplicative factor determined by surface roughness (eq. (4.29)).

Near $\theta = 0$ the coherent component approaches the value for a smooth surface ($\rho = 1$) and the incoherent component approaches zero. As θ increased, the coherent component decreased in accordance with equation

(4.29). Also as θ increased the incoherent component increased until $\Delta\Phi \approx 0.4\pi$ radians, at which point it saturated and thereafter decreased slowly. For $\Delta\Phi \approx 0.4\pi$ radians and larger, the experimental values of the coherent component were larger than is calculated by equation (4.29).

Power contained in the incoherent field was found to be at most about 30 percent of that in the coherent field. The measurements were made for water surfaces with roughness between smooth and those for which $\Delta h \sin\theta = \lambda/3$, where Δh is root-mean-square wave height. The measurements were made at wavelengths of 5.3, 4.2, and 0.86 cm for transmitting and receiving vertical polarization and for transmitting and receiving horizontal polarization. Factors by which the coherent fields were reduced because of roughness and ratios observed in power for the incoherent and coherent fields did not indicate a dependence on polarization.

Beard, Katz, and Spetner (1956) introduced a model for forward scattering that consists of a coherent field and an incoherent field. The intensity of fluctuations of the incoherent field was assumed to be Rayleigh distributed and the phase was assumed to be uniformly distributed. Thus, the model gives a total forward scattered field that is Ricean distributed (sec. 2.8). The analyses of Beard (1961) indicate that the distributions for the experimental data are in fact Ricean. S.P. Zehner, W.M. O'Dowd, Jr., and F.B. Dyer (1974) report that S and X band forward scattered signals are also Ricean distributed for receivers that are only 1/2 to 5 feet above the mean sea. They report a three times larger ratio of incoherent-to-coherent power (for a given roughness) than does Beard, presumably due to the closeness of their antennas to the water.[d] However, it seems the model originally proposed by Beard, Katz, and Spetner (1956) is representative of the physical mechanisms involved.

4.10 Shadowing

The shadowing effect caused by mountains is illustrated in figure 1-1. The mountains are characterized by bright signals from the near slopes and the absence (not illuminated by the radar) of signals from the far slopes. For the sea, it is likely that effects of shadowing become appreciable for angles of incidence closer to horizontal than the angle formed by the crest height and sea wavelength (Court 1955). Several authors [Beckmann 1965; Brockelman and Hagfors 1966; Shaw and Beckmann 1966; Smith 1967] have considered, on theoretical bases, the reduction in scattered power expected to be caused by the shadowing of troughs by crests. It seems that effects of shadowing are more appreciable for upwind or downwind direc-

[d] From personal communications with F.B. Dyer, $\Delta h \sin\theta/\lambda$ was 0.03 or less for these measurements.

tions than for crosswind. At near grazing incidence angles, when shadowing is appreciable, it is expected that the tops of the crests contribute strongly toward defining the effective reflecting surface; at near vertical incidence angles the troughs are expected to contribute significantly to creating the effective reflecting surface. Therefore, the magnitude of the reflected field will depend on incidence angle, and the location of the effective reflecting surface will also be a function of that angle. The subject of shadowing is not yet well understood.

4.11 Depolarization

On the basis of theoretical work, Beckmann showed (Beckmann and Spizzichino 1963, chap.8) that a wave reflected in the plane of incidence is not depolarized if the incident wave is polarized either purely vertically or horizontally. Further, on the basis of theoretical studies by Beckmann it was shown that a horizontally or vertically polarized wave is strongly depolarized if it is scattered out of the plane of incidence. This presumably would concern the diffusely scattered component of the field because irregularities of the nature that would tend to cause diffusely scattered components will contribute because of their nondirectional nature from points on the earth's surface located off the axis of the propagation path.

Spizzichino (Beckmann and Spizzichino 1963, p. 344) describes a propagation experiment for which purely horizontal and purely vertical polarizations were transmitted. The received cross polarized scattered fields were at least 25 dB below that of the polarization transmitted. The experiments were performed at *X* band and over water, but surface roughness was not indicated. The equipment was capable of searching for direction of depolarized scattering and the measurements indicated that the depolarized signals arrived from directions other than from the point of specular reflection. According to Spizzichino, the diffusely scattered field was highly depolarized (ratio estimated as a few decibels).

There are few experimental data available on the depolarization of forward scattered fields and the effects of surface roughness. Beard[e] has reported observations on a one-way experiment over water at *X* band for which the depolarized linear components were down from the transmitted components by more than 20 dB. It is well known that echoes from targets above the sea sometimes (depending on wavelength, incidence angle, and sea roughness) exhibit sharp minima because of interference. Existence of the sharp minima (corresponding to those of figure 4-13) require a reflected field component polarized like, and magnitude comparable to, the direct field. Therefore, the minima are of themselves evidence that depolarization due to reflection off the sea is frequently small.

[e]Private communication by C.I. Beard. See N.I. Durlach (1965, p. 57).

One-way X-band experiments for which circular polarization was transmitted have been reported by I.M. Hunter and T.B.A. Senior (1966). The transmitting antenna was at a fixed site approximately 35 feet above water and receivers were mounted aboard an airplane. Measurements were made for angles of incidence θ of 1 1/2° and less as the plane flew radially outward to the radar horizon. Most flights were made for heights between 100 and 300 feet for which angles of incidence with respect to the sea were 3/4° or less.

During initial measurements, simultaneous recordings were made on signals received at the aircraft with antennas for horizontal, vertical, right-circular, and left-circular polarizations. The records of three of the channels were found to be identical in all major features; the fourth channel, sensitive to circular polarization orthogonal to that transmitted, consistently showed a signal level at least 20 dB below the others. The weak-signal channel was subsequently eliminated in favor of an aircraft-height record. These results also imply that for forward scatter from the sea, depolarization for linear polarization is small.

Measurements were made under prevailing open sea conditions, and the authors include a "typical" plot of field strength versus range that indicates wind speed of 20 knots and sea state of 3-5. The fact that the peaks and nulls for the three signals of comparable level coincide in range indicates that the phase of the effective reflection coefficient for vertical and horizontal polarizations are almost equal, as would be expected if the sea surface were smooth.

Echo from Targets That are Above a Physically Rough Earth

A physically rough surface may appear in an electromagnetic sense to be either smooth, rough, or mixed. Rough surfaces are not well understood from an electromagnetic point of view, and detailed data presently available on one- and two-way reflections from the earth are quite limited. Nonetheless, the information that is available is of significant value for predicting possible effects of the earth's surface.

4.12 Variation of Echo Power with Range

To a good approximation, average echo power from an extended target (one extended over several interference lobes) that is located above a reflecting plane will vary as R^{-4} at close ranges and as R^{-8} at more distant ranges. An example of this effect for ships is depicted in figure 4-16. For certain geometries, effects of the earth's curvature will contribute to the power drop-off with range and then the variation will be even faster than R^{-8}. Also, trapping (sec. 2.12) can cause echo strength to drop less rapidly with range than for a standard atmosphere.

From figures 4-4 and 4-7 it is seen that Γ_H for a smooth sea is close to -1 for angles of a few degrees or less for microwaves. Under these conditions, the R^{-8} zone will exist for values of $\Delta R/\lambda$ small enough for equation (4.16) to be valid. From equation (4.15) the greatest range for a maximum in $|F|$ occurs when ΔR equals $\lambda/2$; clearly equation (4.16) is valid only if ΔR is much less than $\lambda/2$. If $\Delta R \leq \lambda/8$, equation (4.16) will be a good approximation. Since $\Delta R = 2h \sin\theta$, the condition $\Delta R \leq \lambda/8$ can be expressed as

$$\sin\theta \leq \lambda/16h \tag{4.31}$$

where h is effective target height.

To determine if a sea surface is sufficiently smooth for the R^{-8} zone to exist, equation (4.13) can be used with equation (4.29):

$$\Delta\Phi \leq 4\pi\,\Delta h \sin\theta/\lambda = (\pi/4)\,(\Delta h/h)$$

If h is at least four times Δh (standard deviation of wave height), then $\Delta\Phi < \pi/16$. Clearly the condition $h/\Delta h$ equals 4 is a conservative estimate for target height; for example, the "significant" wave height is $4\,\Delta h$ (see sec. 2.11).

From equation (4.29) and the condition $\Delta\Phi \leq \pi/16$, it can be seen that

$$\langle \mathcal{R}_s^2 \rangle = e^{-(\Delta\Phi)^2} \approx 1 - (\Delta\Phi)^2$$

is almost unity. Therefore, the R^{-8} zone can exist whether or not the sea surface is physically rough, and calculations on effect of the sea for the R^{-8} zone can be made as if the sea, independent of roughness, were a perfectly smooth surface.

From equation (4.31) it can be seen that, for the R^{-8} zone to be observed, θ will be less than a few degrees because target height will ordinarily exceed the mean sea surface by a factor of many wavelengths. For example, if effective target height is only 3.6 times transmitter wavelength, the R^{-8} zone will exist for values of θ not exceeding 1°.

4.13 Range at Which Idealized R^{-4} and R^{-8} Curves Intersect

It may be seen from equation (4.10) that echo power varies as R^{-4} if the radar parameters and cross section are fixed and $|F|$ is constant. In principle $|F|$ will be a function of range (because it depends on incidence angle) but there are many practical conditions under which $|F|$ varies so slowly that echo strength varies essentially as R^{-4}. Note that changes in ρ with θ

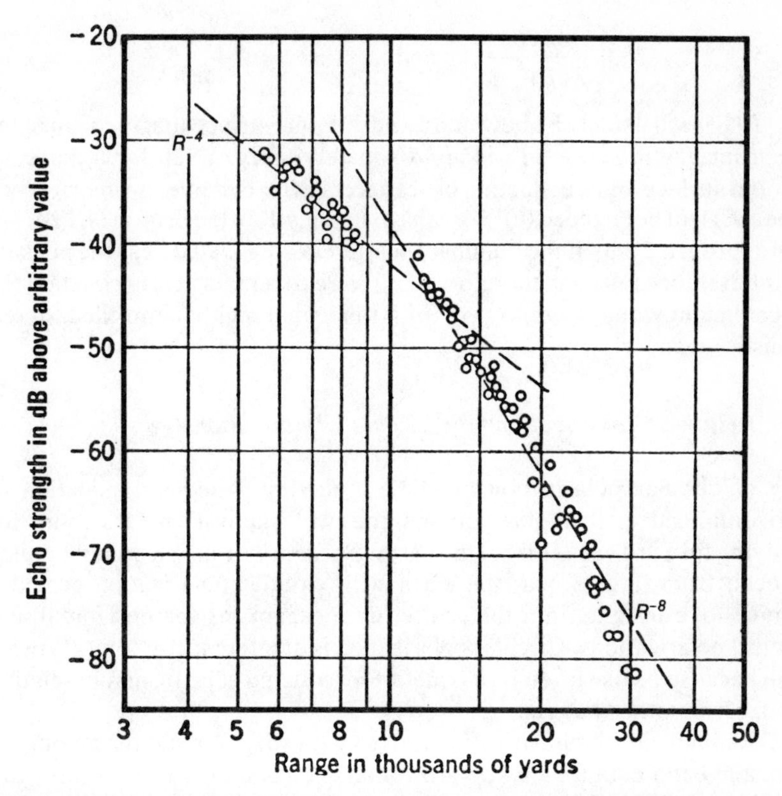

Source: D.E. Kerr, *Propagation of Short Radio Waves.* Copyright 1951, McGraw-Hill Book Company, Inc. Used with permission of McGraw-Hill Book Company.

Figure 4-16. Typical Plot of Ship Echo as a Function of Range. Radar height is 125 feet; $\lambda = 30$ cm. The straight lines correspond to variation of echo strength with R^{-4} and R^{-8}.

are often such as to cause $|F|$ to change so slowly that echo strength will vary no faster than R^{-4} in the "R^{-4}" zone.

Since $|F|$ must be equal at the cross-over of the idealized "R^{-4}" and "R^{-8}" zones, the following is a valid approximation

$$|F|_{R^{-4}} = [(2\pi/\lambda)|\Delta R|_{R=R_t}]$$

where R_t is defined as the cross-over or transition range. Recall that, in general,

$$\Delta R = 2h \sin\theta = 2h(H/R)$$

Therefore,

$$R_t = \frac{4\pi Hh}{\lambda |F|_{R^{-4}}} \qquad \text{and} \qquad \sin\theta_t = \frac{\lambda |F|_{R^{-4}}}{4\pi h} \tag{4.32}$$

$|F|_{R^{-4}}$ will usually be between 1 and 2. For a surface that is rough in the electromagnetic sense, $|F|$ is approximately unity. Even for a perfectly smooth surface and a collection of scatterers extended over an interference lobe, $|F|$ will not exceed $(6)^{1/4}$ or approximately 1.56 (see eq.on (4.19)). $|F|$ will approach 2 only if the dominant scatterers are located near the peak on an interference lobe and therefore if $|F|$ were to remain nearly constant the target height would have to vary with range in a highly controlled (deterministic) manner.

4.14 Relative Cross Sections for Circular Polarizations

Use of circular polarization embodies the simultaneous application of horizontal and vertical polarizations, the two "natural" polarizations for studying forward scatter from the earth. Because of this, an understanding of echo from targets near the earth with circular polarization provides comparative insights into the scattering mechanisms for horizontal and vertical polarizations. Circular polarization is of interest also for its "own" sake because its use often is of benefit for reducing echo from rain relative to that from other targets.

Although circular polarization is used with some radars, the amount of available echo data for that polarization is considerably less than for the two linear polarizations. It is well known (Skolnik 1962, pp. 547-51) that if the same circular polarization antenna is used for transmitting and receiving, echoes from targets are *usually* enchanced relative to those from rain. This is because the reflected energy from targets is generally mixed between two oppositely rotating polarized waves, but the energy reflected from rain, per se, is almost entirely of the circularity that will not be received. M. Gent, I.M. Hunter, and N.P. Robinson (1963) have published results on studies of rain rejection at X band and 35 GHz. R. McFee and T.M. Maher (1959) have presented estimates of the reduction in rain clutter cancellation caused by reflections from a smooth sea for a specific depression angle. G.R. Curry (1965) has reported that the echo for clouds at UHF and L-band is also substantially reduced by using circular polarization over water. It is not clear, however, whether or not the sea had an effect on the echo from the clouds.

Theoretically, all the information needed for describing the two-way signal can be obtained from complete polarization and statistical data for the one-way signal. However, all the information needed does not exist, but if it did a rigorous and general solution for radar would be very complicated. From the previous section it seems reasonable to assume that the effects of depolarization in the forward scattered fields are negligible. This assumption permits equations (4.23) to be used. Even those simplified equations require knowledge that is not available, namely, the relationships between the instantaneous values of amplitude and phase of F_{HH} and F_{VV}.

For a flat, smooth earth F_{HH} and F_{VV} each approach the value minus one for small depression angles. In this section, measurements on the ratio σ_{12}/σ_{11} for a three-sided corner made at sea with small depression angles will be reviewed. The results indicate, at least for the prevailing sea conditions, that it is possible to use the currently limited knowledge of forward scatter to calculate trends to be expected under practical conditions. It should be noted that the sea surface and radar parameters were such that a strong specularly reflected component probably existed for all but the shortest ranges.

During 1966 and 1967, sample measurements were made at X band by personnel of the Engineering Experiment Station, Georgia Institute of Technology, on the ratio σ_{12}/σ_{11} at a site overlooking the Atlantic Ocean near Boca Raton, Florida. Although the measurements on a three-sided corner were not extensive, results indicate that the echo for these targets is primarily of opposite sense (σ_{11} small with respect to σ_{12}) for the various sea states available.

Measurements were made for an antenna height of 80 feet and ranges used were in excess of one mile. For these conditions, the targets were usually at or below the first lobe in the interference pattern ($\Delta R \leq \lambda/2$).

One day was spent measuring the echo from a three-sided corner that was aboard a small, deep-sea fishing boat. Wind speed was 12 mph, peak-to-trough wave height was crudely estimated as being 2 feet. Measurements of σ_{12}/σ_{11} were made as a function of range for the corner mounted on the bow and stern. Of course, the target height varied because of swell; under calm conditions heights would have been approximately 7 and 4 feet for the bow and stern positions.

Measurements were made with two receivers and a calibrated attenuator by observing the same and opposite sense echoes on alternate range sweeps displayed on a dual trace A-scope. The ratio fluctuated rapidly by several decibels but at a rate less than the radar repetition frequency.

For the lower target height, the median of σ_{12}/σ_{11} was estimated to be 20 dB for the measured range interval of 1 to 5 nautical miles. For the higher target height and a range interval of 2½ to 10 nautical miles, the median of σ_{12}/σ_{11} was also estimated as 20 dB. However, for ranges between 1 and 2½ nautical miles, the echo strength dropped substantially and fluctuated over wide limits of amplitude; and the median of σ_{12}/σ_{11} was smaller than 20 dB. For these short ranges, the target was near the first null in the interference pattern and at times the echo was probably due solely to the boat.

The reflector was a conventional three-sided, aluminum corner, 24 inches along each seam, not specifically fabricated for this purpose. It was several years old and showed considerable signs of weathering as a result of being in a salty atmosphere. Therefore, it is not unlikely that the median ratio of 20 dB obtained was limited by the corner itself. However, these data leave little doubt that virtually all the power received from the "odd-bounce" target near water is opposite to the circular polarization transmitted for the prevailing values of $\Delta h \sin \theta/\lambda$.

4.15 A Cloud of Scatterers

Effects of a smooth earth on a large collection of nondepolarizing scatterers (raindrops) are considered in section 4.8. For such scatterers, equations (4.19) and (4.27) can be used to calculate cross sections. It is assumed that the random scatterers are uniformly distributed above a smooth earth and extend to heights that overlap several interference lobes.

Equations (4.19) and (4.27) have also been used (Long and Zehner, 1970) to calculate average cross section of rain above a rough sea even though they are strictly valid only for a smooth and flat surface. The reflection coefficients of these equations were replaced by calculated average values and therefore an approximation used is that the effects of the fluctuating forward scattered fields can be represented by their steady components. Another approximation used, based on section 4.11, is that the forward scattered horizontally and vertically polarized fields are only negligibly depolarized. Although neither of the approximations are strictly valid, they do help to provide insight into a complex problem for which available polarization and statistical data are insufficient.

Radar echo from rain will, of course, fluctuate because of the changing positions of the raindrops. For rain above a sea, the echo will have fluctuations also because of changes in the amplitude and phase of the various polarized components of a forward scattered field. Theoretically, all the information needed for describing the two-way signal can be obtained from complete polarization and statistical data for the one-way signal. A rigorous solution of the radar problem would be very complicated and it cannot be made until more complete information exists for the one-way case.

To provide insight into effects of a rough sea, the mean-square value of the specular scattering coefficient ($\langle \mathscr{R}_s^2 \rangle$ of eq. (4.29)) was used to obtain the ρ's for (4.19) and (4.27). In other words, the values of ρ used were computed from the relationship

$$\rho = \langle \mathscr{R}_s^2 \rangle^{\frac{1}{2}} \rho_0$$

The results of calculating the effects of surface roughness in this way are shown in figures 4-17 and 4-18.

From figure 4-18, it is apparent that for depression angles greater than, say, 10° and for $\Delta h/\lambda = 1$ (from sec. 2.11, wave height four to five times transmitted wavelength) the cross sections are essentially those for free space (i.e., $\sigma_{HH} = \sigma'_{HH}$, $\sigma_{VV} = \sigma'_{VV}$, $\sigma_{12} = \sigma'_{12}$, and $\sigma_{11} = 0$). However, as the depression angle approaches zero, even for a rough sea, specular reflection increases and the cross sections approach those for the smooth surface (fig. 4-14). Therefore, it is expected that rain echo will be strongly dependent on transmitter and received polarizations, surface roughness, transmitter wavelength, and depression angle.

The approximations used in this section are not strictly valid, for example, effects of the diffuse component of forward scattering are

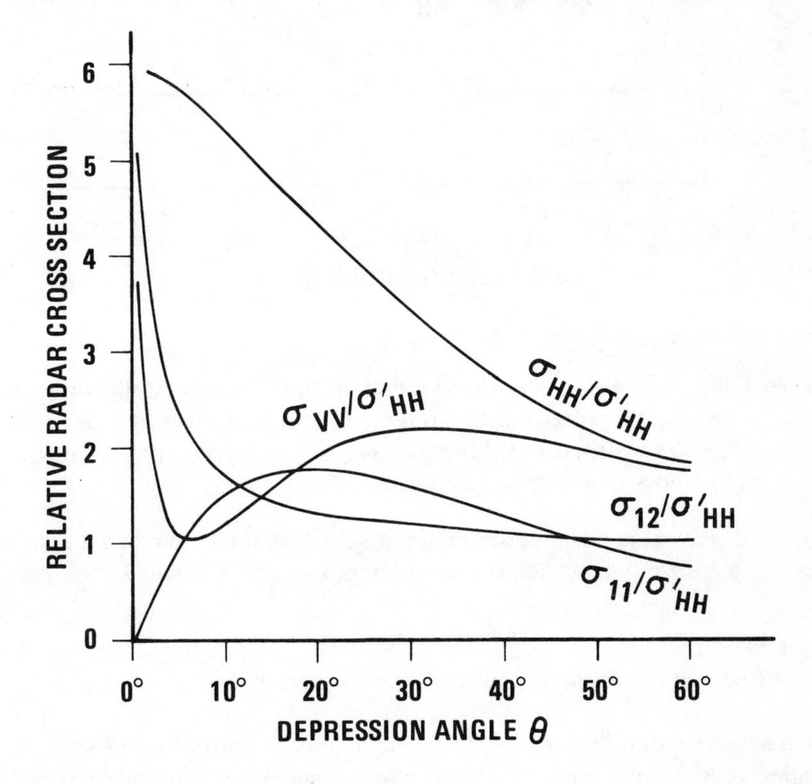

Source: Long and Zehner (1970).

Figure 4-17. Relative Radar Cross Sections for a Large Collection of Nondepolarizing Scatterers. Calculations were made for a transmitted wavelength of 10 cm and an rms surface roughness of 1 cm.

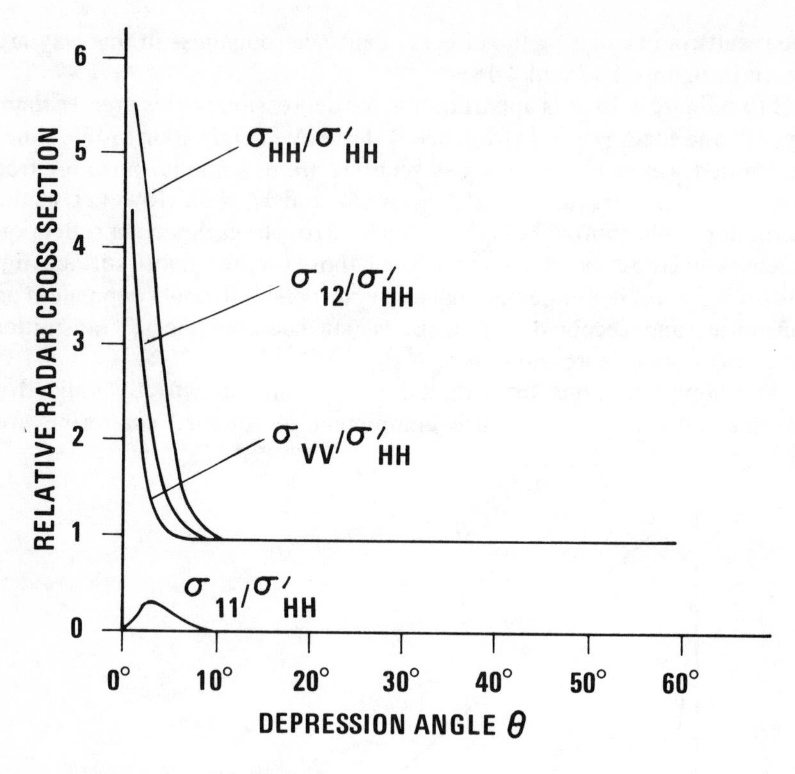

Source: Long and Zehner (1970).

Figure 4-18. Relative Radar Cross Sections for a Large Collection of Nondepolarizing Scatterers. Calculations were made for a transmitted wavelength of 10 cm and an rms surface roughness of 10 cm.

neglected. However, the approximations help with understanding a complex problem for which additional polarization and statistical data are needed.

4.16 Effects of the Diffuse Component on Target Echo

For a perfectly smooth surface, $\mathscr{R}_s$ is unity and $\mathscr{R}_d$ is zero. For rms surface roughness Δh small in terms of the Rayleigh roughness criterion ($\Delta h \sin\theta << \lambda/8$), $\mathscr{R}_s$ exceeds $\mathscr{R}_d$. For Δh large compared to the roughness criterion ($\Delta h \sin\theta >> \lambda/8$), $\mathscr{R}_d$ exceeds $\mathscr{R}_s$. Based on experimental data (sec. 4.9) it seems that $\mathscr{R}_d$ may be as large as 0.4 for $\Delta h \sin\theta = \lambda/8$ and remains essentially constant for rougher surfaces.

Beckmann and Spizzichino (1963, p. 340) assumed that the sum $\langle \mathscr{R}_s^2 \rangle$ +

$\langle \mathcal{R}_d^2 \rangle$ is constant for rough surfaces; that is, they assumed that the total forward scattered power is independent of roughness (for rough surfaces and ρ_0 constant). In other words, it seems that for an electromagnetically rough surface the forward scattered field is predominantly diffuse and is usually considerably smaller than the direct field. Available experimental and theoretical information (see sec. 4.11 and 4.14) indicates that the specularly reflected power is only slightly depolarized and the diffusely scattered power is highly depolarized.

As mentioned in section 4.9, the forward scattered field has, in general, a Ricean amplitude distribution. Target echo may, therefore, contain a discernible Ricean component due to forward scatter in its amplitude distribution.

Even though the diffuse component is usually small, its effect on radar performance is not always negligible. D.K. Barton (1974) has presented a detailed analysis for *HH* and *VV* polarizations for the case of low-angle tracking. His analysis includes effects of the diffuse component. According to Barton, extremely accurate elevation data (one-hundredth beamwidth rms) cannot be expected on targets within two beamwidths of the surface, although that accuracy is available for azimuth measurements if signal fading can be avoided.

Interference Effects on Echo from Land and Sea

4.17 Effects of Forward Scatter on σ° for the Sea

Average echo power for a homogeneous, rough sea has sometimes been observed (secs. 6.4 and 6.14) to vary with range as R^{-3} and R^{-7} at close and more distant ranges, respectively.[f] The R^{-3} dependence is that expected for a collection of scatterers which fill an antenna beam; the R^{-7} dependence has been attributed to an interference effect between direct and reflected electromagnetic waves. From this qualitative explanation (see sec. 6.6) sea echo in the R^{-7} region is considered to be caused by a collection of targets above an effective plane surface—the relatively few waves which exceed some average height play the role of targets.

Equation (6.9) gives the transition range R_c (i.e., the intersection of the idealized R^{-3} and R^{-7} curves) in terms of antenna height H, transmitter wavelength λ, and effective wave height h_e of the sea. Results (sec. 6.14) of comparing experimental data with the equation for transition range

[f]The range dependence observed is not always R^{-3} and R^{-7} (Dyer and Currie 1974). For example, effects of trapping (sec. 2.12, chap. 2) can cause echo strength to drop less rapidly with range than it would for a standard atmosphere.

indicate that h_e is approximately equal to the standard deviation of wave height or, alternatively, the root-mean-squared wave height. Although the scattering model used represents an oversimplification of the actual sea scattering process, the apparent relationship of transition range to wave height is useful for providing limiting performance criteria for the radar designer.

Let σ_{AB} denote the radar cross section for transmitting polarization A and receiving polarization B. The ratio $\sigma^\circ_{VV}/\sigma^\circ_{HH}$ depends on depression angle, wavelength, and sea state, but it is generally recognized that for the calmer seas average σ°_{VV} exceeds average σ°_{HH}. The difference decreases as the sea becomes rougher, and it has been observed that average σ°_{HH} sometimes exceeds average σ°_{VV} by a few decibels. The "classical" interference effect has been used (Katzin, 1957) in an effort to qualitatively account for differences between σ°_{VV} and σ°_{HH}. As already discussed, the effective reflecting plane representing the sea appears rough in the R^{-3} zone and, therefore, it has negligible effect on σ°; in other words, the claim here is that σ°_{VV} and σ°_{HH} would be almost equal for microwaves if the sea scatterers, per se, were insensitive to polarization. A model for explaining polarization differences is discussed in section 6.1.

In summary, the R^{-3} and R^{-7} zones can be explained by the "classical" interference caused by a direct "ray" from the radar to the sea and by an indirect ray reflecting off (and then impinging on the relatively few higher waves) an equivalent reflecting plane that represents the sea. The boundary of these zones is controlled by transmitted wavelength and wave height. For microwaves the reflection coefficient for a smooth sea will be nearly minus one for vertical and for horizontal polarization if θ is very much less than Brewster's angle. Therefore, for microwaves R_c is expected to be essentially independent of polarization.

There are fewer experimental results for uniformly rough land than for sea. However, the classical interference effect has been observed for asphalt and concrete roads and for irrigated farmland (see sec. 6.10).

Radar echo is affected by shadowing. Effects of major shadowers, like trees or mountains, are more obvious than the various shadowing effects of a uniformly rough surface. However, it is easily seen that the effects of shadowing are increased if the radar depression angle is decreased. Shadowing will cause a reduction in reflecting area, reduction in the magnitude of forward scattering, and a change in the height of the effective reflecting plane. Effects of shadowing will depend on viewing direction; for example, shadowing effects of the sea should be more pronounced for upwind and downwind directions than when looking along the troughs. Several authors have considered theoretically the reductions in scattering expected by shadowing (see sec. 4.10).

4.18 Local Interference

Scatterers near a locally flat surface will exhibit effects of interference that are similar to those of the previously discussed classical interference theory. Because of the closeness of the locally flat surface to the scatterer, it is often convenient to consider the combination as the target. Examples include a tree stump on a hillside, a wind ripple on a wave, or a bush protruding from water. Consider figure 4-19 for which the interference formulations previously outlined can often be used although the locally flat surface is only finite in extent.

Using figure 4-19, equation (4.12) for the point scatterer S that is distance z above the locally flat surface becomes[g]

$$F = 1 + \rho e^{-j[(2\pi/\lambda)\,\Delta R + \phi]} = 1 + \rho e^{-j[(2\pi/\lambda)\,2z\sin\alpha + \phi]} \qquad (4.33)$$

In order to create effects of constructive or destructive interference, it will

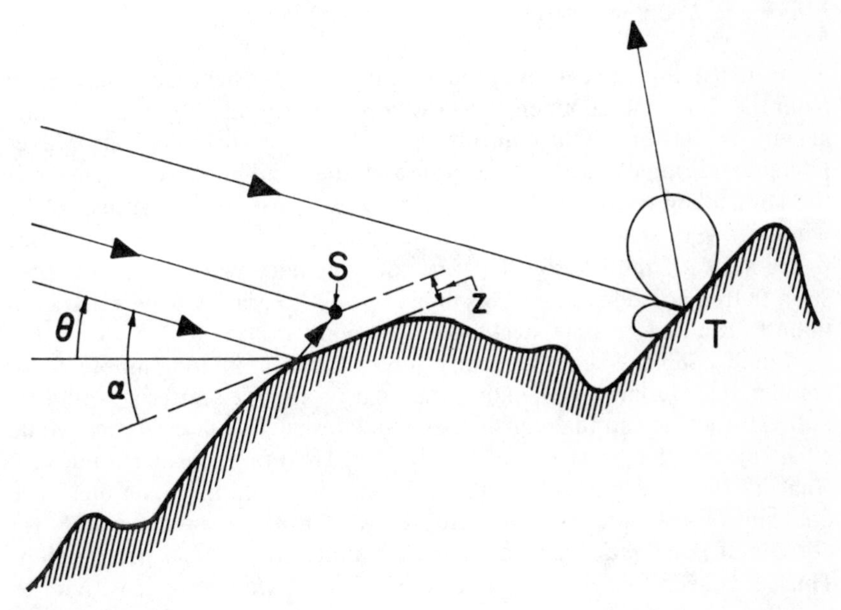

Source: Long (1974).

Figure 4-19. Geometry Relating to Two Types of Scattering.

---

[g]From this equation it is apparent that, for horizontal polarization ($\rho \approx 1$, $\phi \approx 180°$) and with z much smaller than λ, the field at S will be small even if α is as large as 30°.

be assumed that the Rayleigh roughness criterion is satisfied; that is, $z \sin \alpha < \lambda/8$. The height is small under these conditions; for example, if α is 30° or less, height z would be $\lambda/4$.

Local interference can be thought of as a target property that can help explain differences in cross sections observed for different polarizations. For example, ρ and ϕ are such that $|F|$ for water usually will be much less than unity for horizontal polarization and greater than unity for vertical polarization. Therefore, it is seen that the local interference effect will cause sensitivity to polarization and can exist irrespective of the existence of the "classical" interference effect, which is also a polarization sensitive phenomenon. Note that S and a nearby surface can be considered the target of figure 4-10; both rays of figure 4-19 correspond to "direct rays" of figure 4-19.

4.19 Comparisons between the Classical Interference Effect and Local Interference

Some insight into the causes of the two interference phenomena may result from the following comments. Sea waves are generated by the wind and gravity is the force that controls their characteristics. Capillary waves (small wind ripples) are also generated by the wind but surface tension is the controlling characteristic. Capillary waves are quite sensitive to the wind, are very small, and have wavelengths of about one inch or less (see figure 4-20). If the breeze that generates capillary waves dies out, they soon flatten and disappear. This is in contrast to gravity waves that continue to run and become swells after the wind stops.

Since capillary waves usually have dimensions comparable to or smaller than radar wavelengths, the size of such a scatterer would be sufficiently small so that constructive and destructive interference would exist between the direct and reflected fields. Therefore, the magnitude of F (that is, the field strength relative to what it would be if the dielectric constant of sea water were that of air) will be much less than unity for horizontal polarization and greater than unity for vertical polarization. Thus, it is seen that the local interference effect will cause sensitivity to polarization and can exist in the presence (or absence) of the "classical" interference effect. Note that the target we are considering here is a capillary wave, in the vicinity of point S, for which the field strength is modified by the nearby sea surface.

Examples have already been given to illustrate how the direct and indirect rays can cause the cross section of a point target in figure 4-10 to be sensitive to polarization, depression angle, and range. If figure 4-19 were the target, this target might also be sensitive to polarization. The following

Figure 4-20. Capillary Waves Superimposed on Sea Waves

Source: Sittrop (1975)

example is included to illustrate differences between the classical and local interference effects.

Suppose a radar views the sea at a depression angle of 30° and the peak-to-trough wave height is 10 feet. For this example, F would be approximately unity for either horizontal or vertical polarization. Therefore, if an exceptionally large wave were to appear as a target, σ_{HH} and σ_{VV} would not be different by virtue of the classical interference effect. However, the cross section for a small wind ripple riding on that wave might be sensitive to polarization because of local interference.

References

Ament, W.S., "Toward a Theory of Reflection by a Rough Surface," *Proceedings of the IRE*, vol. 41, pp. 142-46, January 1953.

Barton, D.K., "Low-Angle Radar Tracking," *Proceedings of the IEEE*, vol. 62, pp. 687-704, June 1974.

Beard, C.I., "Coherent and Incoherent Scattering from the Ocean," *IRE Transactions on Antennas and Propagation*, vol. AP-9, pp. 470-83, September 1961.

Beard, C.I., Isadore Katz, and L.M. Spetner, "Phenomenological Model of Microwave Reflection from the Ocean," *IRE Transactions on Antennas and Propagation*, vol. AP-4, pp. 162-67, 1956.

Beckmann, Petr, "Shadowing of Random Rough Surfaces," *IEEE Transactions on Antennas and Propagation*, vol. AP-13, pp. 384-88, May 1965.

Beckmann, Petr and André Spizzichino, *The Scattering of Electromagnetic Waves from Rough Surfaces*, Pergamon Press Ltd., Oxford, England, 1963.

Blake, Lamont, V. *Radar Range Performance Analysis*, D.C. Heath and Company, Lexington, Massachusetts, 1980.

Boring, J.G., E.R. Flynt, M.W. Long, and V.R. Widerquist, *Sea Return Study*, Engineering Experiment Station, Georgia Institute of Technology, Final Report, Contract Nobsr-49063, August 1957.

Bremmer, H., *Terrestrial Radio Waves*, Elsevier Publishing Company, New York, 1949.

Brockelman, R.A. and Tor Hagfors, "Notes on the Effect of Shadowing on the Backscattering of Waves from a Random Rough Surface," *IEEE Transactions on Antennas and Propagation*, vol. AP-14, pp. 621-29, September 1966.

Bullington, Kenneth, "Reflection Coefficients of Irregular Terrain," *Proceedings of the IRE*, vol. 42, pp. 1258-62, August 1954.

Court, G.W.G., "Determination of the Reflection Coefficient of the Sea for

Radar Coverage Calculations by an Optical Analogy Method,'' *Proceedings of the IEE*, vol. 102B, pp. 827-30, November 1955.

Curry, G.R., ''Measurements of UHF and L-Band Radar Clutter in the Central Pacific Ocean,'' *IEEE Transactions on Military Electronics*, vol. MIL-9, pp. 39-44, January 1965.

Durlach, N.I., ''Influence of the Earth's Surface on Radar,'' Lincoln Laboratory Technical Report No. 373, Massachusetts Institute of Technology, January 1965.

Dyer, F.B. and N.C. Currie, ''Some Comments on the Characterization of Radar Sea Echo,'' *Digest of the International IEEE Symposium on Antennas and Propagation*, pp. 323-26, June 10-12, 1974.

Ford, L.H. and R. Oliver, ''An Experimental Investigation of the Reflection and Absorption of Radiation of 9 cm Wavelength,'' *Proceedings of the Physical Society*, vol. 58, pp. 265-80, 1946.

Gent, M., I.M. Hunter, and N.P. Robinson, ''Polarization of Radar Echoes, Including Aircraft, Precipitation, and Terrain,'' *Proceedings of the IEE,* vol. 110, pp. 2139-48, December 1963.

Hunter, I.M. and T.B.A. Senior, ''Experimental Studies of Sea-Surface Effects on Low-Angle Radars,'' *Proceedings of the IEE*, vol. 113, pp. 1731-40, November 1966.

Katz, Isadore, ''Radar Reflectivity of the Ocean for Circular Polarization,'' *IEEE Transactions on Antennas and Propagation,* vol. AP-11, pp. 451-53, July 1963.

Katzin, Martin, ''On the Mechanisms of Radar Sea Clutter,'' *Proceedings of the IRE*, vol. 45, pp. 44-54, January, 1957.

Kerr, D.E., *Propagation of Short Radio Waves*, Massachusetts Institute of Technology, Radiation Laboratory Series, vol. 13, McGraw-Hill Book Company, Inc., New York, New York, 1951.

Long, M.W., ''Radar Propagation Factors for Various Linear and Circular Polarizations,'' *Electronics Letters*, vol. 4, p. 5, January 1968.

Long, M.W., ''On a Two-Scatterer Theory of Sea Echo,'' *IEEE Transactions on Antennas and Propagation*, vol. AP-22, pp. 667-72, September 1974.

Long, M.W. and S.P. Zehner, ''Effects of the Sea on Radar Echo from Rain,'' *IEEE Transactions on Aerospace and Electronic Systems*, vol. AES-6, pp. 821-24, November 1970.

Long, M.W., R.D. Wetherington, J.L. Edwards, and A.B. Abeling, ''Wavelength Dependence of Sea Echo,'' Engineering Experiment Station, Georgia Institute of Technology, Final Report, Contract N62269-3019, July 1965.

McFee, R. and T.M. Maher, ''Effect of Surface Reflections on Rain Cancellation of Circularly Polarized Radars,'' *Transactions of the IRE*, vol. AP-7, pp. 199-201, April 1959.

McGavin, R.E. and L.J. Maloney, ''Study at 1,046 Megacycles Per Second of the Reflection Coefficient of Irregular Terrain at Grazing Angles,'' *Journal of Research of the National Bureau of Standards,* vol. 63D, pp. 235-48, 1959.

Nathanson, F.E., *Radar Design Principles,* chap. 7, McGraw-Hill Book Company, Inc., New York, New York, 1969.

Povejsil, D.J., R.S. Raven, and Peter Waterman, *Airborne Radar,* D. Van Nostrand Company, Inc., Princeton, New Jersey, 1961.

Shaw, Leonard and Petr Beckmann, ''Comments on Shadowing of Random Surfaces,'' *IEEE Transactions on Antennas and Propagation*, vol. AP-14, p. 253, March 1966.

Sherwood, E.M. and E.L. Ginzton, ''Reflection Coefficient of an Irregular Terrain at 10 cm,'' *Proceedings of the IRE*, vol. 43, pp. 877-78, July 1955.

Sittrop, Ir. H., ''X and K_u-Band Radar Backscatter Characteristics of Sea Clutter,'' Part II, Report PHL 1975-09, Physics Laboratory TNO, Netherlands, 1975.

Skolnik, M.I., *Introduction of Radar Systems*, McGraw-Hill Book Company, Inc., New York, New York, 1962.

Smith, B.G., ''Geometrical Shadowing of a Random Rough Surface,'' *IEEE Transactions of Antennas and Propagation*, vol. AP-15, pp. 668-71, September 1967.

Stratton, J.A., *Electromagnetic Theory*, McGraw-Hill Book Company, Inc., New York, New York, 1941.

U.S. Department of Commerce, *A World Atlas of Atmospheric Radio Refractivity,* ESSA Monograph No. 1, 1966.

Zehner, S.P., W.M. O'Dowd, Jr., and F.B. Dyer, ''Forward-Scatter Properties of Microwaves Near the Surface of the Ocean,'' *Digest of the International IEEE Symposium on Antennas and Propagation,* pp. 8-10, June 10-12, 1974.

5 Echo Fluctuations

The echo from a simple target composed of one type of surface will vary if the target orientation or transmitter frequency is changed. If the target is complex in the sense that it consists of a large number of simple targets in relative motion, the echo will fluctuate in a seemingly noise-like manner. This fluctuation is caused by changes between constructive and destructive interference of the waves (received at the antenna) that are reradiated (scattered) from the individual simple targets. Therefore, the fluctuation spectrum for land or sea will, in general, be a composite of at least two spectra: the scattering spectrum or spectra of the individual scatterers and the spectrum or spectra caused by the group or ensemble of scatterers.

It is expected that the rates of fluctuation (frequency spectrum) will increase if the transmitter frequency is increased. This is because the scattering lobes (diffraction patterns) for simple scatterers and the scattering from an ensemble of scatterers are strongly influenced by relative electrical phases that are determined by distances expressed in terms of wavelength.

It is obvious that an extended target such as rain will include a large number of elementary scatterers (raindrops) and that the echo phases, as determined by the radial distances from the radar, will tend to be random but uniformly distributed. That is, the number of echoing centers with any given phase will tend to be the same as that with any other phase. Although the raindrops may be moving with the wind, it is only the relative phases that affect the echo amplitude. The relative motion (possibly altered by the wind) will change the relative phases from the scattering centers, causing the resultant signal to fluctuate. An alternative way of considering rates of fluctuation is to think in terms of differences between doppler frequencies from the individual targets. A conventional noncoherent radar will not detect the absolute doppler frequency (mass motion of raindrops), but it can be shown that differences in dopplers will cause fluctuations in amplitude that are detected.

A simple relationship exists between relative velocity and spectrum width for an ensemble of identical isotropic scatterers for spectra and velocity distributions that have Gaussian shapes. According to J.L. Lawson and G.E. Uhlenbeck (1950, p. 140), if $\overline{V}$ is defined so that one-half of the scatterers have relative velocities in the direction of the radar lying between $-\overline{V}$ and $+\overline{V}$, then

$$\overline{V} = 0.2\,\lambda\, f_{1/2} \tag{5.1}$$

where λ is the wavelength and $f_{1/2}$ is the frequency (spectrum width) at which the power spectrum is down to one-half of its maximum value.

The shape of the continuum of fluctuations that constitute radar echo can be described either by the power density spectrum or by the time autocorrelation function. The autocorrelation function $R(\tau)$ of a stationary (that is, temporally homogeneous) function is defined as

$$R(\tau) = \lim_{T\to\infty} \frac{1}{2T}\int_{-T}^{T} X(t)\cdot X(t+\tau)\,dt \tag{5.2}$$

The power density spectrum

$$P(f) = \lim_{T\to\infty} \frac{1}{2T}\left|\int_{-T}^{T} X(t)\cdot e^{-j2\pi ft}\,dt\right|^2 \tag{5.3}$$

is the Fourier transform of the correlation function, and conversely (see chap. 2). Therefore, the total power contained in the spectrum is equal to $R(\tau)$ at $\tau = 0$:

$$R(0) = \int_{-\infty}^{\infty} P(f)\cdot df = 2\int_{0}^{\infty} P(f)\,df \tag{5.4}$$

In addition to rate of change, the range of amplitude excursion is also of importance to the radar designer. Probability functions are described in chapter 2 and the Appendix. Frequently used distributions for describing signal fluctuations include the Rayleigh, Ricean (Raleigh plus a constant), lognormal and Weibull distributions.

Echo amplitude is expected to vary so that its probability density function is Rayleigh if caused by many randomly moving scatterers of about equal size. Then, as described in section 2.8, the probability density function for power is exponential and, specifically, the probability of echo power being between a level P and an infinitesimally larger level $P + dP$ is given by

$$W(P)\,dP = \frac{1}{\overline{P}}\, e^{-P/\overline{P}} dP \tag{5.5}$$

where $\overline{P}$ is the average power. Therefore, a collection of scatterers (source of echo) that is *physically* distributed so as to obey equation (5.5) for power is one for which its amplitude is described by the Raleigh density function. Such a collection of scatterers can appropriately be called Rayleigh or Rayleigh distributed, but this author and others have mistakenly called $W(P)$ of equation (5.5) a Rayleigh density function.

Sometimes the received signal from terrain includes an essentially constant echo (slow compared to fast fluctuations) in addition to a Rayleigh distributed fluctuation. Such a distribution is called a Ricean distribution and can be written as

$$W(P)\,dP = (1 + m^2)e^{-m^2}e^{-P(1+m^2)/\overline{P}}J_o(2im\sqrt{1 + m^2}\sqrt{P/\overline{P}})\,dP/\overline{P} \qquad (5.6)$$

where J_o is the Bessel function described by the series

$$J_o(x) = 1 - \frac{x^2}{2^2} + \frac{x^4}{2^2 \cdot 4^2} - \frac{x^6}{2^2 \cdot 4^2 \cdot 6^2} + \ldots$$

In figure 5-3 the Ricean distribution is plotted against $P/\overline{P}$ for several values of m^2 (ratio of the constant power to the random power).

Sometimes terrain echo contains large amplitude components that cause the total distribution to depart appreciably from the Rayleigh or the Ricean. The lognormal distribution, one for which amplitudes expressed in decibels is Gaussian distributed, is often useful for describing the larger fluctuations. The density function for a lognormal distribution is

$$W(Y) = \frac{1}{Ys\sqrt{2\pi}} \exp\left[-\frac{1}{2s^2}\left(\ln\frac{Y}{Y_m}\right)^2\right], \qquad (5.7)$$

where Y = the lognormally distributed variable

Y_m = the median value of Y $\quad s$ = the standard deviation of $\ln(Y/Y_m)$

The Weibull distribution has properties that lie between those of the Rayleigh and the lognormal distributions. A comparison between the Weibull and the lognormal is given in the Appendix. The Weibull density function can be expressed as

$$W(P) = \frac{1}{\overline{P}}\, bP^{(b-1)} \exp\left(-\frac{P^b}{\overline{P}}\right) \qquad (5.8)$$

It is of interest to notice that for b equals unity, equation (5.8) reduces to the exponential density function of equation (5.5). Therefore, for b equals unity, Weibull statistics can be used to describe the echo statistics of Rayleigh clutter.

Ground Echo Fluctuations

5.1 Nature of Ground Echoes

Echo from land is caused by plants, trees, rocks, hills, and even bare ground. For some purposes, land scatterers can be grouped into two general classes: (1) those that are moved by wind, and (2) the relatively fixed objects such as tree trunks, rocks, and bare hills. If there are *moving* scatterers—grass, flowers, leaves, twigs and possibly branches—there are usually many within an illuminated area. Therefore, it seems that the moving scatterers should have electromagnetic properties similar to those of a large collection of random scatterers.

The classifications *fixed* and *moving* are broad generalities and, in fact, whether a scatterer is of one class or the other depends on extent of motion expressed in terms of wavelength. For example, received reradiation from a tree branch that is moving back and forth through one centimeter goes through nearly all possible electrical phases for millimeter wavelengths, but the phase is nearly constant for decimeter wavelengths. Therefore, other factors being fixed, the percentage of echo power from fixed and moving scatterers illuminated by the radar depends on wavelength. In other words, the rapidity of echo fluctuations and the statistical distributions of the amplitudes are wavelength dependent.

Figures 5-1 and 5-2 are examples of pen recorder playouts of parallel and cross-polarized echoes from deciduous and pine trees reported by R.D. Hayes et al. (1958). The data represent echo from a fixed range. In addition to illustrating the amplitude variation of return obtained for range-gated samples, the figures show an increase in fluctuation rate that generally accompanies an increase in wind speed. The radar (Hayes, Currie, and Long 1957) operates at 9,375 MHz with 0.25 microsecond pulses, and has equal azimuth and elevation beamwidths of 1.86°. Notice that, in general, neither P_{HH} and P_{HV} nor P_{VV} and P_{VH} fluctuate together. A similar lack of correlation in fluctuations is observed for trees when transmitting and receiving circular polarizations. The interdependence of echo for various polarizations is discussed in chapter 7.

5.2 *Amplitude Distributions for Terrain*

Theoretically, for a large number of independently moving scatterers, the probability density function of received power is of the Rayleigh type. Radar observations on heavy foliage in high winds have yielded the Rayleigh statistical character. In general, echo from ground contains a constant (or nearly constant) component caused by the "fixed" targets plus the fluctuating echoes that are sometimes describable by the Rayleigh (exponential) distribution. Therefore, ground echo amplitude distributions are expected to be peaked at a value near the amplitude of the constant component caused by the fixed scatterers (Ricean distribution). The next three paragraphs, almost[a] verbatim from D.E. Kerr (1951, p. 582), illustrate this point.

The probability distribution for the echo has been given in equation (5.6) as a function of m^2, the ratio of the intensity of the steady echo to the average power of the echo from the moving scatterers. This ratio will be called the steady-to-random ratio. For $m^2 = 0$, the distribution reduces to the familiar exponential for random scatterers alone. When m^2 is small, that is, less than 1, there is little difference between equation (5.6) and equation (5.5). As m^2 increases, a maximum appears in the function. When $m^2 >> 1$, the distribution can be approximated by a narrow Gaussian curve centered about the intensity of the steady echo.

[a]From D.E. Kerr (ed.), *Propagation of Short Radio Waves*. Copyright 1951 by McGraw-Hill Book Company, Inc. Used with permission of McGraw-Hill Book Company. The figure numbers, the equation numbers, and the symbol $W(P)$ as used here were selected for consistency with this book.

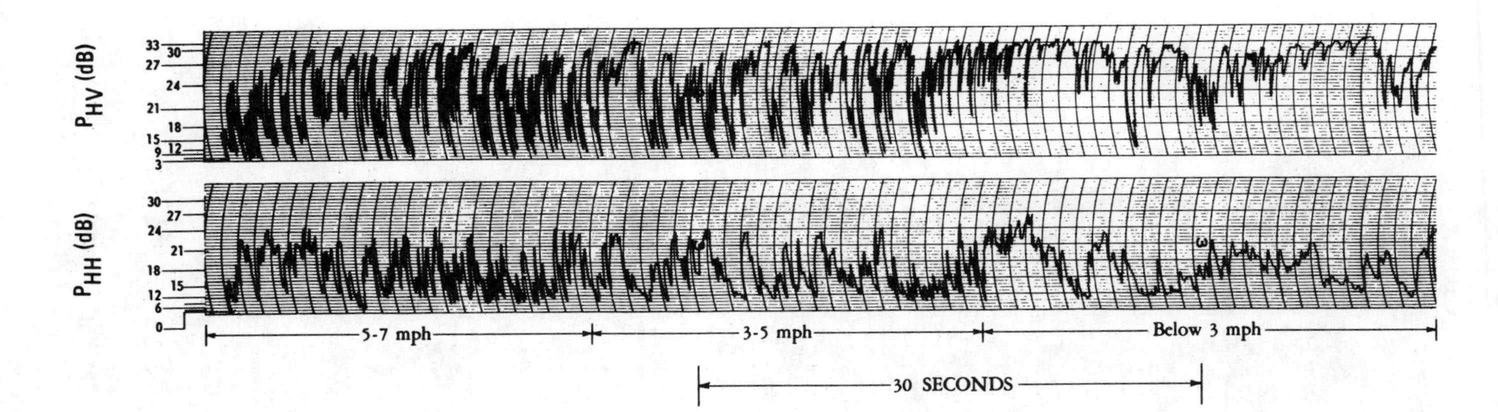

Source: Hayes et al. (1958).

Figure 5-1. Deciduous Tree Echo for *HH* and *HV* Polarizations (Windspeed 0-7 mph).

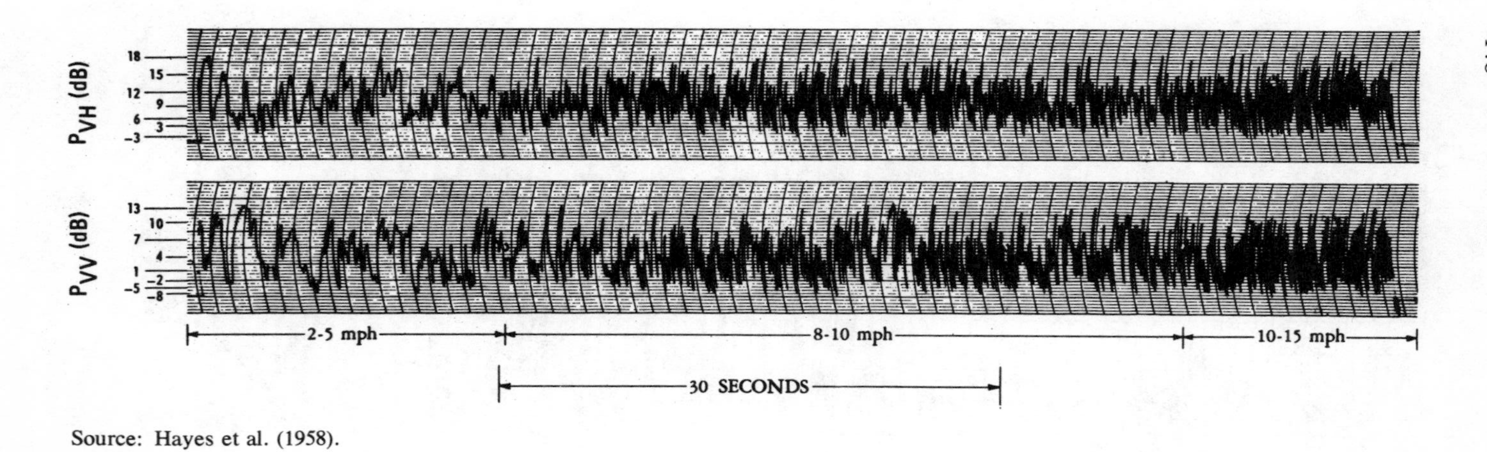

Source: Hayes et al. (1958).

Figure 5-2. Pine Tree Echo for *VV* and *VH* Polarizations (Windspeed 2-15 mph).

All steps in this sequence, in the development of $W(P)\,dP$, can be illustrated with observed ground-clutter distributions. At very high wind speeds almost everything moves in the wind and there are very few "steady" targets. As the wind velocity decreases, for the same average intensity of ground clutter, the steady-to-random ratio would be expected to increase.

The echo from a heavily wooded hill at 9.2 cm has been measured for gusty winds of about 50 mph, and the density function is indistinguishable from the simple exponential curve. A less extreme case is illustrated in figure 5-3a, which shows the distribution of the echo from a densely wooded section at a wind speed of 25 mph. The flattening of the distribution for low values of signal intensity is evident, and the histogram has been fitted with a theoretical curve of $m^2 = 0.8$. In figure 5-3b is shown the distribution of the echo from similar terrain but at a wind speed of 10 mph. There is now a maximum in the distribution at slightly less than the average value of the echo intensity, and the histogram is best fitted by a theoretical curve with $m^2 = 5.2$. The final stage in the sequence is illustrated in figure 5-3c. The distribution shown there is also for a wind speed of 10 mph, but the target consisted chiefly of rocks and sparse vegetation. All of the echoes are closely clustered around the average value, and the theoretical curve best fitting the experimental distribution corresponds to $m^2 = 30$.

R.D. Hayes and J.R. Walsh, Jr. (1959) have reported measurements on vegetation with an *X*-band radar (Hayes, Currie, and Long 1957) and they found after repeated measurements that distributions for heavy winds were Rayleigh distributed with the various polarizations available: *HH*, *VV*, *HV*, *VH*, *RL*, *RR*, *LR*, and *LL*. Angles of incidence or depression were not defined for these measurements because the radar was ground based in terrain that can be described, generally, as rough. The distributions were sometimes peaked under conditions with no wind. Figure 5-4 is typical of the distributions observed by Hayes and Walsh, and figure 5-5 shows a peaking that was sometimes observed.

T. Linell (1963) reported results of a two-year *X*-band measurements program on the short-term amplitude distributions for the mid-80 percent of the pulses (the strongest and the weakest 10 percent of the pulses were excluded) from cultivated terrain and a forest. The distributions, obtained for pulses from contiguous radar range cells while the antenna beam scanned an azimuth sector, were reported to be approximately lognormal. The radar was mounted on a 100-foot-high water-works tower and this permitted clutter to be investigated at small depression angles. Major system parameters are given in table 5-1.

Figure 5-6 shows cumulative distributions reported by Linell for cultivated terrain. For each curve, the difference in the abscissa points at the dashed lines is twice the standard deviation. For a Rayleigh distribution (see fig. 5-21), the standard deviation is 5.6 dB. In figure 5-6 the data for March when the ground was covered with 10 cm of snow has a standard deviation of about 7 dB, but without snow cover the curves seem to be indistinguishable from a Rayleigh distribution.

For smaller depression angles, the measured standard deviations are

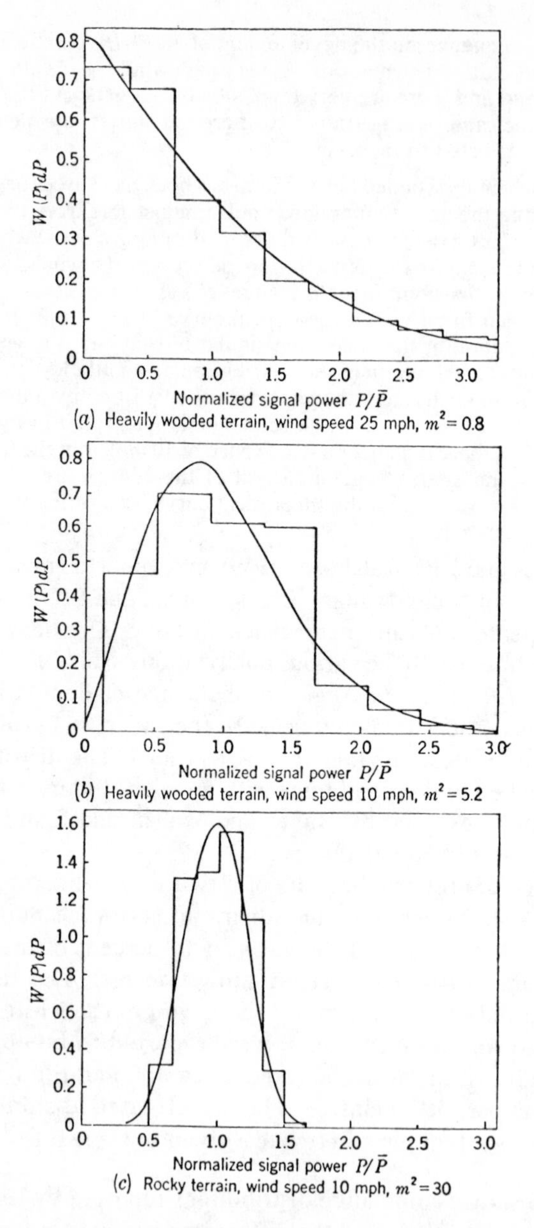

Source: D.E. Kerr, *Propagation of Short Radio Waves*. Copyright 1951, McGraw-Hill Book Company, Inc. Used by permission of McGraw-Hill Book Company.

Figure 5-3. Several Probability Distributions for Ground Clutter at 9.2 cm. Experimental data are shown by histograms. The continuous curves are Ricean distributions.

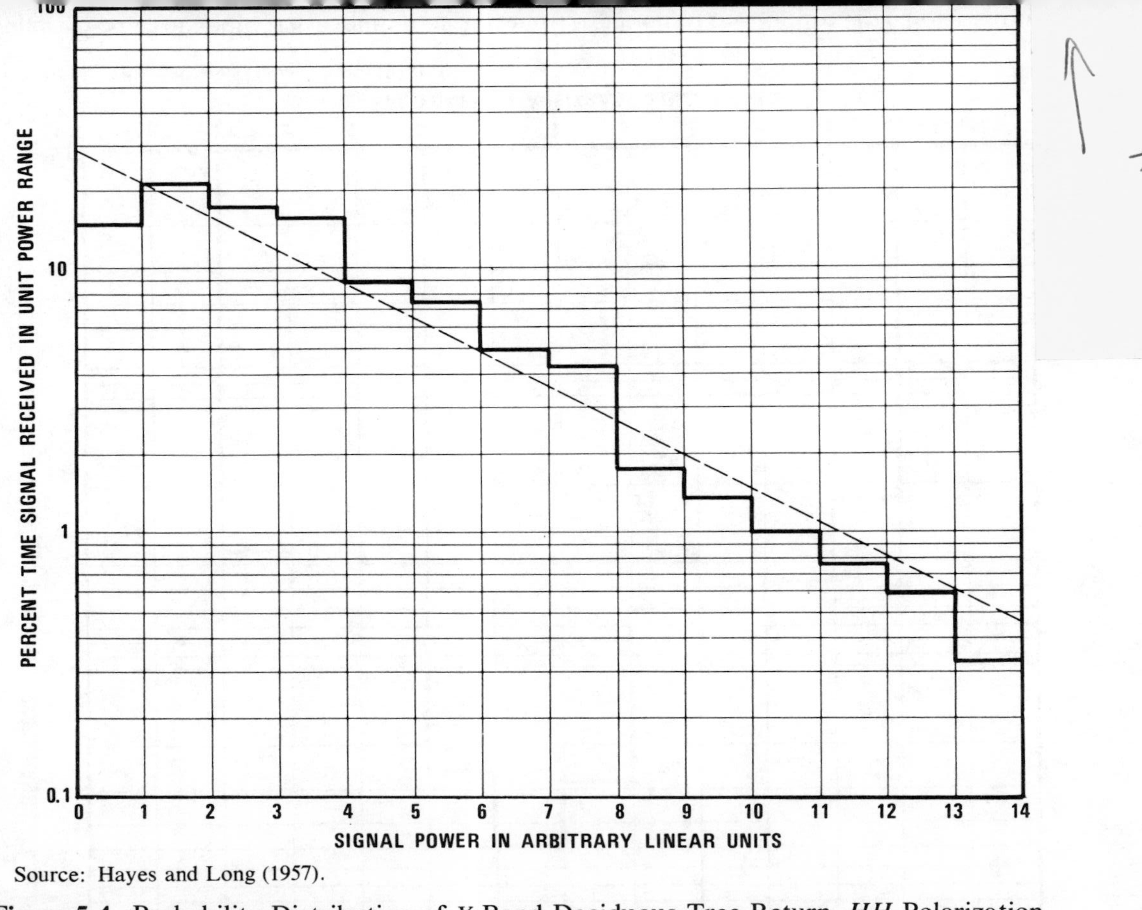

Source: Hayes and Long (1957).

Figure 5-4. Probability Distribution of *X*-Band Deciduous Tree Return. *HH* Polarization.

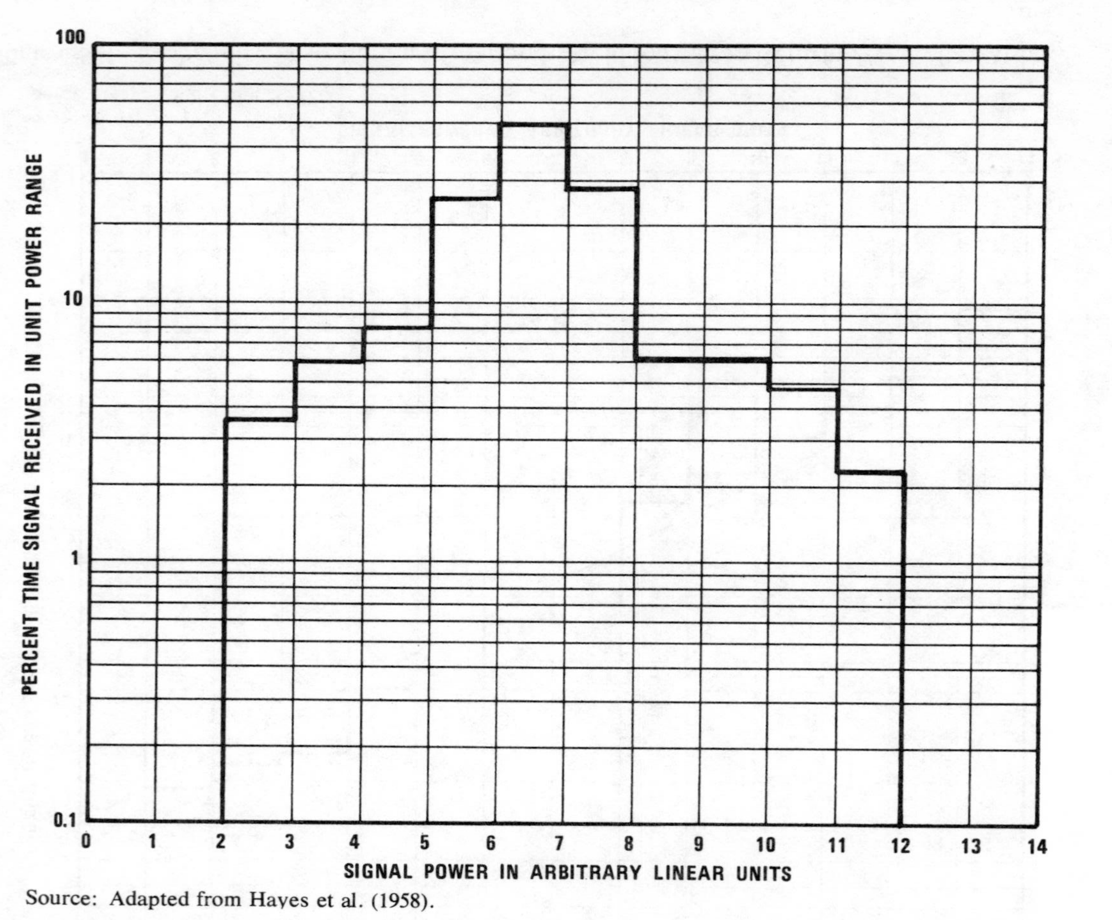

Source: Adapted from Hayes et al. (1958).

Figure 5-5. Probability Distribution of X-Band Deciduous Tree Return. VH Polarization.

Table 5-1
System Parameters Used by Linell

Horizontal beamwidth	1.4°
Vertical beamwidth	30°
Range resolution	25 meters
Polarization	changeable, horizontal or vertical
Calibration	standard target

much larger than for a Rayleigh curve. This is immediately obvious by viewing figure 5-7. In this figure, for a depression angle of 1.25°, the standard deviations are between 12 and 16 dB. Additional information on standard deviations reported by Linell is given in a discussion of differences between averages and medians in section 6.8. From table 6-2, standard deviations measured for a forest at a 0.7° depression angle are between 15 and 17 dB.

In summary, Linell's data at 5° for cultivated terrain without snow seem to be Rayleigh distributed. However, at smaller depression angles, the forest and cultivated terrain data have large standard deviations and can be approximated by either lognormal or Weibull (see Appendix) distributions. Linell observed no discernable difference in the distributions for *HH* and *VV* polarizations.

N.W. Guinard et al. (1967) studied echo distributions for *P*, *L*, *C*, and *X* bands measured simultaneously with airborne equipment for horizontal and vertical polarizations for depression angles between 5° and 60°. Normalized radar cross section was reported in terms of the 10, 50 (median), and 90 percentile values of its distribution. Two types of statistical processes were determined for the cross-section data: the Rayleigh distribution between 10 and 90 percentile values, and one that departed radically from this distribution. The one that departed radically from the Rayleigh distribution was identified as a Ricean distribution produced by a strong point target plus random "Rayleigh" scatterers. In other words, statistical characteristics obtained with airborne equipment at four radar bands and for horizontal and vertical polarizations agree with that reported much earlier by Kerr.

More recently, measurements over Arizona on deserts, mountains, farmlands, and urban areas have been made at *X*, *C*, *L*, and *P* bands for *HH* and *VV* echoes (Daley et al. 1968). Depression angles of 5° to 60° were used. The radar output was in units of decibels as a function of time. G.R. Valenzuela and M.B. Laing (1972) reported that the variances (standard deviations squared) of these observations were rarely less than 20 dB^2, and the values were seldom as large as 60 dB^2. The values of variance tended to

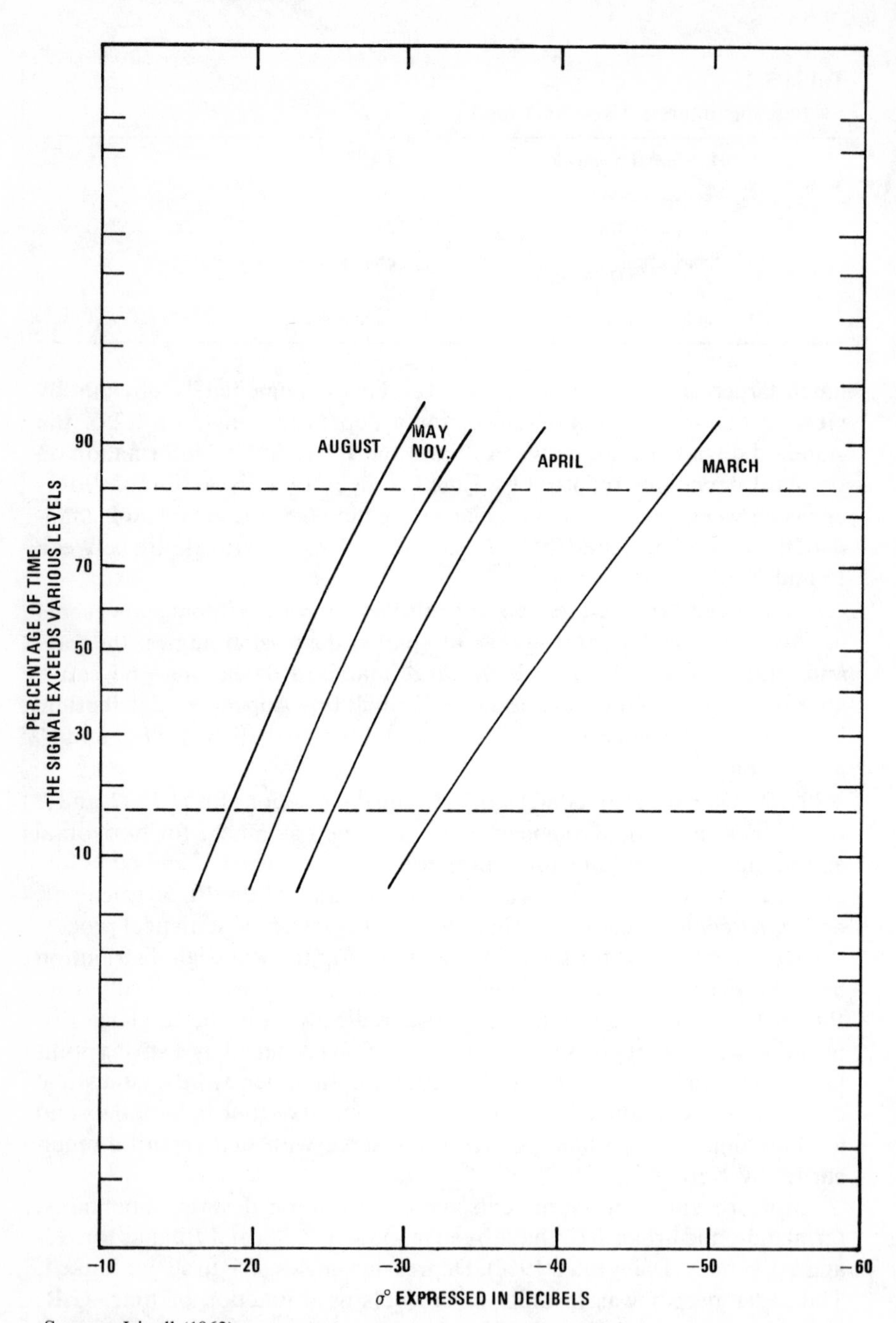

Source: Linell (1963).

Figure 5-6. σ° for Cultivated Terrain at Different Times of the Year. Depression Angle 5°.

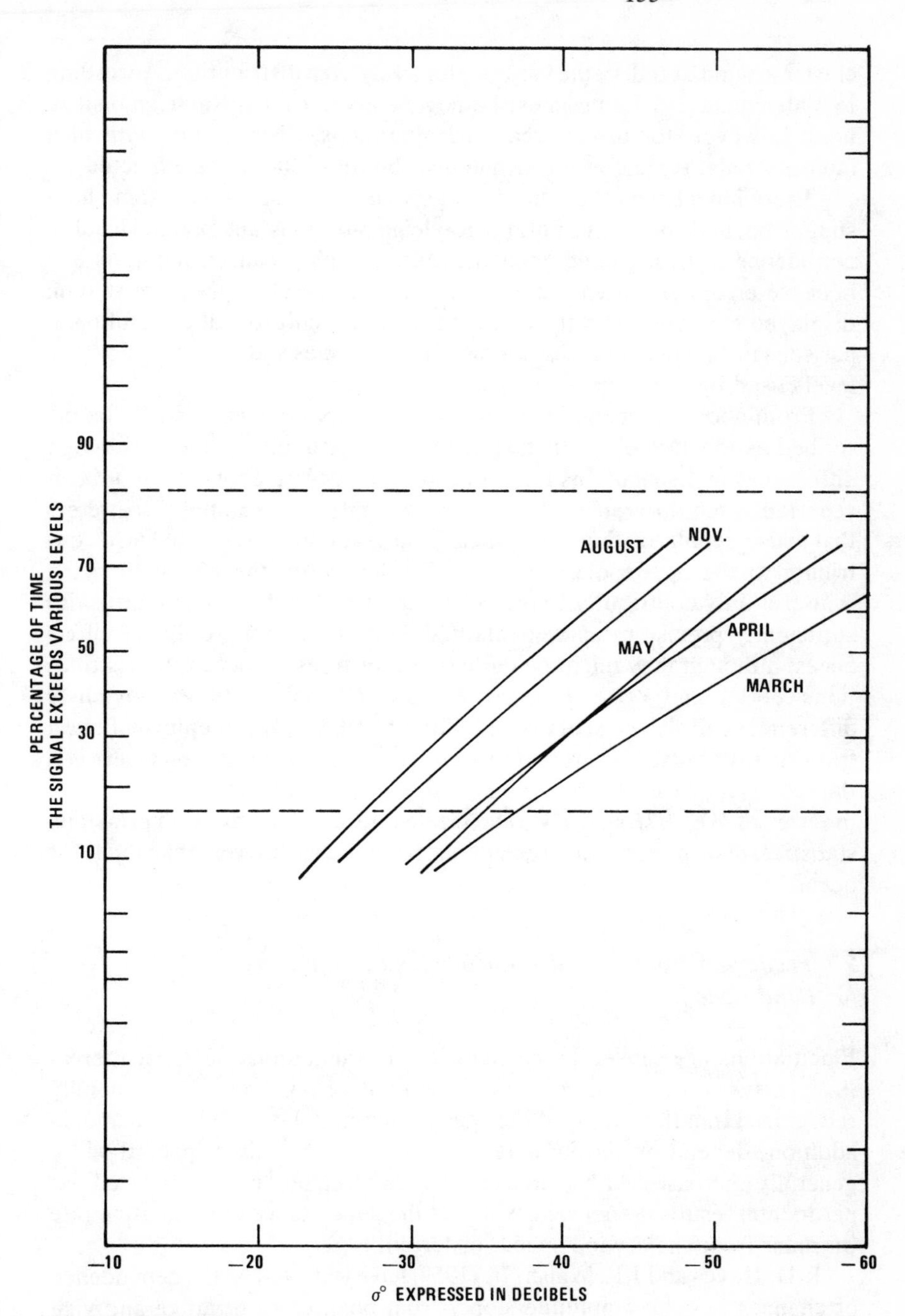

Source: Linell (1963).

Figure 5-7. $\sigma°$ for Cultivated Terrain at Different Times of the Year. Depression Angle 1.25°.

cluster around 31.0 dB^2, the variance for a Rayleigh distribution. According to Valenzuela and Laing, most homogeneous terrain is Rayleigh distributed; however, for urban areas and mountains, distributions with high intensity tails, typical of lognormal distributions, should be expected.

There have been other studies of ground clutter statistics—some have supported, and some have not, the Rayleigh plus constant model. Usually, confidence in fitting experimental results to any idealized model is low because errors are often not well known and therefore they are seldom displayed explicitly with the data. It is also difficult to make useful comparisons in data because of differences in the ranges and intervals of power levels used by the various investigators.

From above, it seems that amplitude distributions can usually be described as the sum of a constant echo plus fluctuating echoes. Although differences in distributions for land caused by polarization have not been reported, such differences do exist in principle. For example, scatterers that cause depolarization when linear polarization is transmitted also contribute to the nondepolarized echo. By definition, the nondepolarizing scatterer only contributes to the polarization transmitted. Therefore, with sufficiently precise measurements it should be possible to detect differences, although they might be quite small for trees, between the distributions for *VV* and *VH* or between *HH* and *HV* polarizations. Obviously differences can also exist between *VV* and *HH* data; for example, multipath transmission caused by ground reflections will cause a difference between the echoes for various polarizations. Circular-polarized echo is describable in terms of *VV*, *HH*, and *HV* echo (see eq. (3.2)). Therefore, in principle, statistical distributions for circular polarizations are different than those for linear.

5.3 Frequency Spectra and Autocorrelation Functions for Land Echo

Fluctuations are caused by changes in the orientations of the scatterers (leaves, twigs, etc.) and changes in the relative positions. It is generally recognized that the widths of the spectra increase with wind speed and, in addition, depend on terrain details to some extent. Also, spectral width generally increases with increases in radar frequency, but reported experimental results do not clearly depict the dependence of fluctuation rate on radar frequency, wind speed, or terrain type.

R.D. Hayes and J.R. Walsh, Jr. (1959) have reported on the dependence of changes in echo amplitude slope (from positive to negative and vice versa) per unit time as a function of wind speed for speeds up to 13 mph. Their measurements were made with *HH*, *HV*, *VV*, and *VH* polarizations and for each polarization there was an abrupt increase in fluctuation rate (of slope sign change) near 10 mph. According to table 5-2, leaves and twigs are in constant motion for winds of 8 to 12 mph. Although not directly related, it is interesting to notice from table 5-4 that m^2, the ratio of steady-to-random echo power, drops from 1.0 to 0.2 for wind speeds that increase from 22 to 30 mph. Large branches and most trees are in motion at 30 mph (table 5-2).

Figures 5-1 and 5-2 do not reveal a dependence of frequency spectra on polarization. As in the case of amplitude distributions, the author does not know of any clearly describable differences of frequency spectra for trees as a function of polarization.

Ground clutter often has characteristics similar to those of random scatterers. Theoretically, the power frequency spectra of echo for a large number of randomly moving scatterers is Gaussian in shape. Therefore, the desire exists to describe spectra from terrain solely by this simple function. According to discussions in this section that follow, Gaussian functions do not always accurately describe the power spectra for terrain. The power spectra caused by terrain are very complex and much remains to be learned about this subject.

E.J. Barlow (1949) reported spectra for wooded hills, sea echo, rain clouds, and chaff measured at a frequency of 1 GHz. According to Barlow, clutter power spectra can be represented by Gaussian-shaped curves of the form

$$P(f) = \exp\left[-a(f/f_o)^2\right]$$

where f_o is the radar transmitter frequency and a is a parameter dependent on radar target type. Values given by Barlow for the parameter a are included in table 5-3. Several spectra for 1 GHz are included in figure 5-8.

Performance predicted for an X-band noncoherent pulse-doppler radar, under the assumption that trees provide a Gaussian-shaped power spectrum, has not been achieved in practice (Fishbein, Graveline, and Rittenbach 1967). Because of this, the spectrum versus wind speed was examined in order to provide improved video processing equipment for a ground-based radar. According to the authors, a simple expression that gives good agreement with their measured power spectra for deciduous foliage is

$$P(f) = \frac{1}{1 + (f/f_c)^3} \tag{5.9}$$

where $f_c = 1.33e^{0.1356v}$ and v is wind speed in knots. The measurements were made with horizontal polarization.

In order to compare the predicted spectra of Barlow with that of Fishbein, Graveline, and Rittenbach, assume that the radar frequency is 10 GHz (multiply the abscissa of figure 5-8 by 10). Then for a 20-mph wind, the half-power points of the two spectra would occur at frequencies between 15

Table 5-2
Qualitative Effects of Wind Speed[a]

International Description	*Speed Miles Per Hour*	*Specifications*
Calm	less than 1	Smoke rises straight up. Trees and bushes do not move. A lake looks as smooth as a mirror.
Light air	1 to 3	Wind direction shown by drift of smoke but not by wind vane.
Light breeze	4 to 7	Wind felt on face. Leaves rustle. Wind vanes begin to move.
Gentle breeze	8 to 12	Leaves and small twigs in constant motion. Wind extends light flags.
Moderate breeze	13 to 18	Dust, loose paper, and small branches are moved.
Fresh breeze	19 to 24	Small trees in leaf begin to sway.
Strong breeze	25 to 31	Large branches in motion. Whistling heard in wires.
Near gale	32 to 38	Whole trees in motion. Inconvenience felt in walking against wind.
Gale	39 to 46	Twigs and small branches break off trees. Walking is impeded.
Strong gale	47 to 54	Slight structural damage occurs. Slate blown off roofs.
Storm	55 to 63	Seldom experienced inland. Trees broken or uprooted. Considerable structural damage inflicted.
Violent storm	64 to 72	Very rarely experienced on land. Widespread damage occurs.
Hurricane	73 and above	Excessive damage and destruction.

[a]The specifications in this table were developed circa 1805 to provide an over-land scale similar to that devised by Beaufort (see table 2-1) for the sea. See, for example, Nathaniel Bowditch (1966, app. R).

and 20 Hz. From this it would seem that spectra widths are proportional to radar frequency, as is expected for a cloud of randomly positioned scatterers. From the results that follow, however, it is clear that the spectra of ground clutter are often more complicated than is depicted by the single Gaussian-shaped curve.

Figure 5-9 gives measurement results reported by Fishbein, Graveline, and Rittenbach (1967) for trees in a 12-knot wind at *X* band. The figure also includes a Gaussian-shaped spectrum and one calculated with equation (5.9). It is clear that the Gaussian curve drops off with frequency more rapidly at the higher frequencies than do the measured results.

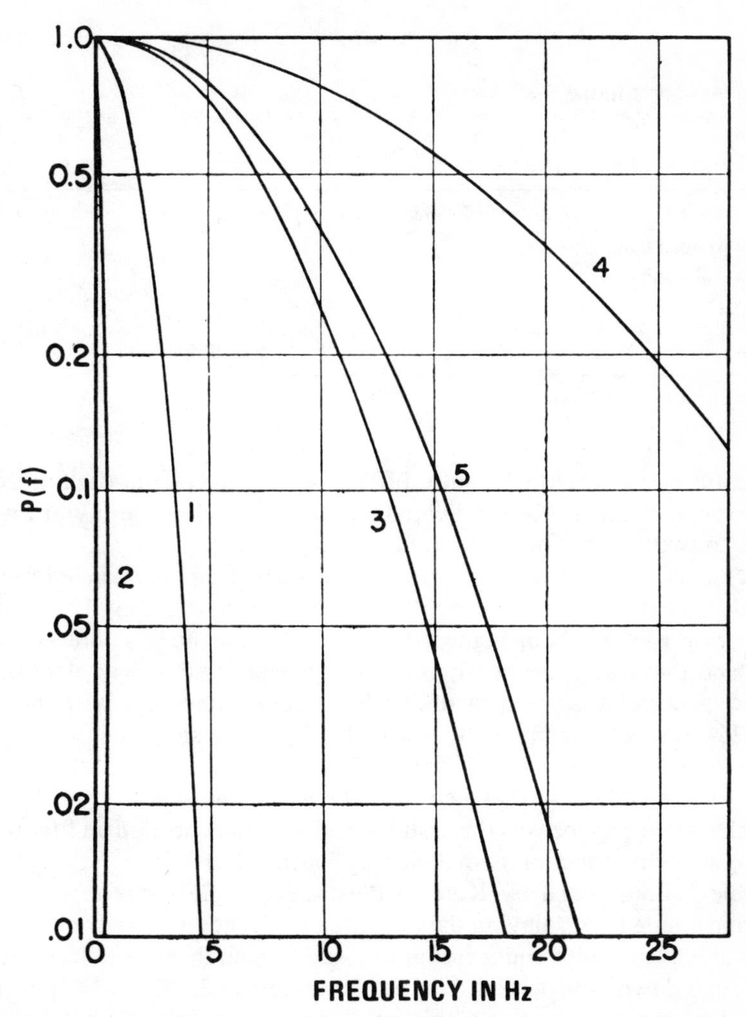

Source:Barlow (1949).

Figure 5-8. Frequency Spectra of Various Types of Complex Targets at 1 GHz. See table 5-3 to identify curves.

Kerr (1951, pp. 571-87) gives results on the measurements of chaff (metallic dipoles) and ground clutter. In the case of chaff measurements it was reported that, in addition to the fast doppler-beat fluctuations characteristic of a large number of point scatterers, a slow secular variation was almost always observed. In the case of ground clutter Kerr (1951, p. 585) reported that the width of the fluctuation spectrum increased with radar frequency, but not linearly as would be expected if the fluctuations were due entirely to the velocity distribution of the scatterers. Sometimes the fluctuation spectra resembled those of randomly moving scatterers, while

Table 5-3
Parameters for Figure 5-8[a]

Target	*a*	*Curve in Figure 5-8*
Heavily wooded hills, 20-mph wind blowing	2.3×10^{17}	1
Sparsely wooded hills, calm day	3.9×10^{19}	2
Sea echo, windy day	1.4×10^{16}	3
Rain clouds	2.8×10^{15}	4
Window "jamming"	1.0×10^{16}	5

[a]From E.J. Barlow (1949).

at other times the measured results differed significantly from the theoretically expected values. Kerr indicated that this anomalous behavior could be due to two sources of fluctuations instead of one.

A comparison between fluctuation spectra reported for wavelengths of 3.2 and 1.25 cm is given in figure 5-10. The target area was heavily wooded terrain with gale winds or higher. These curves are roughly Gaussian in shape and the widths are nearly inversely proportional to wavelength.

The spectrum was also measured at 9.2 cm simultaneously with the 3.2- and 1.25-cm spectra mentioned above. However, the shape of the 9.2-cm spectrum (listed as 50 mph wind in table 5-4) was reported as markedly out of line with the others so that a comparison was impossible. According to Kerr, the strange shape was the result of a kink in the correlation function. (That correlation function is included in figure 5-13.)

Table 5-4 was used by Kerr to illustrate complexities observed in spectrum shape by displaying data at three frequency points for wooded terrain at 9.2 cm. The frequencies are those at which the power frequency spectrum is down to 80 percent (−1 dB), 50 percent (−3 dB), and 10 percent (−10 dB), respectively, of its maximum value. In addition, the wind speed and the steady-to-random power ratio m^2 are listed.

Notice that the data are ranked by frequency at the 10-percent power points and this ranking properly orders wind speed for the strong winds reported. However, notice that the order of frequencies is not maintained for the other frequency columns; thus, table 5-4 illustrates that spectra shape, as well as spectra width, is a complex function of wind speed. Therefore, the simple Gaussian shape will not fit all of these data.

Described below are the results made with ground-based 6.3-GHz and 35-GHz radars that used the same video instrumentation and analysis equipment (Ivey, Long and Widerquist 1956). The objective of the investigation was to determine a way that the time autocorrelation function of tree

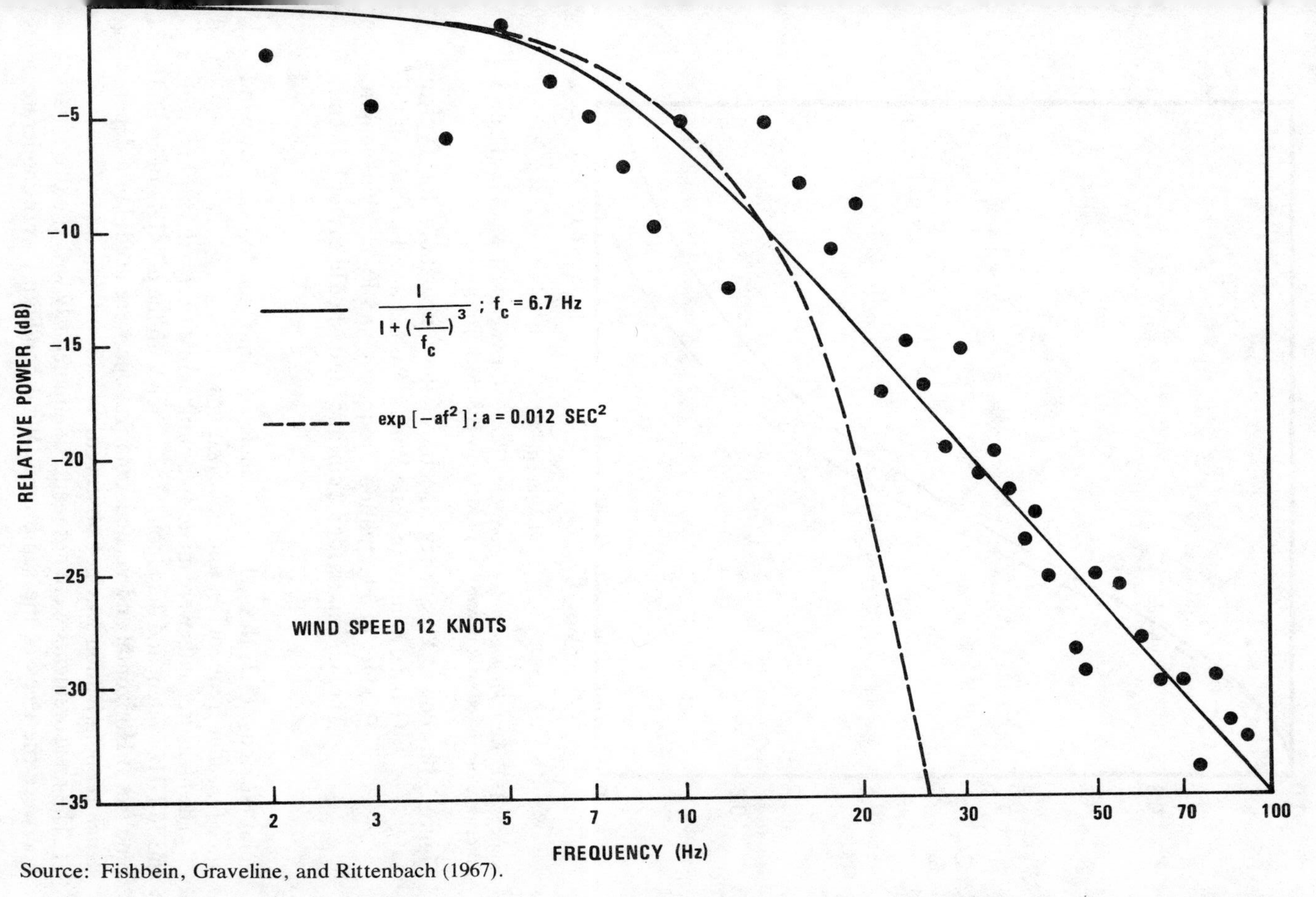

Source: Fishbein, Graveline, and Rittenbach (1967).

Figure 5-9. Power Spectrum Obtained with an *X*-Band Radar. Each solid circle represents a measurement.

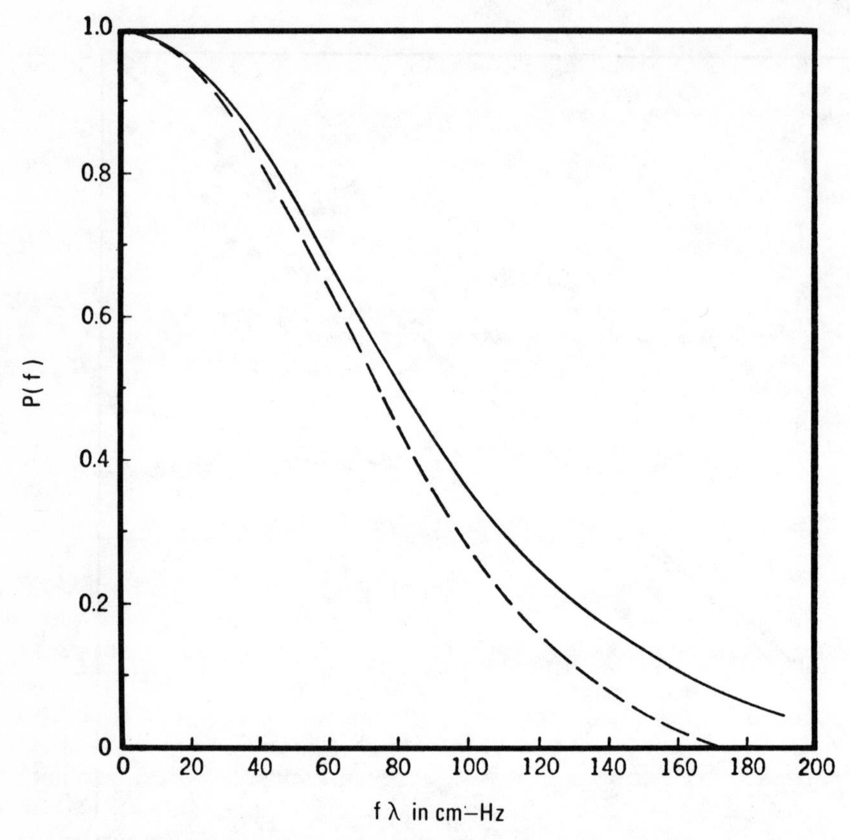

Source: D.E. Kerr, *Propagation of Short Radio Waves*. Copyright 1951, McGraw-Hill Book Company, Inc. Used with permission of McGraw-Hill Book Company.

Figure 5-10. Power Frequency Spectra of the Fluctuations for Typical Ground Clutter in Gale Winds, Plotted as a Function of the Product of Fluctuation Frequency and Radar Wavelength. The solid line is for 3.2 cm, and the dashed line is for 1.25 cm.

echo (see eqs. (5.2) through (5.4)) could be simulated for various wind speeds, radar frequencies, and polarizations.

The measurements were made on a number of different days—first with the 35-GHz system and deciduous trees, then with the 6.3-GHz radar and pine trees. Horizontal and vertical polarizations were used, but a dependence on polarization was not observed.

The measurements were not made simultaneously with the two radars nor were the trees of the same type. Wind conditions were moderate,

Table 5-4
Examples of the Fluctuation Spectra for Wooded Terrain. Wavelength 9.2 cm.

Wind, mph	m^2	$f_{0.8}$, *Hz*	$f_{0.5}$, *Hz*	$f_{0.1}$, *Hz*
17	1.3	1.44	0.77	1.67
10	5.2	0.65	1.10	2.07
23	1.0	0.72	1.27	2.7
22	1.0	1.9	3.1	5.9
23	0.8	1.9	3.3	8.6
30	0.2	3.3	5.9	10.1
50	0	2.1	3.9	14.8

Source: From D.E. Kerr, *Propagation of Short Radio Waves* Copyright 1951, McGraw-Hill Book Company, Inc. Used with permission of McGraw-Hill Book Company.

describable only as those prevailing over the several days involved. As would be expected, a wide range of decorrelation times were obtained; but as a group, the 35-GHz correlation functions had considerably shorter decorrelation times (wider bandwidths) than the 6.3-GHz group. There were certain features regarding shape (as distinct from width) that were observed to be common to the correlation functions for each radar frequency (and target area) group, and this preliminary investigation also revealed certain features common to both groups. As a result of these preliminary studies, the authors expressed the opinion that the dominant characteristics of the autocorrelation function for tree echo may be represented by the sum of two autocorrelation functions corresponding to those of white noise passed through two different types of integrating networks. The radars, on loan for the ground studies, were never again available for continuing the investigation.

Averaged correlation functions obtained from the deciduous-tree echo at 35 GHz and from the pine-tree echo at 6.3 GHz are shown in figure 5-11. These two curves were obtained by averaging the two groups of curves for which the individual members were derived from the original correlation functions by the following process. If the original correlation function was asymptotic to some value other than zero, this asymptote was subtracted. The difference curve was then renormalized in both amplitude and time so that it passed through the points $[\tau = 0, R(\tau) = 1]$ and $[\tau = \tau_e, R(\tau) = 1/e]$ and was asymptotic to zero. The value of time delay required for the difference curve to fall to a value $1/e$ of its original value was designated τ_e and recorded. Disregarding all other variables except the radar frequency (6.3 or 35 GHz), the normalized difference curves were averaged to obtain the correlation functions displayed.

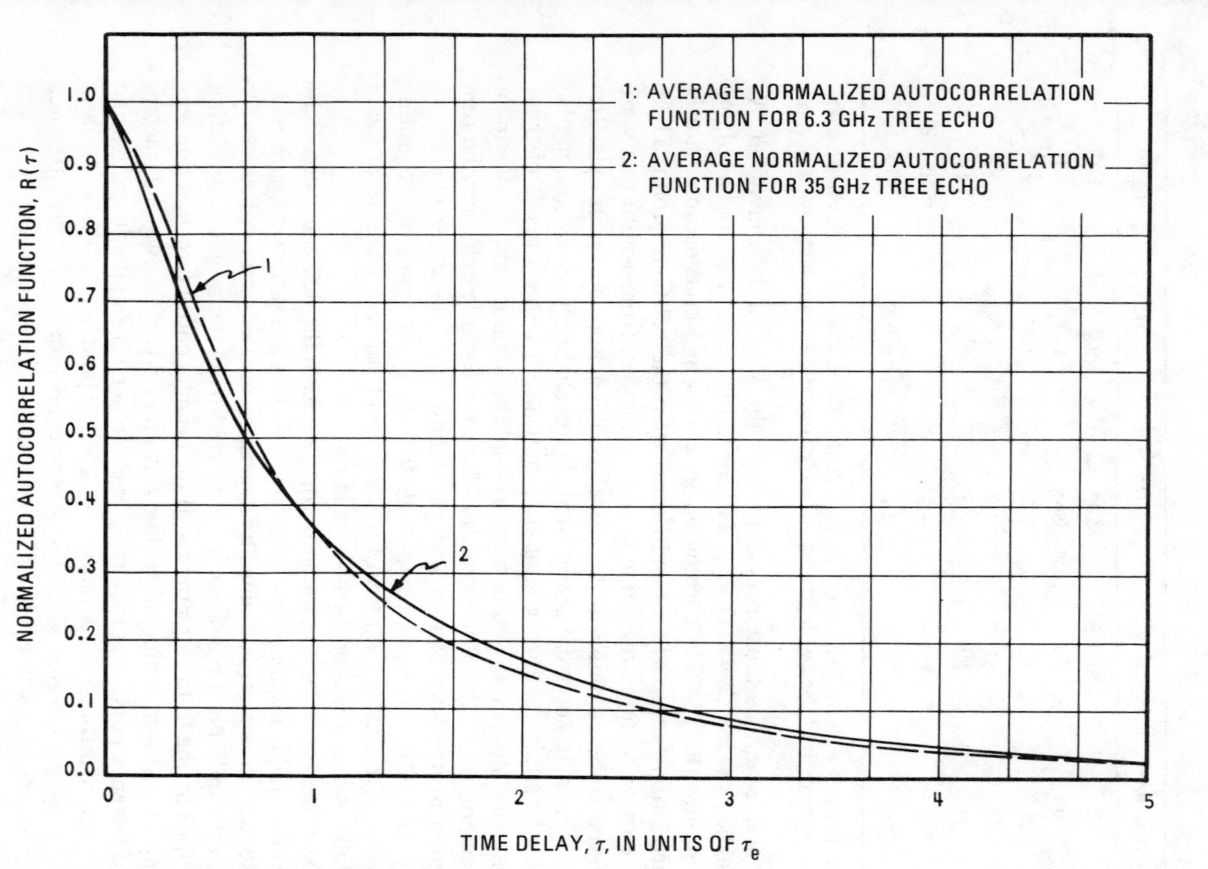

Source: Ivey, Long, and Widerquist (1956).

Figure 5-11. Experimental Autocorrelation Functions for 6.3-GHz and 35-GHz Tree Echo.

For 35-GHz echo, recordings of three to five minutes duration were made on four days. These recordings yielded a total of 13 runs and for each run data were obtained for horizontally and vertically polarized transmissions; that is, 26 individual correlation functions were computed. These correlation functions were reasonably smooth and, after the initial rapid decrease, usually approached a small asymptote. The range of τ_e recorded for the individual curves was between 12 and 155 milliseconds with the majority falling between 25 and 80 milliseconds.

For 6.3-GHz echo, runs with durations from three to five minutes were recorded on one day. Horizontal and vertical transmissions were used, and a total of 17 correlation functions were computed. The value of τ_e for the individual curves varied between 0.07 and 0.41 second with the majority lying between 0.07 and 0.13 second. The individual correlation functions were smooth and in all cases the curves were asymptotic to zero.

According to the authors, statistical examinations of the behavior of individual curves of a group (6.3 or 35 GHz) indicated that there was no marked difference between an individual curve and the group average. They also reported that, in addition to there being a visually obvious difference between the two averaged correlation functions shown in figure 5-10, a difference was demonstrated in a statistical sense. However, as described below, each of the averaged curves can be expressed as the sum of a Gaussian and an exponentially shaped autocorrelation function.

Figure 5-12 shows the 35-GHz curve fitted with an exponential curve for values of τ greater than one and the extension of this exponential back to $\tau = 0$. By subtracting the exponential curve B from the original curve A, a residue C was obtained. This same process was applied to the 6.3-GHz averaged correlation function.

From this analysis and the available data, tree echo fluctuations seem to be describable as the sum of two independent fluctuations that have different widths and shapes. The power in a spectrum is proportional (see eq. (5.4)) to the correlation function at $\tau = 0$. Therefore, from the average 35-GHz curve, the spectrum comprising the highest frequencies (shorter times) contains 30 percent (see caption in fig. 5-12) of the total power, and the other spectrum contains 70 percent of the power.

The authors found that the residue C for 35 GHz (fig. 5-12) can be closely approximated by a Gaussian function that drops to half its peak value at a time $\tau = 0.4\ \tau_e$. The exponential function B, representing a correlation function that drops off more slowly, reaches half its peak value at a time $\tau = \tau_e$. Therefore, the autocorrelation function for 35 GHz is considered here to be caused by separate and independent scattering mechanisms so that the frequency spectra can be represented by the sum of spectra that are the Fourier transforms of $R'(\tau)$ and $R''(\tau)$ as follows

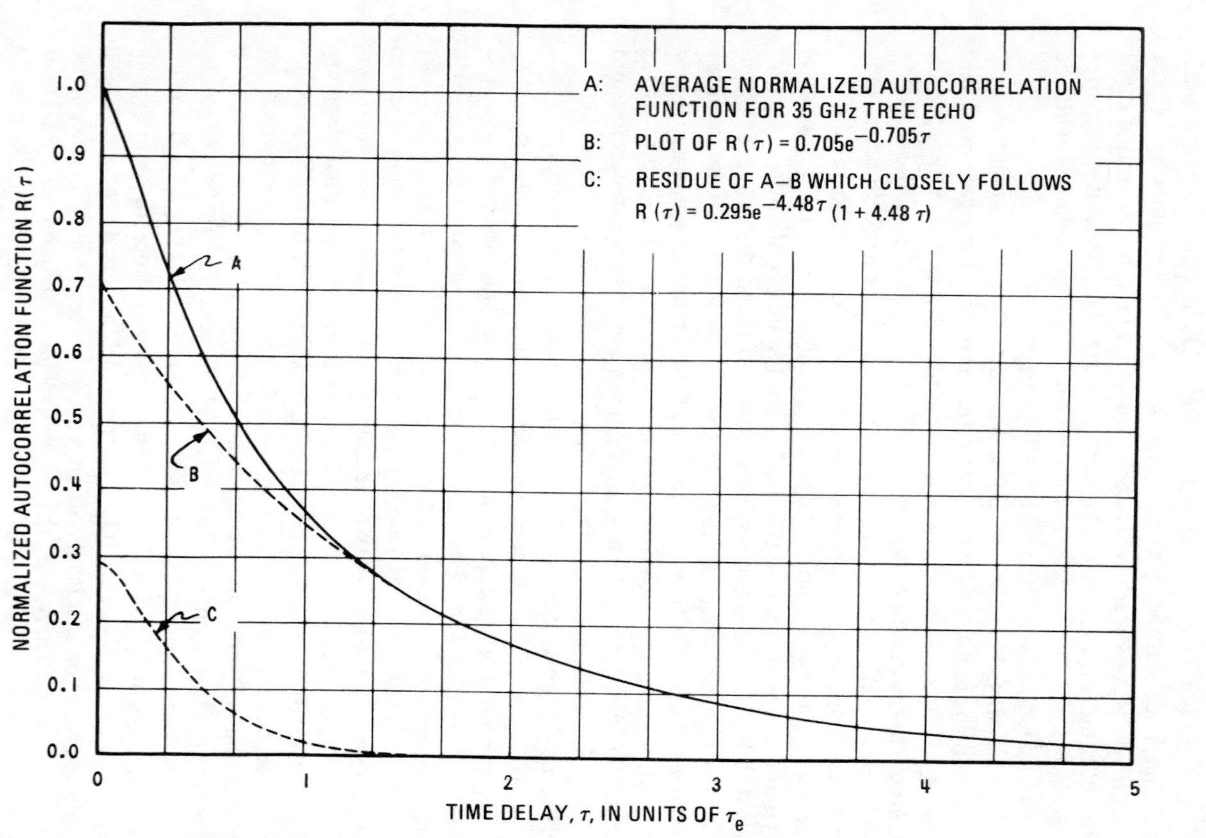

Source: Ivey, Long, and Widerquist (1956).

Figure 5-12. Components of Experimental Autocorrelation Function for 35-GHz Tree Echo.

$$R(\tau) = R'(\tau) + R''(\tau) = 0.7e^{-\beta\tau} + 0.3e^{-\alpha^2\tau^2} \qquad (5.10)$$

where β and α are such that $R'(\tau)$ and $R''(\tau)$ reach their half-amplitude points at times of τ_e and $0.4\tau_e$, respectively.

It was also found that the autocorrelation function at 6.3 GHz could be expressed as in equation (5.10), where in this case $R'(\tau)$ reaches its half-amplitude point before that point is reached for $R''(\tau)$. Specifically, $R'(\tau)$ and $R''(\tau)$ for 6.3 GHz were reported to have half-amplitude points at times $0.7\tau_e$ and $0.9\tau_e$, respectively.

From J.L. Lawson and G.E. Uhlenbeck (1950, p. 42), if

$$R'(\tau) = e^{-\beta\tau} \quad \text{then} \quad P'(f) = \frac{4\beta}{\beta^2 + (2\pi f)^2} \qquad (5.11)$$

and if

$$R''(\tau) = e^{-\alpha^2\tau^2} \quad \text{then} \quad P''(f) = \frac{2\sqrt{\pi}}{\alpha} e^{-\pi^2 f^2/\alpha^2} \qquad (5.12)$$

It can be shown from equations (5.11) and (5.12) that the half-power frequencies $f_{1/2}$ are

$$f'_{1/2} = 0.11(1/\tau'_0) \qquad (5.13)$$

and

$$f''_{1/2} = 0.22(1/\tau''_0) \qquad (5.14)$$

where τ'_0 and τ''_0 are the time delays at which $R'(\tau)$ and $R''(\tau)$, respectively, reach one-half the peak amplitudes, $R'(0)$ and $R''(0)$.

Fluctuation spectra are, of course, strongly dependent on wind speed and the measurements at 6.3 GHz and 35 GHz were not made simultaneously. The trees were not even the same type of trees. Perhaps, however, some insight can be obtained by making comparisons in spectra, by using the most frequently measured values of decorrelation time τ_e.

Recall that $f''_{1/2}$ is the half-power point for the Gaussian spectrum, and it is the Gaussian spectrum that is expected for a cloud of randomly positioned point scatterers. Based on the values of $f''_{1/2}$ in table 5-5 it seems likely, even though ranges of wind speed are unknown, that $f''_{1/2}$ for 6.3 GHz and 35 GHz might differ by a factor 5.6 under similar wind conditions. This factor is the ratio of radar frequencies (35/6.3) and it is the ratio expected for Gaussian spectra. It may be, for example, that the $P''(f)$ spectra are caused by the random motion of leaves moving with the breeze and the $P'(f)$ spectra are caused by the movement of tree branches and trunks. These are, of course, conjectures with no known basis of fact.

As already mentioned, Kerr observed that the general shape of spectra

Table 5-5
Spectra Bandwidth as a Function of Decorrelation Time

Radar Frequency	*Range for Majority of* τ_e	τ_0'	$f_{1/2}'$	τ_0''	$f_{1/2}''$
6.3 GHz	0.07-0.13 sec.	$0.7\ \tau_e$	1.2-2.2 Hz	$0.9\ \tau_e$	1.9-3.5 Hz
35 GHz	0.025-0.080 sec.	τ_e	1.4-4.4 Hz	$0.4\ \tau_e$	6.9-22 Hz

is a complex function of wind speed. In particular, he noted (Kerr 1951, p. 587) that the autocorrelation function for trees at 9.2 cm and a 50 mph wind had an unusual kink that might be caused by two sources of fluctuation. That autocorrelation function is reproduced in figure 5-13 and is labeled experimental.

J.L. Wong, I.S. Reed, and Z.A. Kaprielian (1967) obtained the autocorrelation function labeled theoretical in figure 5-13 by an analysis of many randomly moving dipole scatterers that takes into account the effects of scatterer rotation. Figure 5-14 contains a power spectrum obtained by Wong, Reed, and Kaprielian by computing the Fourier transform of the theoretically obtained correlation function of figure 5-13. The calculated spectrum (for which the autocorrelation function closely matches the experimental one reported by Kerr) is caused by changes in relative positions of the scatterers plus scatterer rotation. The relative magnitudes of the spectral widths of the two sources of fluctuation therefore control the composite spectral shape. The analysis by Wong, Reed, and Kaprielian is significant because it shows that scatterers with two sources of fluctuation can give a spectrum that replicates the experimental result discussed by Kerr.

The discussion in this section has been on spectral width caused by changes in positions and orientations of the scatterers. Measured spectra are broadened by motion of the radar bean (Ridenour 1947, p. 658). For example, a major source of broadening for an airborne radar is due to radial velocity (and Doppler frequency) being a function of position within the resolution cell. Therefore, to determine intrinsic spectra (effects of platform motion removed) it is usually necessary to make detailed analyses to estimate effects of platform motion. M.B. Laing (1971) includes an analysis of the effects of aircraft motion in a report on the intrinsic spectra of sea echo (see sec. 5.13).

Visual Observations of Sea Echo

Sections 5.4 through 5.7 discuss observations of sea echo as revealed by

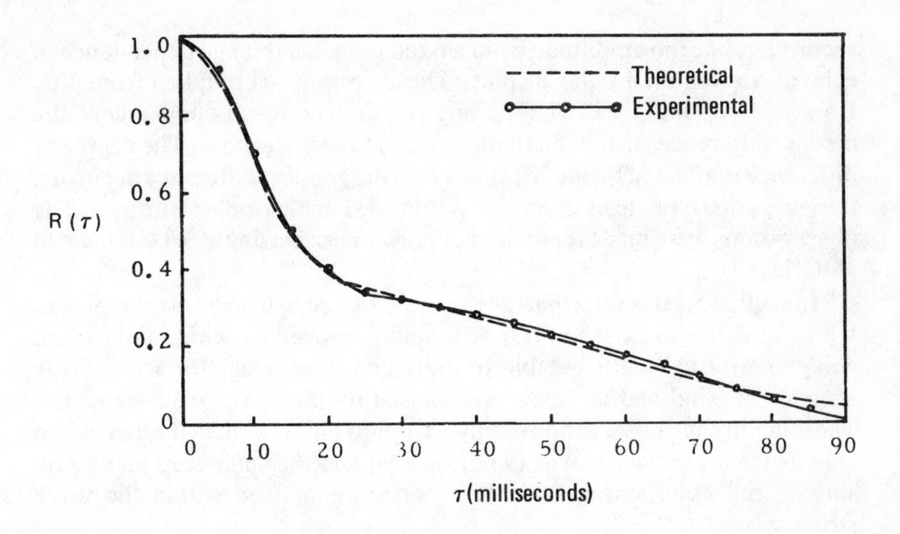

Source: Wong, Reed, and Kaprielian (1967).

Figure 5-13. Correlation Function for Ground Clutter at 9.2 cm. (Heavily wooded terrain at wind speeds of 50 mph.)

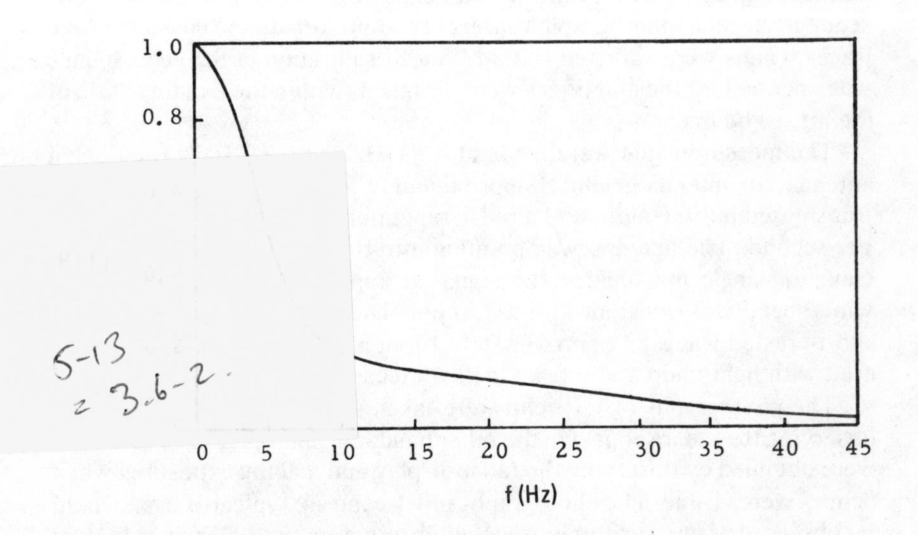

Source: Wong, Reed, and Kaprielian (1967).

Figure 5-14. Power Frequency Spectrum for Ground Clutter at 9.2 cm. (Heavily wooded terrain at wind speeds of 50 mph.)

recordings of echo amplitude from a fixed range, and range dependence of echo as viewed on a radar display. These sections were taken from J.G. Boring et al. (1957) and M.W. Long (1974). The results underscore the marked differences in the fluctuations of *HH* and *VV* echoes. The degree of difference is affected by the sea state, with the greatest difference occurring during periods of dead calm or a calm sea with rippled surface. The observations were made for incidence angles near grazing at 6.3 GHz and at 35 GHz.

In section 5.7 it is seen that (a) the noise-like echo usually attributable to *VV* echo can also exist for *HH* echo and is caused by wind, and (b) the spiky echo usually attributable to *HH* echo does occur for *VV* echo in absence of wind and is, therefore, caused by the gross wave structure. These and other observations support the hypothesis that the noise-like echo is from capillary waves (wind ripples) and the spikes are caused by smooth reflecting surfaces (facets) that are contained within the wave structure.

5.4 Characteristics Revealed by an A-Scope Display

Differences between *VV* and *HH* echo are illustrated in figure 5-15. These were obtained by photographing an *A*-scope having a sweep repetition frequency of 3,000 per second with an exposure time of 0.1 second. Consequently, each photograph represents approximately 300 successive traces. Gains were adjusted so that "hard" saturation of the echo signals was uncommon; the amplifiers were designed, within the existing state of the art, to be linear.

The measurements were made at 6.3 GHz using a 2 1/4° pencil-beam antenna, an antenna height of approximately 52 feet, a 0.19-microsecond transmitted pulse length, and a pulse repetition frequency of 3,035 pulses per second. The antenna was pointing into the wind and waves with an elevation angle maximizing the signal at approximately midrange. The wind speed was constant at 8 1/2 knots. The sea was characterized by short-crested waves of approximately 1 foot average height and was covered with light chop and a few small whitecaps.

The photographs of *VV* echo were taken at random intervals without observing the radar scope. On the other hand, the photographs of *HH* echo were obtained by observing the radar display and making exposures when echoes were visible. The photographs of *VV* echo are typical of what would be observed at any randomly selected observation time. A causal look at figure 5-15 gives an entirely false impression about the rate of occurrence of spikes in the *HH* echo, since much of the time the scope trace was unbro-

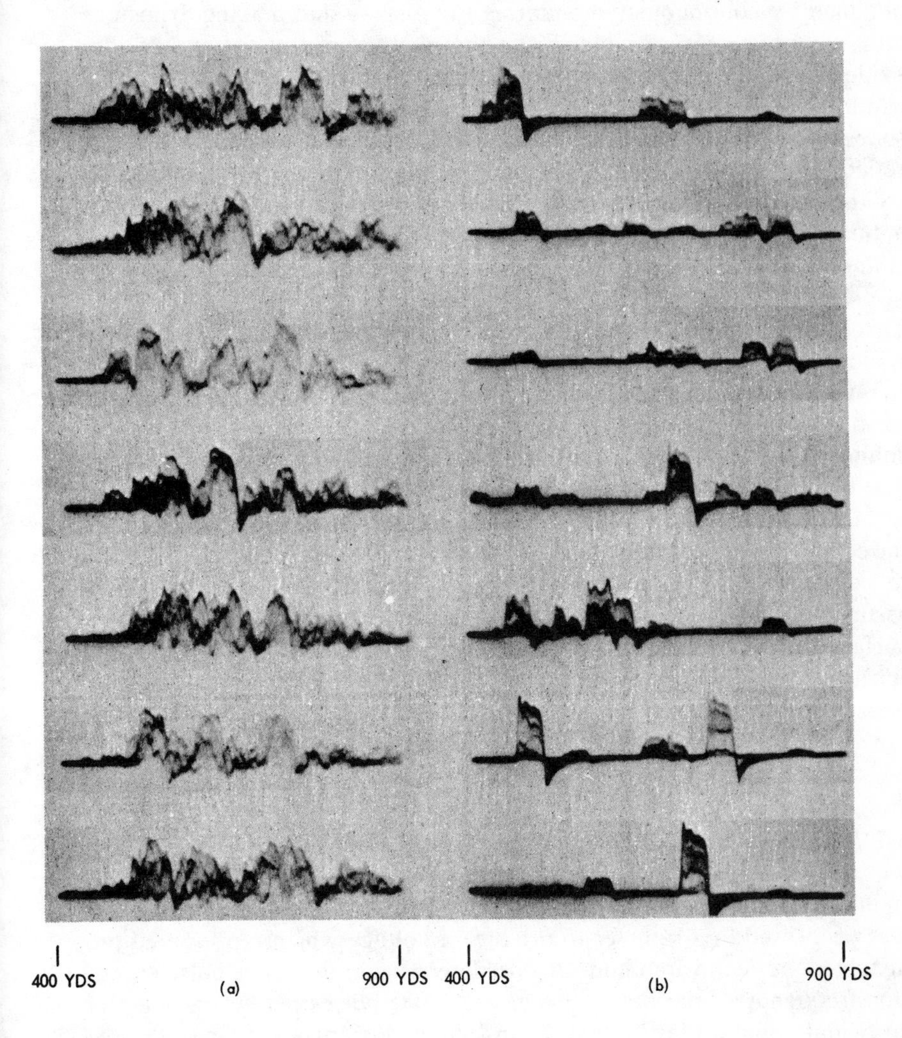

Source: Long (1974).

Figure 5-15. A-Scope Photographs Showing Range Dependence.
(a) *VV* Polarization (photographs taken randomly).
(b) *HH* Polarization (photographs taken only when echo was visible). Each photograph contains 300 successive range traces.

ken or had a barely perceptible echo present. These photographs were obtained by continuously monitoring the display and making exposures only when echo was visible. From the photographs in figure 5-15 it is evident that *VV* echo has no well-defined characteristics. On the other hand, the *HH* echo appears as a single well-defined echo having the same shape as the transmitted pulse—thus appearing as if caused by a single reflector. The overshoot on the trailing edge of the echo is due to an equipment limitation. This pronounced difference between the echoes obtained for the two polarizations was commonly observed for low sea states.

The behavior of echoes was also examined on a pulse-to-pulse basis. This was accomplished by successively photographing individual range traces that were triggered at a trace repetition frequency of 30 per second. In reviewing the film for *VV* echo, it was found that the continued presence of an individual echo (fixed range) over several traces was not particularly common. However, from strip-film recordings for *HH* echo it could be seen that individual echoes appeared and disappeared in a somewhat periodic fashion. The period was about 1/4 or 1/3 second. This fading in and out of the echo is similar to that obtained when a simple target, such as a flat plate or a cylinder, is moved to change the aspect from which it is viewed. If it is assumed that the echo is produced by a single target, such as a curved surface on the surface of water, then it is the reflection pattern of the surface that is being observed as a function of the surface orientation with respect to the radar.

5.5 Results from Fixed Range Sampling

In the equipment used for fixed range sampling, the output of each receiver was processed in a sampler to produce a voltage which represented the instantaneous echo amplitude at a predetermined range. The pulse repetition frequency of the radar system was 3,035 pulses per second—a rate sufficiently high to insure that the instantaneous cross section of the sea would not change by a measurable amount between successive pulses. The wave form, which represents the instantaneous cross section, was integrated in a simple *r-c* filter having an adjustable time constant. The output of the integrator was plotted with a paper chart recorder.

1. Short Integration Time. A time plot of the results obtained by integrating the instantaneous *VV* echo from a fixed range, by use of a low-pass filter having a time constant of 0.06 second, is presented in figure 5-16. The echo fluctuated in a random fashion even though the high-frequency components had been removed.

The results obtained when *HH* echo was processed in a similar manner are also displayed in figure 5-16. Again, the spiky echo produced when utilizing horizontal transmission is obvious, and it is evident that for a considerable portion of the time the echo amplitude was essentially zero. When the amplitudes of the fluctuations for the *VV* and *HH* echoes are compared, it is seen that the dynamic range of the *VV* echo is much smaller than that of the *HH* echo, and at all times the *VV* echo amplitude remains well above zero. On the other hand, the *HH* echo was essentially zero for a large portion of the time and actually reached saturation for occasional large spikes. The dynamic range of the system and calibration of the recorder was approximately the same for the two polarizations.

2. Medium and Long Integration Times. If a longer integration time constant is employed, the residual fluctuations displayed in the previous plots are largely removed, and the *VV* and *HH* echoes continue to behave in a different manner. Fluctuations of instantaneous *VV* and *HH* echo after averaging with a time constant of 0.6 second have been compared (Boring et al. 1957, p. 16) with plots of instantaneous wave heights. The wave-height data were obtained with a beach erosion board guage (measures wave height in 0.2-foot steps). Although each wave-height recording was made at the same time as its associated echo recording, the location at which wave heights were measured was separated by several hundred yards from the illuminated area.

In general, the *VV* echo plots for a 0.6-second time constant fluctuated in a manner similar to the ocean waves. For *HH* echo, the spiky characteristic was still present but not as obvious as when a shorter integration time was employed. It seemed from these results that the occurrence of successive spikes in the *HH* echo was associated with the sea waves, but the *HH* echo did not follow the sea-wave pattern nearly as well as the *VV* echo. Examination of the wave height revealed a tendency for some peaks or crests of waves to be appreciably larger than the average of the crests. The correlation between the frequency of occurrence of these high waves and the larger echoes in the *HH* echo plot was fairly good.

This wave-like fluctuation does not always occur for *VV* echo; the presence or absence of it can be definitely associated with the wave characteristics of the sea surface. For 6.3 GHz the fluctuation was more noticeable for angles of arrival larger than 1.5°, usually being observed when the sea waves were well defined and fairly long-crested with periods of 5 or more seconds. Obviously, the conditions under which it is observed are dependent upon the size of the illuminated area. The wave-like fluctua-

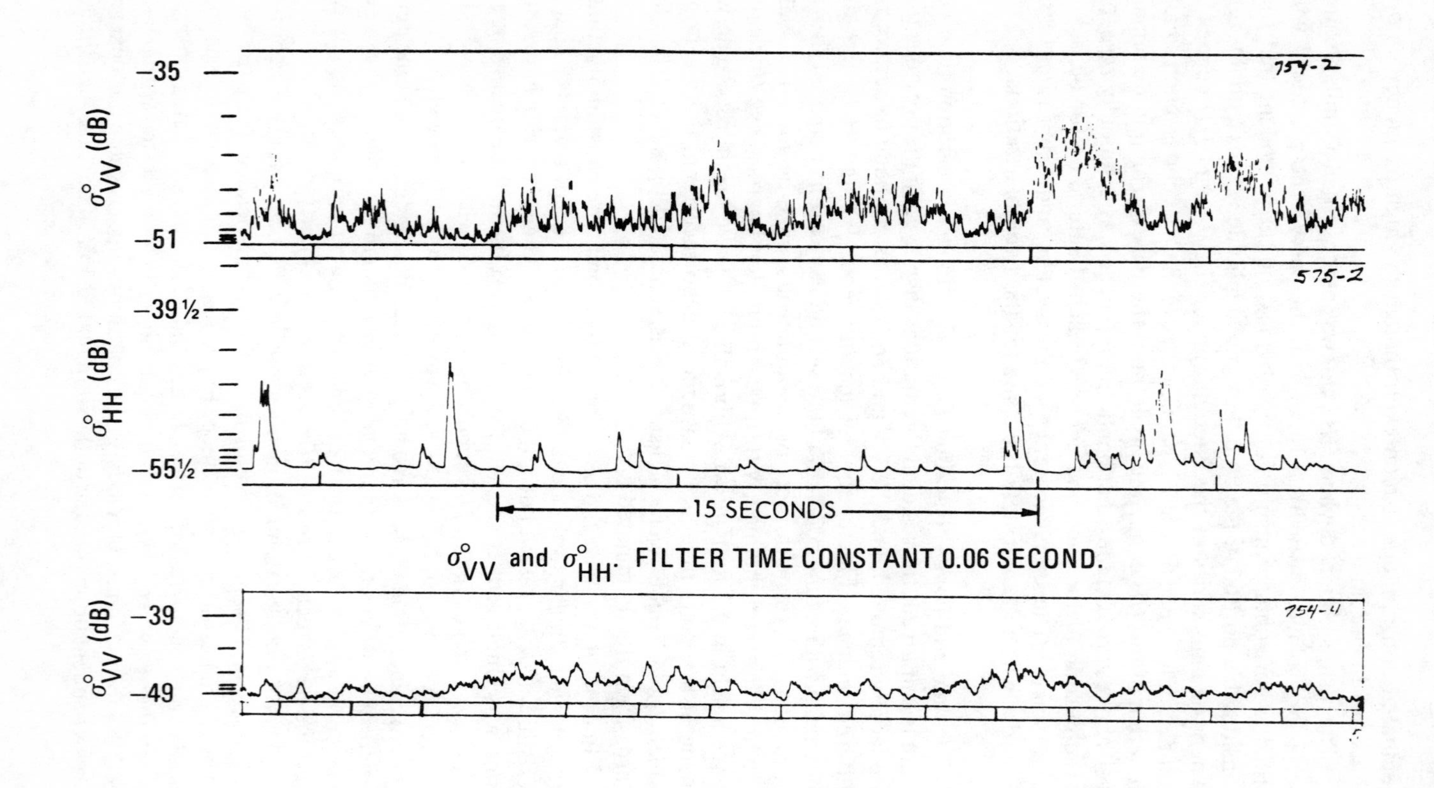

−35
σ°VV (dB)
754-2
−51
575-2
−39½
σ°HH (dB)
−55½
15 SECONDS
σ°VV and σ°HH. FILTER TIME CONSTANT 0.06 SECOND.
σ°VV (dB)
−39
754-1
−49

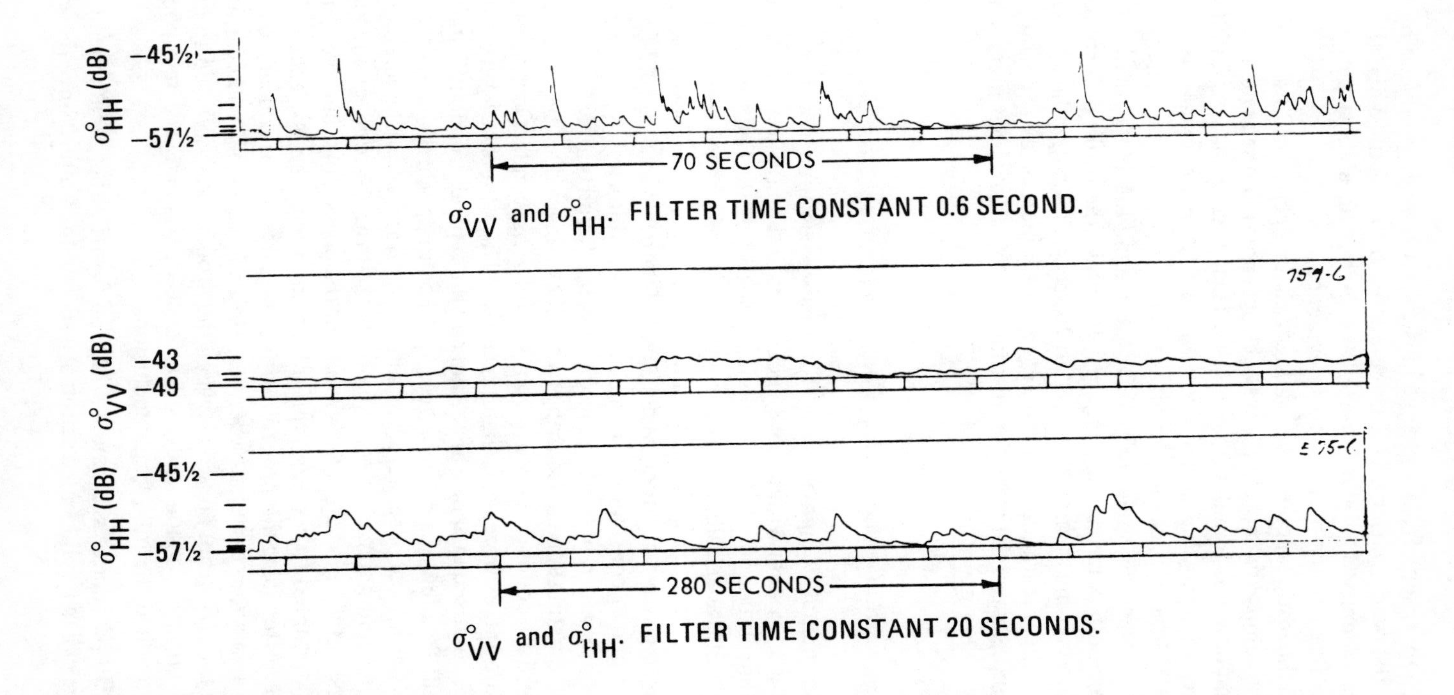

Source: Boring et al. (1957).

Figure 5-16. σ°_{VV} and σ°_{HH} at 6.3 GHz Showing the Effects of Integration. Wind speed 6 knots, average wave height 1 1/2 feet, and arrival angle 2.3°.

tion probably results from a change in the arrival angle produced by the tilting of the water surface as a wave passes by a point. When the sea surface is composed of short-crested waves having a short period, the wave-like fluctuations may not be in evidence.

When the instantaneous echo is integrated for a period of 20 seconds, all rapid fluctuations and most of the wave effects are smoothed out. With this integration time it is found, for angles of arrival greater than 1.5°, that *VV* echo quite commonly has an average value that is fairly constant—within 1 or 2 dB. *HH* echo will usually exhibit a greater range of fluctuation than *VV* echo. As the arrival angle decreases, the magnitude of fluctuations increases for both polarizations. Figure 5-16 shows the result of increasing the integration time from 0.06 second to 20 seconds. During the course of recording, the integration time was successively increased. The wind and sea conditions were the same for all charts in this figure, but none of the recordings were made simultaneously.

5.6 Subjective Radar and Optical Comparison

The experimental equipment used for this observation consisted of a 35-GHz radar system with a 5/8° pencil-beam antenna and a 0.23-microsecond transmitted pulse, and an optical telescope attached to and collimated with the radar antenna. A depression angle of 3.8° was used, and the antenna was pointed into the wind and waves. Wind speed was 14 knots, and the sea consisted of 3-foot swells covered with waves 1- to 1 1/2-feet in average height. A considerable number of small whitecaps were present.

From observation of the *A*-scope display, it is found that *HH* echoes are discrete fluctuating spikes occurring (within the ranges displayed) at intervals of 3 to 10 seconds and lasting from 1/5 to 1 1/2 seconds. Occasionally, two or more spikes were seen simultaneously. The *VV* echoes were range extensive, confused in appearance, and had the typical "random" characteristic; but frequently a recognizable spike echo appeared.

The experimental procedure was simple. An observer at the telescope, in telephone communication with an observer at the *A*-scope, announced the presence of a whitecap in the vicinity of the cross hairs, and the radar observer attempted to correlate this information with the presence of an echo spike. After using this procedure for 10 minutes, it was concluded that there was no obvious correlation between the occurrence of a whitecap and a spike in the *HH* echo. However, when the process was reversed so that the radar observer announced the presence of a spike, the telescopic observer reached the following conclusions:

1. About 50 percent of the time a whitecap forms either simultaneously with the call of a spike or a fraction of a second thereafter. Since there is

a definite time delay between the actual development of a spike and the vocalized message, it appears that the formation of a whitecap appears after the echo spike develops.

2. There is a correlation between the duration of the spike and the size and duration of the whitecap.
3. Thirty-five to 40 percent of the time a spike will be called when the wave structure has a very peaked crest, as if a whitecap were about to form but did not develop.
4. The remainder of the time there will be no peculiarity of the sea surface associated with the occurrence of the spike.
5. No whitecap will be observed in the absence of a spike.

Based on these observations, it appears that *HH* echo for this particular sea state is predominantly produced by single-peaked wave crests that become unstable and break to form a whitecap about half the time.

When the process was repeated for *VV* echoes, it was found that the correlation was not as good, possibly due to confusion of the radar observer by the large amount of random echo present. Large spikes, having amplitudes significantly greater than the average echo, were associated with the occurrence of a whitecap or a peaked wave.

5.7 Anomalies

Under certain conditions *VV* and *HH* echoes were observed to exhibit the properties normally attributed to their counterpart; that is, observations revealed a *spiky-like VV* echo and a *random-like HH* echo. Also, as the sea state and wind speed increased, there was a tendency for the two to become similar.

On one occasion a smooth, flat, glassy sea having no visible wave structure was excited by a rather abrupt increase in wind speed from a dead calm. The *HH* echo observed was random in nature and typical of normal *VV* echo. The average radar cross section was extremely small. As chop and waves built up, the echo became spiky with the magnitude of the spikes large enough to mask the weak random component.

On another occasion a fairly choppy sea had been generated by a fresh sea wind. The wind stopped abruptly. In a very short period the water surface was devoid of small ripples and had a glassy surface even though the gross structure was unchanged. In this case, the *VV* echo consisted primarily of spikes. When the wind speed increased (from a different direction), the glassy surface disappeared and the spikes were lost in the general random return.

Sea Echo Statistics and Spectra

Sea echo is the vector sum of scattering from the sea surface within the illuminated area. The movement of the scatterers (waves, ripples, etc.) causes a change in the relative phases of their separate echoes and a resulting change in the total echo (the vector sum of the component echoes). In figure 5-15, photographs of sea echo as displayed on a radar A-scope were given for small depression angles and for transmitting and receiving horizontally or vertically polarized waves. The echo is often noise-like, as would be expected from a collection of randomly moving scatterers, but sometimes the echo appears to resolve itself into distinct point target-like echoes with regions of very low signal occurring between these echoes. Similar results were also reported by F.C. Macdonald (1957). The distinct "spikyness" tends to increase with decreasing depression angle and decreasing area of illumination (i.e., higher radar resolution). Also, this spikyness is more pronounced with horizontal than with vertical polarization and there is evidence that the occurrence of the spikes is related to the occurrence of well-defined, steep-created ocean waves.

There are considerably more data on fluctuations in time than on fluctuations in range. In general, it has been found that time fluctuations tend to resolve themselves into the following types: (1) a fast random fluctuation caused by constructive and destructive interference in the echoes from many independently moving scatterers; (2) a slower fluctuation caused by variations in the amplitude or slope of individual ocean waves passing through the illuminated area, and possibly also a periodic fluctuation caused by the recurrence of ocean waves in the illuminated area; and (3) very slow fluctuations or secular changes caused by gross or long-term modifications of the sea.

5.8 Amplitude Distributions

Sea echo as viewed on an oscilloscope is rapid and irregular and this gives it a family resemblance to ordinary receiver noise. The range or extent of the fluctuations is usually specified by the probability distribution for the instantaneous intensity. Many of the density functions of received power reported for sea echo can be approximated by exponentials as given by the Rayleigh model (eq. (5.6)). However, approximation is probably never valid over the total power range of a density function. An exponential distribution is represented on semilogarithmic graph paper as a straight line with the slope being determined by the average power (see fig. 5-17).

Some experimental curves, predominatly for small depression angles and small illuminated areas, have a break or breaks in the slope suggesting

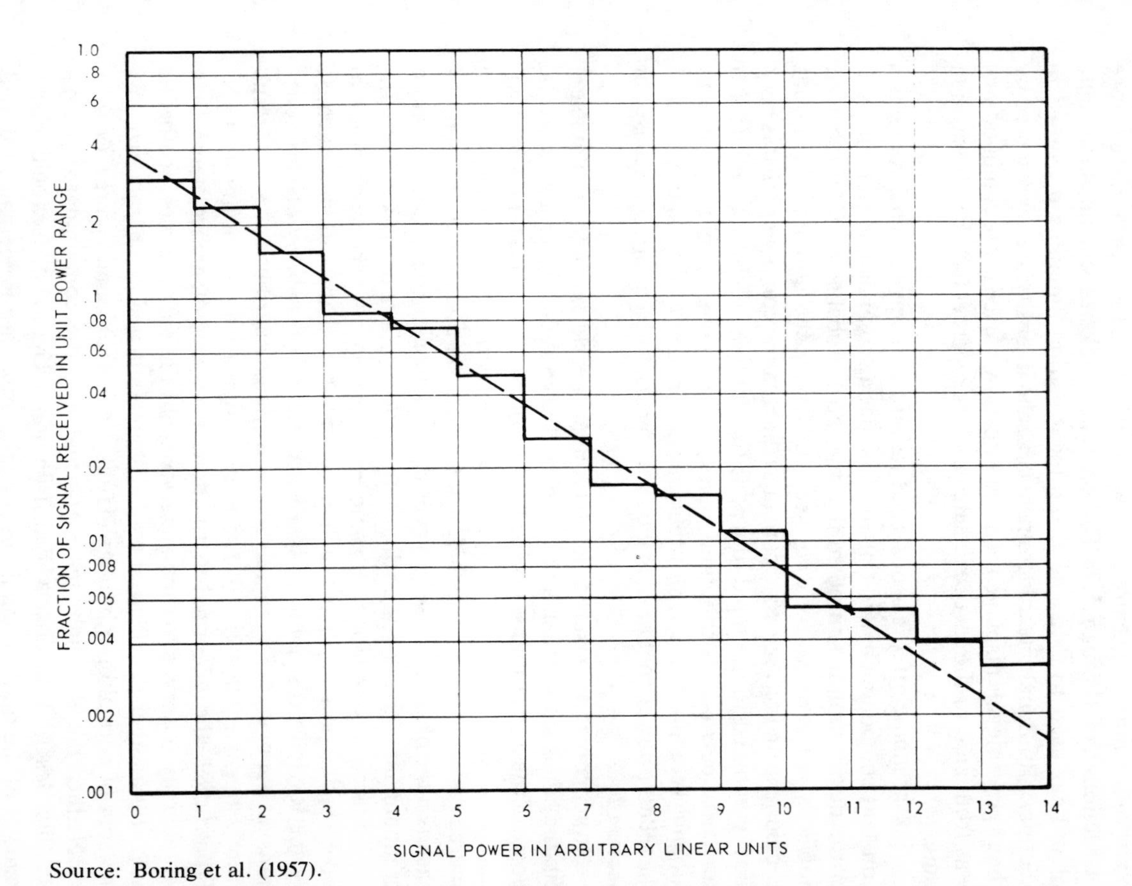

Source: Boring et al. (1957).

Figure 5-17. A Probability Distribution for 6.3-GHz Sea Echo. *VV* Polarization.

an additional source or sources of fluctuation. Figures 5-17 through 5-20 show the appearance of probability distributions for pulse-to-pulse fluctuations in power as a function of polarization at 6.3 GHz with 10-minute observation times. Measurements were made with the same radar as those in sections 5.4 through 5.7. The data were taken from magnetic tape recordings used to calculate the autocorrelation functions discussed in section 5.10. Equal linear increments of echo intensity are represented by the 14 power levels displayed. The range of power used in the computations was from zero to the level exceeded only one percent of the observation time.

For ground clutter it is observed that if the received signal contains a constant component in addition to a fluctuating component, then the peak of the distribution is shifted so that the most probable value of received power is not zero. It does not seem that sea echo contains a constant component. In fact, sea echo contains a fluctuating component that is very slow compared to the rapid (doppler) fluctuations, thereby resulting in a similarity between distributions that contain a constant plus a Rayleigh (doppler) component. The slow fluctuations are depicted by figure 5-16 and the relative power contained in the slow and fast fluctuations is discussed in section 5.12.

In general, the measured distributions are nearly exponential (Rayleigh) but they drop slower than an exponential curve at the higher power levels. Some of the distributions, primarily those obtained for horizontal polarization, have a break in the slope suggesting two separate (independent) sources of fluctuation. This helps to support the finding that fluctuations based on the passing of ocean waves through the illuminated area are more predominant with horizontal than with vertical polarization (see sec. 5.5).

As already mentioned, in some cases sea echo has been observed to have a higher probability of reaching larger values than would be indicated by the Rayleigh distribution; that is, the actual probability density functions do not decrease as rapidly as the Rayleigh distribution. S.F. George (1968) suggested that the distribution is lognormal. According to G.N. Trunk (1969), the lognormal distribution has, in general, higher "tails" than are actually observed. Trunk suggested that the data are matched better with still another distribution (called "contaminated"), characterized as the sum of normal distributions (Trunk 1969; Trunk and George 1970).

Neither the Rayleigh, the lognormal, nor the contaminated normal actually describe all observed distributions. The echo, it would seem, consists of the rapid fluctuations (dopplers) that are Rayleigh distributed plus a component that is fluctuating slowly, and for which distributions have not been reported. For time intervals short as compared to the average period of the slow fluctuations, the distribution on this basis is the Ricean; that is, a Rayleigh plus constant (Norton et al. 1955). Distributions

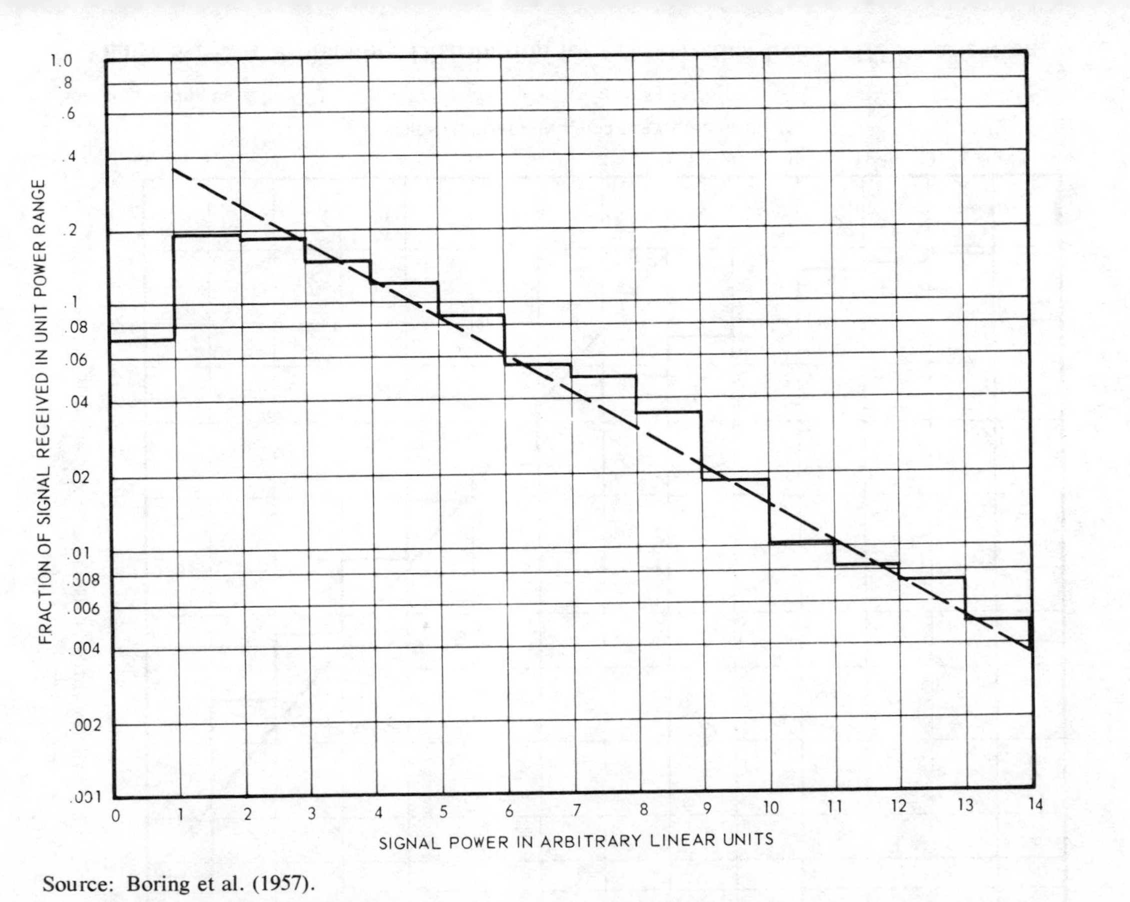

Source: Boring et al. (1957).

Figure 5-18. A Probability Distribution for 6.3-GHz Sea Echo. *VV* Polarization.

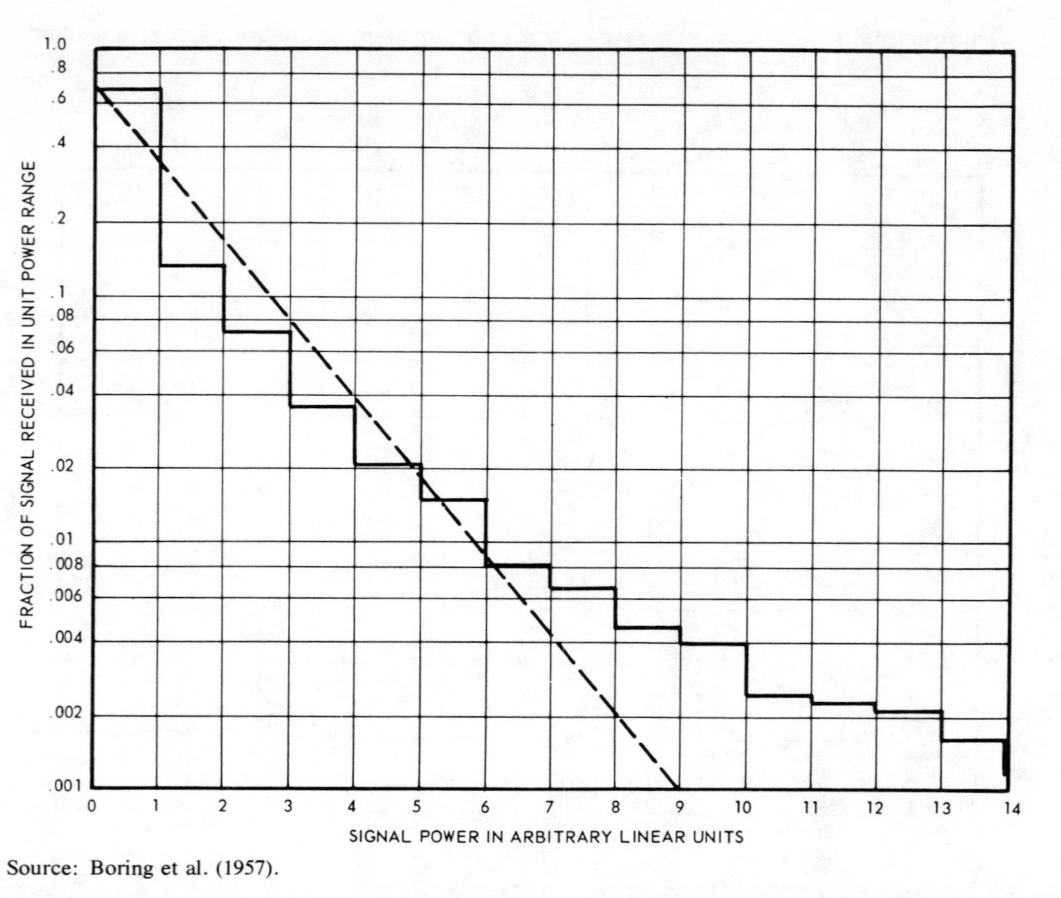

Source: Boring et al. (1957).

Figure 5-19. A Probability Distribution for 6.3-GHz Sea Echo. *HH* Polarization.

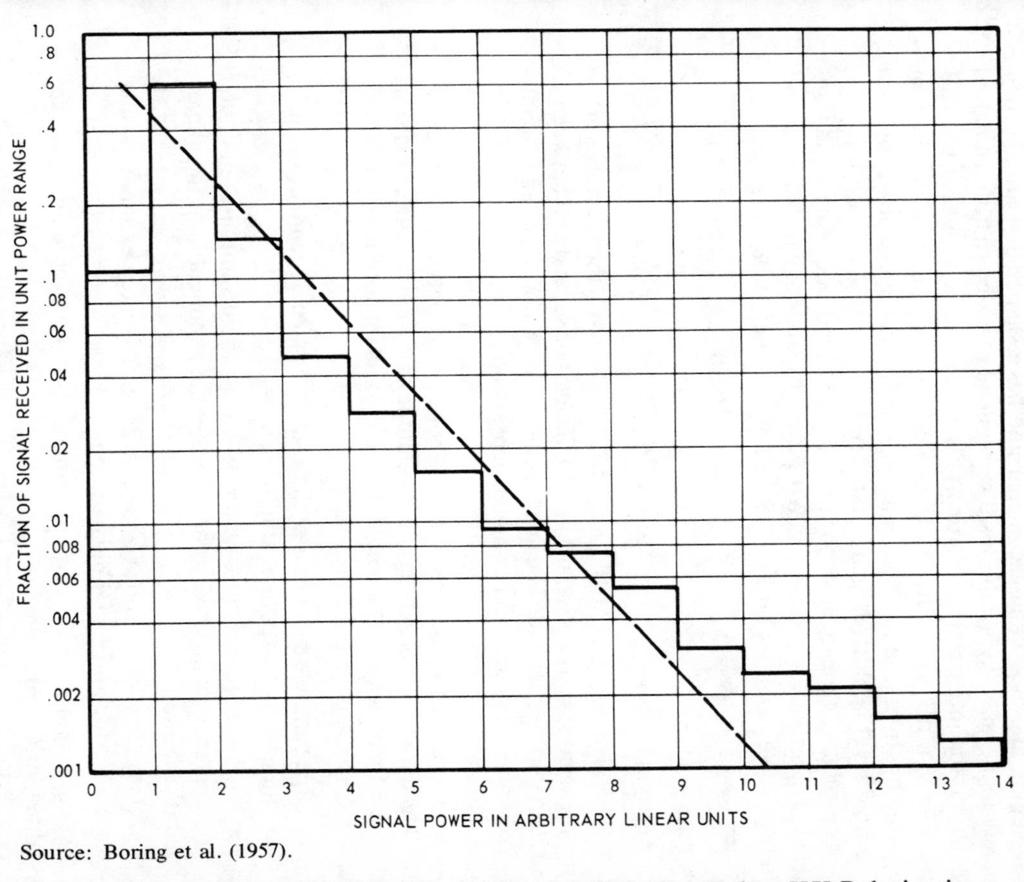

Source: Boring et al. (1957).

Figure 5-20. A Probability Distribution for 6.3-GHz Sea Echo. *HH* Polarization.

for longer time periods will depend on details of the distributions for both the slow and the fast fluctuations.

Recently there has been a considerable effort on measuring amplitude distributions both in the United States and in the United Kingdom (Bishop 1970). W.K. Rivers (1970) reported that statistical distributions at a 3-mm wavelength and at *X* band are similar and have shapes that are approximately lognormal. Typically, his observations revealed that the standard deviations for horizontal and vertical polarizations were about 7.0 ± 0.6 dB and 4.7± 0.6 dB, respectively. Figure 5-21 is a cumulative distribution reported by Rivers. In this figure, the probability that the echo lies below a threshold level is plotted against that level.

The radars were located on the sea coast, and operating conditions included depressions angles of 5° to 0.7°, wind speeds of 2 to 14 knots, and average wave heights of 0.5 to 2.5 feet. Key radar parameters were:

Wavelength (frequency)	3 mm (95 GHz)	3.2 cm (9.4 GHz)
Pulse length	0.1 μs	0.4 μs
Azimuthal beamwidth	0.38°	0.8°

G.R. Valenzuela and M.B. Laing (1971) reported on the variances of sea echo for wind speeds between 5 and 10 meters per second and between 15 and 20 meters per second. The measurements were at *X*, *C*, *L*, and *P* bands for *HH* and *VV* polarizations, and the depression angles were 5° to 30°. The radar output was in units of decibels as a function of time and the values of variance (standard deviation squared) reported were between 16 dB^2 and 38 dB^2. The authors reported that the amplitude distributions of sea echo are intermediate between exponential (Rayleigh) and lognormal. Other results given by Valenzuela and Laing follow:

1. For large sample sizes (greater than about 1,000 independent samples) sea clutter is not exponential nor lognormal.
2. For small sample sizes (less than 200 independent samples) sea-clutter statistics may be approximated by either the exponential or the lognormal distribution.
3. Sea clutter for vertical polarization, in general, is more exponential than sea clutter for horizontal polarization.
4. Sea clutter for *P* and *L* band, in general, is more exponential than sea clutter for *C* and *X* band.
5. The number of independent samples is roughly proportional to radar frequency.
6. Sea clutter for calm seas is more exponential than sea clutter for rougher seas.

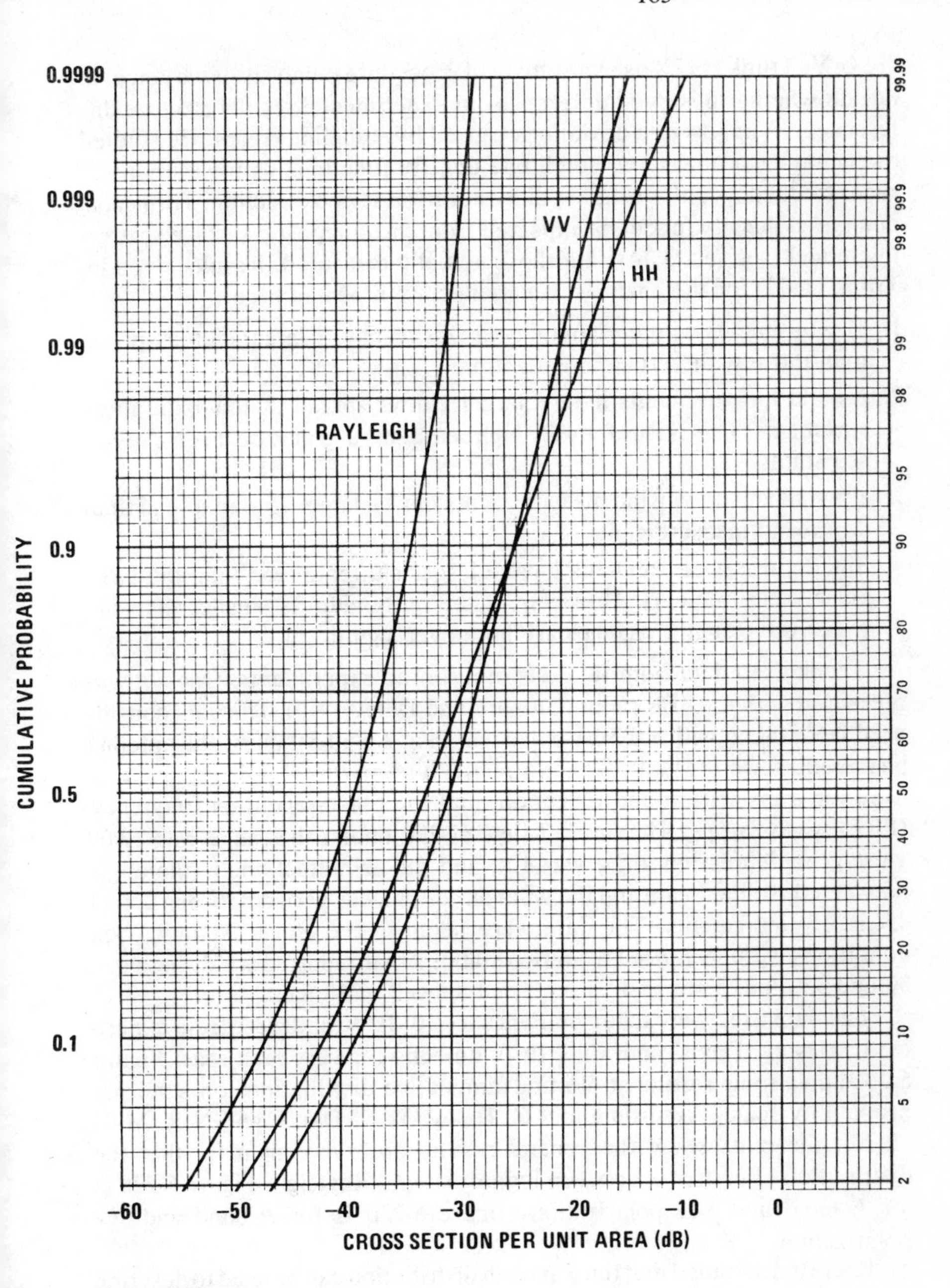

Source: Rivers (1970).

Figure 5-21. Cumulative Distribution for *X*-Band Sea Echo.

G.V. Trunk (1972) used a statistical procedure (analysis of variance) to decide whether observed differences in experimental results are true differences, as opposed to deviations caused by sampling errors. He studied the behavior of measured amplitude density functions with variations in radar frequency, pulse length, polarization, and wind direction. Trunk also compared density functions expected from the composite surface scattering model (see sec. 3.14) with those actually measured. Results given by Trunk (1972) for the analysis of variance are repeated here.

1. Measurements with horizontal polarization have a larger clutter spread than those with vertical polarization.
2. *L*-band measurements have a larger clutter spread than *X*-band measurements. This is true for very high sea states and may be true for lesser sea states.
3. Upwind and downwind measurements have a larger clutter spread than crosswind measurements.
4. Small pulse measurements have a larger clutter spread than the large pulse measurements. However, if the pulsewidth is smaller than the water wavelength, changes are no longer significant.[b]

Trunk used data from two sets of airborne experiments. One set was measured with the 4FR radar (Guinard and Daley 1970). The data considered were taken with a 0.5-μs pulse length at *L* and *X* bands; the antenna beamwidths were 5° and 5.5°, respectively; horizontal polarization was used for transmission and reception; and the pulse-repetition frequency (PRF) was 683 pps. The aircraft flew at 200 knots; the range was 2,000 yards; the depression angle was 10°; and the azimuthal angle (the angle between the radar beam and the wind direction) was varied between 0° and 45° in 15° increments. The data were taken in the North Atlantic. Sea conditions were measured from two stations that reported winds between 30 and 35 mph, a sea of 13.1 feet, and a swell of 18 feet.

The data used were taken on February 10, 1969, and were from experiments (Daley, Davis, and Mills 1970) on echo for very high seas. Figure 5-22 shows some results of plotting cumulative distributions on a normal probability scale. According to J.C. Daley, W.T. Davis, and N.R. Mills (1970), the plots show the general results observed; namely, that the distribution of clutter is between lognormal and Rayleigh. Figure 5-22a is for *L* band and *HH* polarization—figure 5-22b is for *P* band and *VV* polarization.

It is often assumed that the Rayleigh distribution can be used to describe sea echo if the illuminated patch is very large. However, notice that for the

[b]Trunk acknowledges that water wavelength is usually defined in terms of a completely developed sea, a condition that is rarely reached. However, for a possible contradiction to this conclusion for short pulse lengths, see F.E. Nathanson (1969, p. 254).

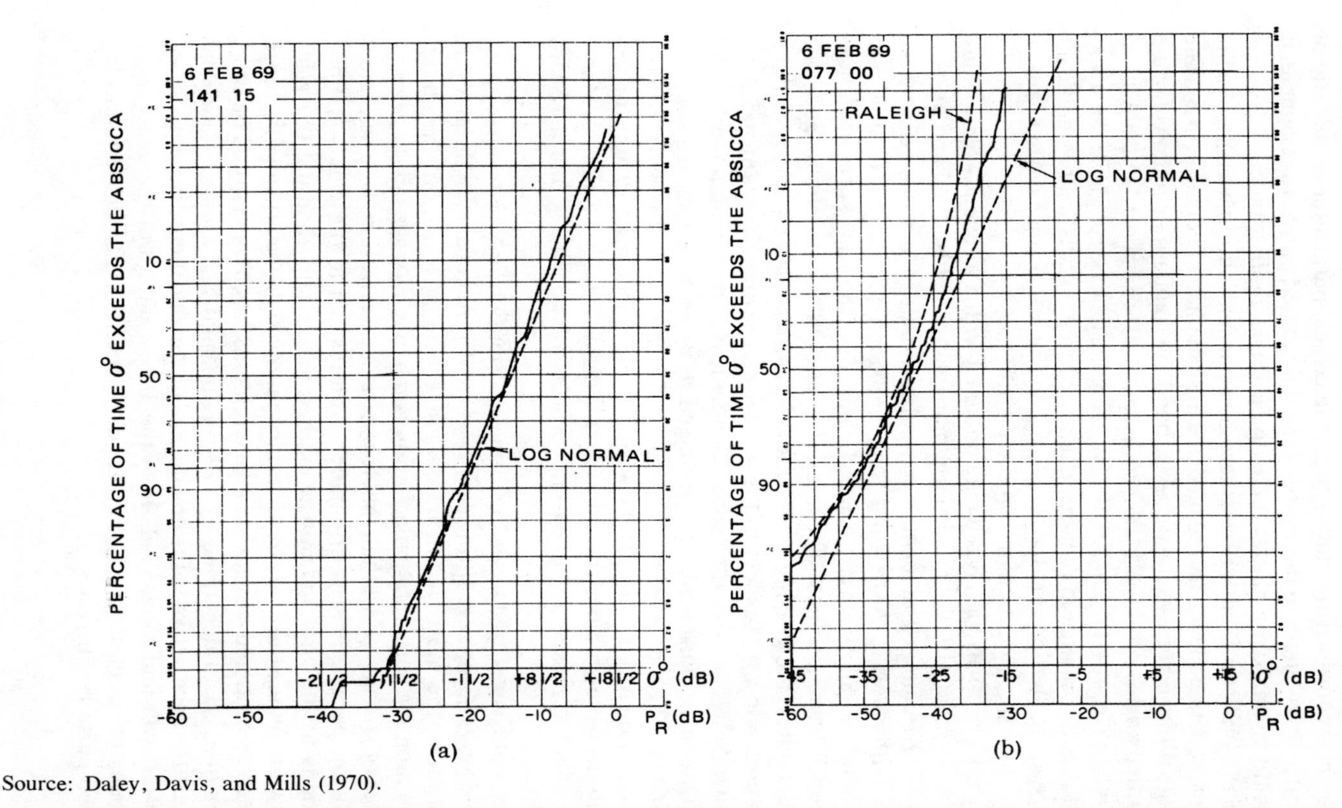

Source: Daley, Davis, and Mills (1970).

Figure 5-22. Samples of Cumulative Probability Distributions for Echo Power. (a) is for *L* band and *HH* polarization; (b) is for *P* band and *VV* polarization.

experiment of Daley et al. the patch is large and the distributions are not Rayleigh. According to Trunk (1972), the critical parameter is the ratio of the water wavelength to the width of the illuminated patch; for example, if this ratio is small, the clutter tends to be Rayleigh distributed.

In the study by Trunk (1972), the lognormal distribution was fitted to the data by equating the measured median to the median of equation (5.8). The average fitting error was 0.8 dB. At *L* band the standard deviations for the azimuthal angles of 0° to 45° ranged from 5.4 dB to 6.3 dB, and at *X* band they ranged from 4.4 dB to 5.0 dB. By using the analysis of variance, it was found that the change in radar frequency caused a statistically significant change in the distribution, but the change in azimuthal angle did not cause a statistically significant change in the distribution.

Trunk (1972) also analyzed data obtained with a frequency-agile, high-resolution radar (FHR), an airborne noncoherent pulsed *X*-band system capable of frequency diversity on a pulse-to-pulse basis. The data were taken with short *S*(20-*ns*) and long *L*(100-*ns*) pulses, vertical *V* and horizontal *H* polarizations, a 0.5° beamwidth, and the pulse repetition frequency was 2,560 pps. The aircraft flew at 180 knots, the range was 2 nautical miles, the grazing angle was 4.7°, and the aircraft was flown in a given direction with respect to the wind (i.e., upwind *U*, downwind *D*, or crosswind *C*) for a period of about 20 minutes. After 2 minutes either the polarization or pulsewidth was changed in order to obtain a variety of conditions. The data used were taken on March 11 and 12, 1969, when the FHR was flown 200 miles off the Virginia Capes (Schmidt 1970). The sea conditions were obtained from the Fleet Weather Facility, Suitland, Maryland, which reported 25- to 31-knot winds, 8-foot waves, and a 12-foot swell.

The data were fitted to the lognormal distribution and the results appear in table 5-6. The average fitting error was 0.8 dB. According to the analysis of variance, the most significant parameter was polarization: the clutter distribution for horizontal polarization had a much longer tail (larger percentage of strong pulses) than that for vertical polarization. The next most significant parameter was orientation with respect to the wind—the distributions for upwind and downwind had longer tails than did those for crosswind. While the analysis of variance showed no significant difference between upwind and downwind measurements, such an analysis of short pulse (20-*ns*) data described by G.V. Trunk and S.J. George (1970) did yield a difference between the upwind and downwind measurements. Also, the 20-*ns* data had longer tails than the 100-*ns* data, but the analysis of variance did not indicate a significant difference between distributions for the two pulse lengths.

Table 5-6
Lognormal Fit of Data Obtained with the *FHR* Radar.[a] **The symbol *s*(*dB*) denotes standard deviation expressed in decibels.**

Date	Time	Identifier	s(dB)
3/11	1344	VUL	4.8
3/11	1346	HUL	6.0
3/11	1411	VDL	4.7
3/11	1413	HDL	7.3
3/12	1026	HDS	7.5
3/12	1028	VDS	5.5
3/12	1049	VCS	4.4
3/12	1051	HCS	6.1
3/12	1125	VCL	4.4
3/12	1127	HCL	5.6
3/12	1154	HUS	5.9
3/12	1156	VUS	5.8

[a]From Trunk (1972).

5.9 Spectra Observed with Noncoherent Radar

The basic cause for sea echo fluctuations is the relative motion of the individual scattering areas on the surface. In this section the frequency spectra of echo from a given patch of the sea as observed with noncoherent radar is reviewed.[c] The spectra have periods comparable to the time between individual waves passing through the radar patch and shorter. They consist of a domain of fast fluctuations with frequencies from near zero to approximately 100 Hz and a second region of moderate fluctuations (and periodicities) with frequencies of the order of 1 Hz.

The first region is ascribed to changing interference effects produced by the relative motion of the individual scattering centers, such as wind ripples. The second region is ascribed to the growth and decay of groups of scattering areas, including the passage of wave crests through the radar viewing area. The relative power contained in the two spectra and the dependence of this relative power on polarization and sea conditions, as interpreted from autocorrelation functions, is discussed in section 5.10. In that section a distinction is made between the periodicities caused by the passage of waves through the sea area under investigation and random fluctuation caused by the presence of these waves within that area.

The fast fluctuation spectrum is usually called the "doppler spectrum."

[c]Section 5.13 contains a discussion on spectra observed with phase coherent radar.

The power spectrum of these fluctuations was reported by Herbert Goldstein (Kerr 1951, pp. 579-81) to be approximately Gaussian in shape, and the measurement results available to him did not indicate a dependence of spectral width on sea roughness. However, more recent results indicate that there is a strong dependence of spectral width on sea surface conditions.

Goldstein describes attempts to measure simultaneously the incoherent doppler spectrum at wavelengths of 1.25 cm, 3.2 cm, and 9.2 cm. None of the measurements were made in less than a few minutes apart, but he conjectured that the product $f_{1/2}\lambda$ ($f_{1/2}$ is the half-power width of the spectrum) seemed to be independent of transmitter wavelength. A value for $f_{1/2}$ of approximately 60 Hz was given for a 3.2-cm wavelength.

If the fluctuations are caused solely by the doppler effect, the product $f_{1/2}\lambda$ will be independent of transmitter wavelength providing the scattering centers (from which relative motion causes the doppler spectrum) are independent of transmitter wavelength. There have been recent reports (see sec. 5.13) that this product, as observed with coherent doppler radar, is not constant throughout the microwave region.

In addition to the fast fluctuations that comprise the doppler spectrum, slower fluctuations are also present. Much of the discussion on subjective observations in sections 5.4 through 5.6 was oriented toward the slow fluctuations. Goldstein (Kerr 1951, pp. 514-18) was aware of these slower fluctuations too. Although it was not possible for Goldstein to specifically identify the causes for the fluctuations, he noted that the slow fluctuations affected measured pulse-to-pulse amplitude distributions. The relationship between the sea surface and the slow fluctuations as observed through analysis of correlation functions is discussed in section 5.11.

G.F. Andrews and H.W. Hiser (1969) have developed a sea state analyzer that measures the average period of the slow fluctuation spectrum and thereby remotely senses the ocean wave period. A 10-cm wavelength, horizontally polarized, one-microsecond pulse radar was used. The data were obtained from 150-meter (depth in range) range gates set at approximately 25 kilometers and a 3°-azimuthal beamwidth was used. Therefore, the radar cell size was 150 by 1,300 meters. In order to obtain a repeatable measurement, the analyzer was designed to calculate an average of 10 periods because ocean wave alternations are neither pure sinusoids nor of precisely uniform duration. Hiser and Andrews (1968) reported that the correlation obtained between the periods of the radar video spectrum and the ocean waves is high, and that the radar spectrum analysis provides a useful way of remotely sensing the period of gravity waves from which wavelength and wave velocity can be readily computed for deep water (Kinsman 1965). A graphic portrayal, taken from H.B. Bigelow and W.T. Edmondson (1947), of the theoretical relationships between period, wavelength, and wave velocity for deep water are given in figure 5-23.

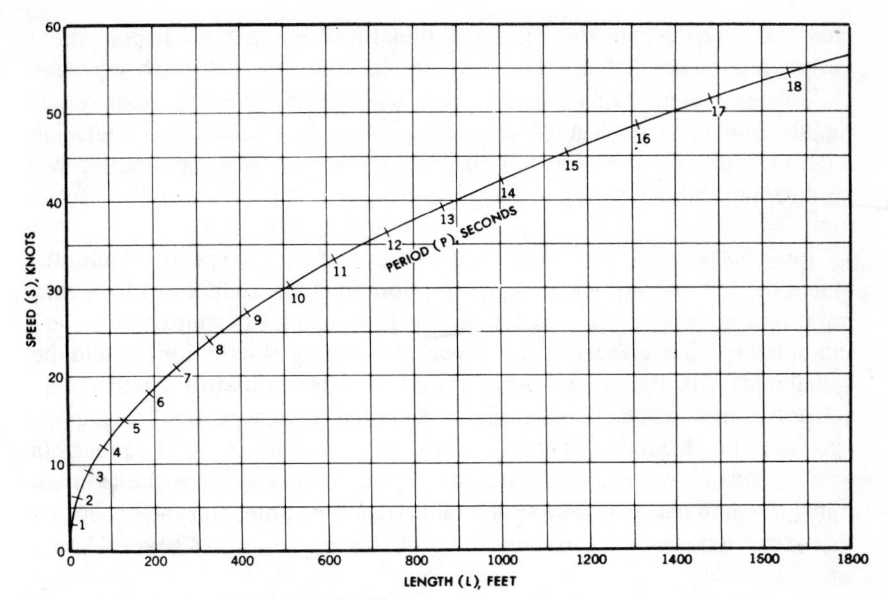

Source: Bigelow and Edmondson (1947).

Figure 5-23. Graphical Presentation of the Theoretical Relationship between Wavelengths, Velocities, and Periods in Deep Water.

5.10 Autocorrelation Functions

Details regarding the shape of the continuum of fluctuations that constitute sea echo can be given either by the power density spectrum or the time autocorrelation function. Of major interest here is the relative power contained in various spectra, and the correlation function is well suited for accomplishing this task (see eq. (5.4)). This section includes results of measurements obtained simultaneously with the observations reported in section 5.5 and the statistical data given in figures 5-17 through 5-20. E.R. Flynt developed the instrumentation for measuring the correlation functions and he also analyzed much of the data discussed here (Boring et al. 1957).

The autocorrelation functions presented in this section were obtained by processing radar data recorded on magnetic tape with a special-purpose analog computer. The recorded signals were derived by gating the radar receiver in range and stretching the resulting pulses to the length of the interpulse period.

From equation (5.2), the autocorrelation function was given as

$$R(\tau) = \lim_{T\to\infty} \frac{1}{2T} \int_{-T}^{T} X(t) \cdot X(t + \tau)\, dt \tag{5.2}$$

Here, $X(t)$ represents the recorded signal and τ is a time-displacement parameter introduced by the computer mechanism. Although equation (5.2) is designated as an integral over an infinite time interval, it converges rapidly enough so that a good approximation can be obtained by integrating a data sample of reasonable length. Only the autocorrelation function will be discussed here and for brevity it will be called the "correlation function."

Each correlation curve was computed from a data sample of 10 minutes in length, and consequently the integration time was 10 minutes for every point on the curve. The long length for data samples permitted sufficient sampling so that effects of the lower frequency fluctuations could be calculated. The signals were monitored in order to assure a reasonable constancy in average signal strength. As noted elsewhere (fig. 5-24), average sea echo strength will change abruptly as a result of rapid changes in wind speed. However, under stable sea and wind conditions the average signal strength may not vary appreciably over long time intervals—periods that are as long as one-half hour and probably for periods of several hours (see fig. 5-25).

Figure 5-26 is a reproduction of an actual autocorrelation function and it illustrates the major features of sea echo data. All parameters of figure 5-26 were observed to be dependent on wave and/or wind conditions, but the analyses revealed few clear-cut patterns of dependences. The broken-line curve is a short segment of the initial portion of the curve drawn on an expanded time scale to illustrate the shape of the initial drop. The amplitude of this component is represented by measurement A. Following the initial rapid drop, a second portion of the correlation curves continues to drop with a much more gradual slope. The solid curve in figure 5-26 includes measurements B and F. Measurement B of width E is analogous to the main lobe and amplitude F with half-period G corresponds to a sidelobe. The amplitude F in figure 5-26 is somewhat larger than the average for all data runs, and in many runs no periodic component was apparent.

The amplitude C of a very slow random component was probably caused by drifts in the equipment, but it may have been caused by very slow changes in the sea environment. The slight changes that cause C occur within the 10-minute observation intervals, but at a rate that is slow compared with the 6-second time displacement generally used in computing the correlation curves.

Measurements D and E are the time delays for which the respective curves drop to half their peak amplitudes. Measurement D is about 10 milliseconds and measurement E and the half-period G are roughly 1 second. Since the time scales (and thus the spectral widths) corresponding to the solid and dashed curves are vastly different, it is reasonable to assume that these curves are generated by separate, uncorrelated

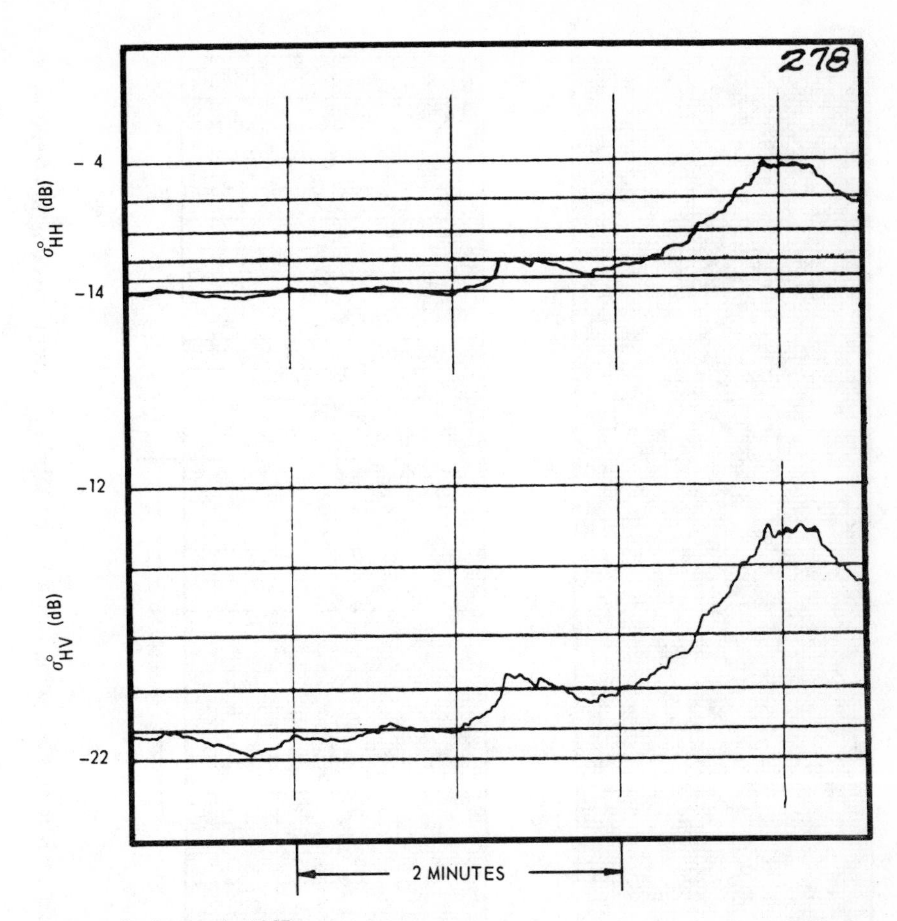

Source: Boring et al. (1957).

Figure 5-24. Signal Strength Referred to Arbitrary Reference Versus Time to Illustrate Rapid Increase in $\sigma°$ Produced by Abrupt Increase in Wind Speed. Filter time constant 20 seconds.

mechanisms. For mathematical functions that are uncorrelated, the correlation function of their sum is the sum of the separate correlation functions. Therefore, the dashed curve is interpreted as representing a scattering mechanism different from those that produce the solid curve.

Of major interest here is the relative power for the fast fluctuations corresponding to measurement A and the slower fluctuations corresponding to measurements B and F. Measurements A and B correspond to the $R(O)$ of equation (5.4) for the fast and slow fluctuations, respectively. The reference level for measuring the amplitudes of the individual curves is lost in the process of normalizing the correlation function, but only the ratio

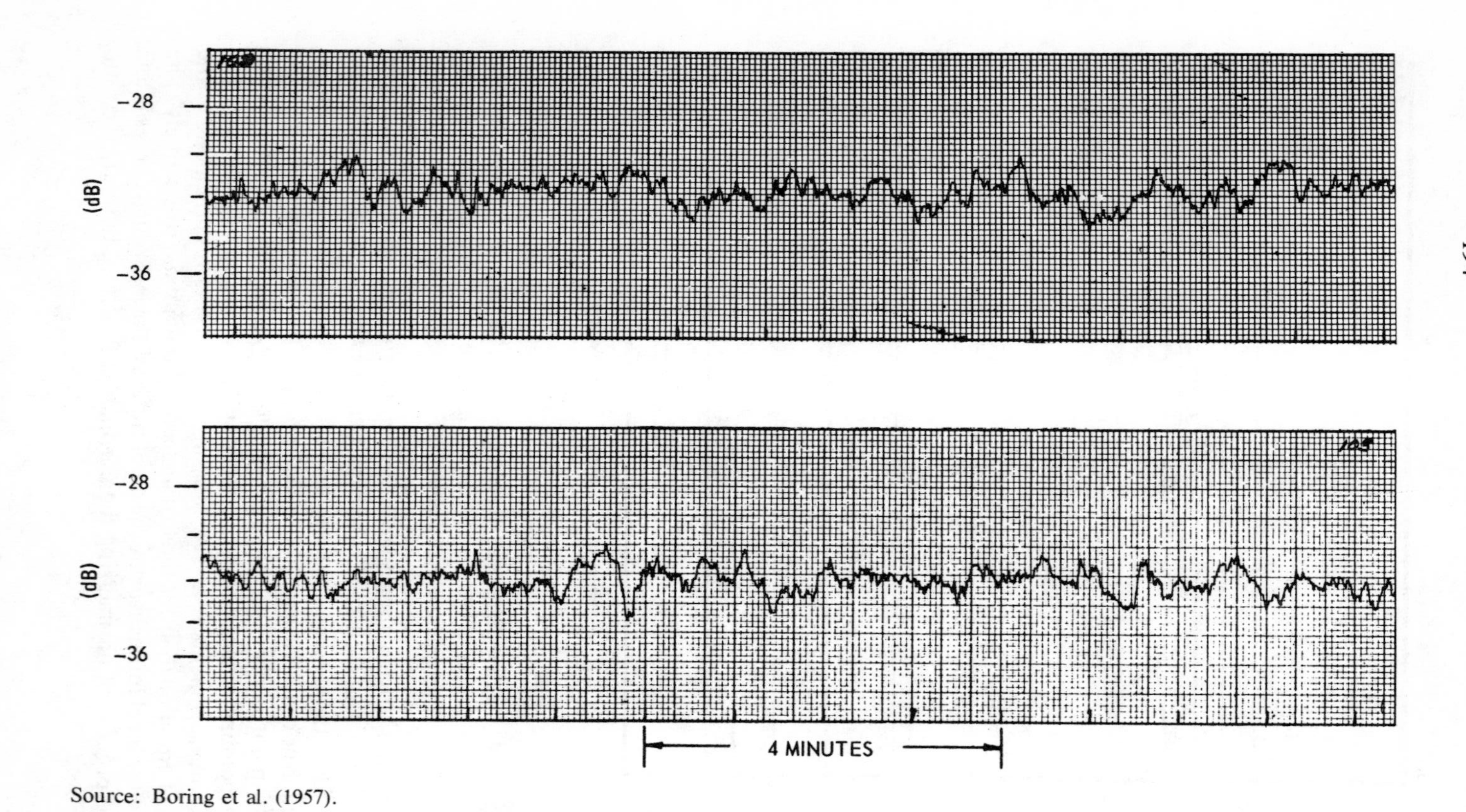

Source: Boring et al. (1957).

Figure 5-25. A Continuous Record of σ°_{VV} for 26 Minutes Showing Long Time Stability. Filter time constant 6 seconds.

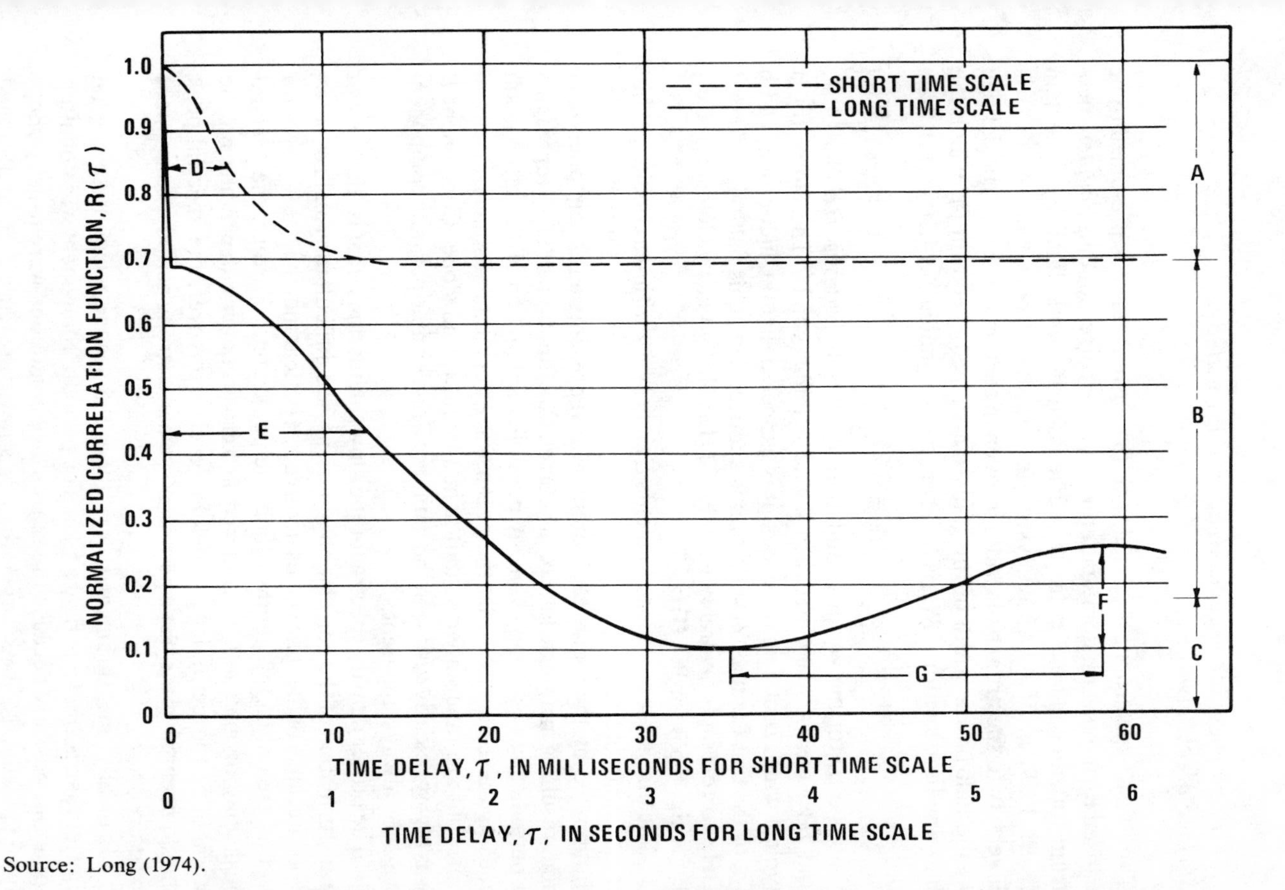

Source: Long (1974).

Figure 5-26. Illustration of the Characteristics Measured on Correlation Functions.

B/A is of primary interest to the present discussion. A strong dependence on transmit and receive polarizations was observed for this ratio.

5.11 Frequency Spectra and Relationships with Sea Surface Mechanisms

The "fast" spectrum associated with the dashed curve is presumed to be caused by moving random scatterers, and the corresponding *doppler* spectrum for noncoherent radar was discussed in some detail by Goldstein (Kerr 1959, p. 580). An analysis showing that decorrelation time D is inversely to spectrum bandwidth is given in section 5.3. Let $f_{1/2}$ be defined as the half-power width of the Gaussian-shaped spectrum for which the half-amplitude width of $R(\tau)$ is D. Then, from equation (5.14), it is seen that

$$f_{1/2} = 0.22/D \tag{5.15}$$

Values of D for the various combinations of polarizations (*HH, VV, VH,* and *HV*) were usually between 4 and 8 milliseconds. Therefore, for a typical width D of 6 milliseconds, it is expected that the half-power width $f_{1/2}$ of the fast fluctuation spectrum is about 35 Hz. This half-width for 6.3 GHz corresponds favorably with the 60 Hz value given by Goldstein (Kerr 1951, fig. 6-55) for 9.4 GHz.

The time scales for measurements E and G vary with sea conditions, but are comparable to the periodic passing of ocean waves through the area illuminated by the radar. Measurement B represents a random phenomenon, resulting in a single lobe of the correlation function, while F represents a periodically recurring event. A possible explanation of the mechanism causing these curves is that the ocean waves occur almost periodically in time, but are randomly distributed in amplitude and slope. Component B was always clearly visible in the data, but a discernible periodic component was not always apparent.

The power density spectra for the slow fluctuations, that is, the Fourier transformation of the correlation functions, are difficult to visualize. However, because of the time scales involved, it is clear that the "slow" spectra widths are only a few tenths of a cycle per second. Further, the spectral peak may or may not be centered at zero frequency depending on the strength of the periodicity (sidelobe level). However, the spectral peak should occur near the wave period, $2G$, when a strong periodic component exists.

Typically, measurement E was about 0.5 second which is somewhat shorter than is shown in figure 5-26. The periodic component was not often apparent, possibly because the sea surface conditions were usually choppy. Typically, $2G$ was larger than is shown in figure 5-26 and ranged between 3 and 8 seconds. For one unusual data run, the curve was computed to a time displacement of 40 seconds, resulting in five loops of the curve with very nearly constant amplitude and period. At least for this run, the ocean waves passed through the illuminated area in a truly periodic manner. Therefore, the spectrum for this run should be sharply peaked near $2G$.

Detailed comparisons of wave conditions with the correlation functions have not been undertaken, but it was clear that the slow fluctuations are closely associated with wave structure. For example, on one occasion an observer counted the waves passing a fixed point on the sea surface during a time interval, and the wave period was found to be equal to the period indicated on the correlation curve for that run.

Some measurements of E were made with a 35-GHz radar at the same site and for similar sea and depression angles to those used for 6.3 GHz. Although the measurements were rarely made simultaneously with the two radars, it seemed that E was essentially independent of radar frequency.

The only other known studies on the correlation functions and the frequency spectra for the slow fluctuations were undertaken by the Naval Research Laboratory at the same field site, and were reported by G.F. Myers (1958). He reported results of calculating correlation functions and frequency spectra for the slow fluctuations obtained at 9,375 MHz for *HH* and *VV* echo with a very high resolution pulse radar. The correlation functions and frequency spectra obtained for *VV* echo exhibited the same general appearance of correlation functions and frequency spectra of wave heights measured with a wave gage placed in the sea area under investigation; the results indicated that the energy for *HH* echo is spread out over a wider range of frequencies than it is for *VV* echo. The pulse length and the azimuthal antenna beam were 0.008 μs and 1.2°, respectively, and the depression angles used ranged between 1/4° and 6°.

The studies were chiefly concerned with incoming ocean waves passing a pile-mounted wave gage located 430 yards off shore. The radar cell size used by Myers was very small (5 by 22 feet), but his results with wave spectra corresponded, in general, to observations by V.R. Widerquist (see sec. 5.5): "It seemed from these results that the occurrence of successive spikes in the *HH* echo are associated with the sea waves, but the *HH* echo did not follow the sea wave pattern nearly as well as the *VV* echo." The report by Myers included a number of figures to illustrate changes in the shapes of the correlation functions and spectra depending on sea conditions. A trend of shape versus sea state was not discernible; figures 5-27 and 5-28 were selected in an effort to include representative data.

This discussion of sea wave effects has centered on the use of high resolution radar. More recently, G.F. Andrews and H.W. Hiser (1969) developed a sea state analyzer that measures the average period of the slow

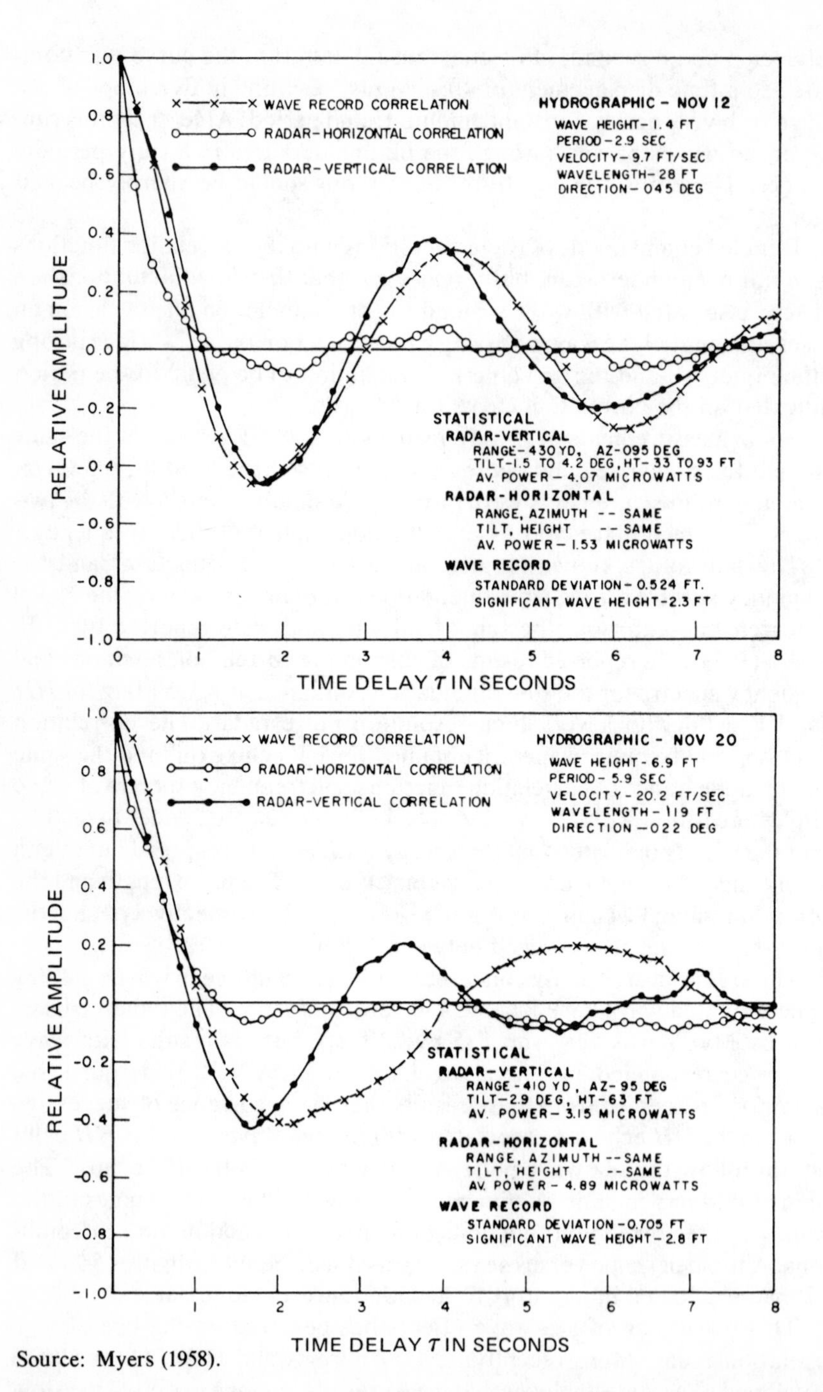

Source: Myers (1958).

Figure 5-27 Variability of the Correlation Coefficients for Different Weather Conditions.

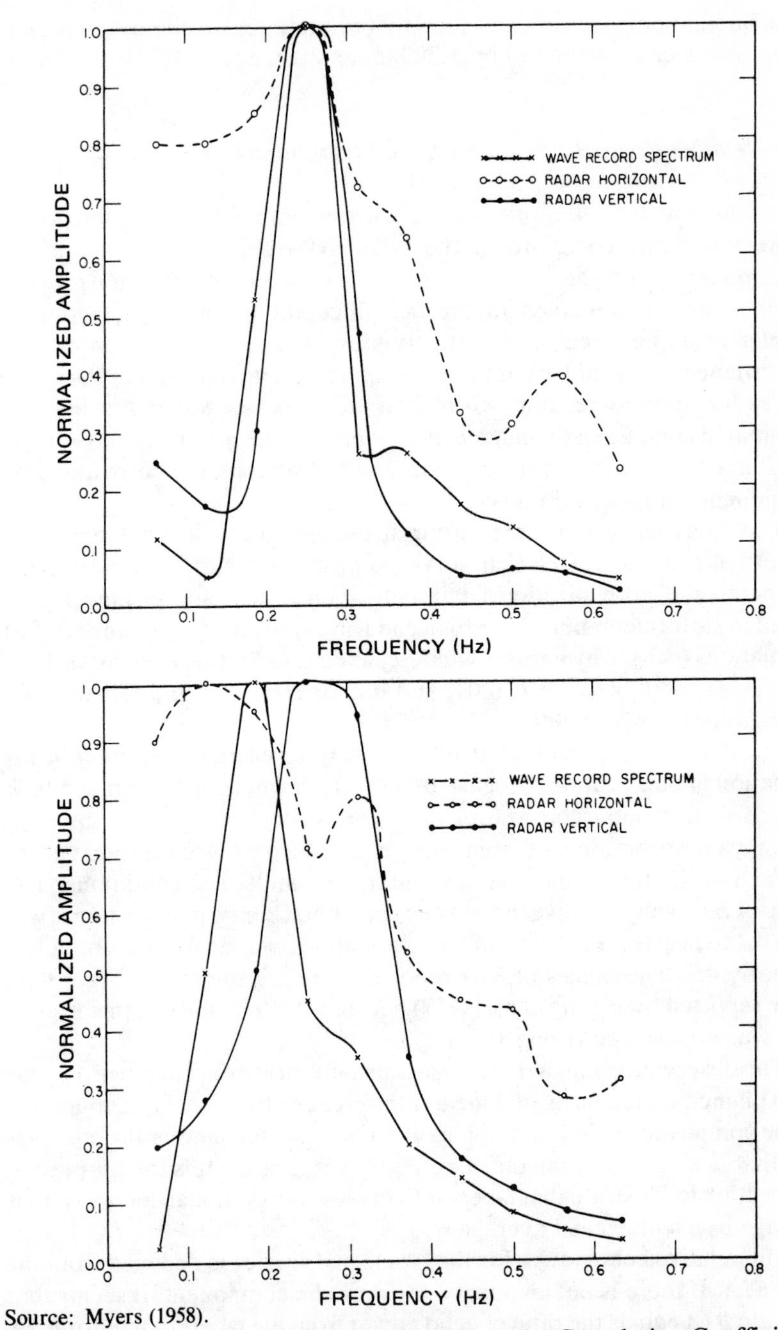

Source: Myers (1958).

Figure 5-28. Power Spectra Calculated from the Correlation Coefficients of Figure 5-27. Measured on November 12 (upper) and November 20 (lower).

radar fluctuation spectrum and thereby remotely senses the ocean period with a radar cell size of 150 by 1,300 meters (see sec. 5.9).

5.12 Relative Power in Fast and Slow Fluctuations

From equation (5.4), the correlation coefficient for $\tau = 0$ equals total power in the spectrum. Therefore, if the radar system were perfectly linear, measurement A on the autocorrelation curve would be directly proportional to power contained in the fast fluctuations (presumably due to doppler beats between independently-moving scattering elements), and measurement B would be directly proportional to power contained in slower fluctuations resulting from passage of ocean waves through the illuminated area. Even though receiver nonlinearities are always present, A and B may be considered as approximate measures of the corresponding components of received power.

The reference level for measuring absolute power is lost in the process of normalizing the correlation curves. For a given curve the ratio B/A, irrespective of normalization, is an indication of the ratio of power contained in slow fluctuations (due to ocean waves) to power contained in fast fluctuations (caused by various wind-dependent scattering elements). Typical values for B/A are 0.7, 0.07, and 0.2 for *HH, VV,* and *HV* (or *VH*) polarizations, respectively.

The B/A ratio is a significant indicator of differences between echoes as a function of polarization, because that ratio is a whole order of magnitude larger for horizontal polarization than for vertical. The values given as typical values were, in fact, mean values taken from 50 correlation curves for each of the three polarizations and under similar sea conditions. The range of B/A values, however, was large; for horizontal polarization it was from 0.1 to nearly 3 and for vertical polarization it varied from essentially 0 to nearly 0.3. The ranges of B/A reported here are somewhat larger than those reported by Boring et al. (1957) because further study of the original data was undertaken (Long 1974).

The observation that average σ_{VV} is usually larger than average σ_{HH} can be explained on the basis of a local interference effect (and quantitatively by the composite model), but ripples alone will not account for the observation that average σ_{HH} sometimes exceeds average σ_{VV}. It is the purpose of this section to illustrate that facets reflect enough additional energy so that average σ_{HH} will exceed average σ_{VV}.

The relative echo power for the facets and ripples is deduced from the ratio B/A. If there is not an observable periodic component, it seems that the ratio B/A equals the ratio of echo power from the facets to that from the ripples. The periodic component has been observed to be associated with the passing of sea waves through the area viewed by the radar—it may actually include contributions from both ripples and facets. Often the periodic component was too weak to be observed, possibly because the sea was usually choppy.

The discussion above can be illustrated by the following equations. Assume that sea echo is caused by two independent mechanisms for which the scattering cross sections are denoted by superscripts h and l. The h and l indicate high and low frequency spectra, respectively. Then

$$\sigma_{HH} = \sigma_{HH}^{h} + \sigma_{HH}^{l}$$

$$\sigma_{VV} = \sigma_{VV}^{h} + \sigma_{VV}^{l}$$

The relative magnitudes of the σ^h and σ^l can be obtained from the relative powers (B/A) obtained from correlation functions:

$$\frac{B_{HH}}{A_{HH}} = \frac{\sigma^{l}_{HH}}{\sigma^{h}_{HH}} \quad \text{and} \quad \frac{B_{VV}}{A_{VV}} = \frac{\sigma^{l}_{VV}}{\sigma^{h}_{VV}}$$

Therefore, it may be seen that

$$\frac{\sigma_{HH}}{\sigma_{VV}} = \frac{\left(1 + \dfrac{B_{HH}}{A_{HH}}\right)\sigma^{h}_{HH}}{\left(1 + \dfrac{B_{VV}}{A_{VV}}\right)\sigma^{h}_{VV}}$$

Under the assumptions above, the σ^h are caused by wind ripples that are in the presence of fields controlled by a local interference effect. Therefore $\sigma^{h}_{HH}/\sigma^{h}_{VV}$ cannot exceed unity. However, values of relative power measured for the low and high frequency spectra (B_{HH}/A_{HH}, B_{VV}/A_{VV}) will permit, consistent with observations, averages of σ_{HH}/σ_{VV} to exceed unity.

In the last several sections, observations are discussed which indicate that reflections from the facets (smooth areas making up sea waves) are the cause for average σ_{HH} to exceed average σ_{VV}. Some of these observations are listed below.

1. Under certain conditions, there is a marked difference between the appearance of *VV* and *HH* echo. Usually when differences exist, the *VV* echo has a noise-like appearance and the *HH* echo has a spiky, target-like appearance.
2. The strength of the noise-like echo increases (or decreases) almost instantaneously with increases (or decreases) in wind speed and it is therefore considered to be caused by wind ripples.
3. When the spiky appearance is apparent, it seems to be associated with the gross wave structure and not the wind ripples.

4. Separate and distinct fluctuation spectra can be identified from time autocorrelation functions. The ratio of power contained in the slow fluctuations to that in the fast ones is usually substantially greater for *HH* than for *VV* polarization.

5.13 Phase Coherent Doppler Spectra

In a phase-coherent doppler radar, the echoes are compared with the transmitted signal so that receiver output frequencies that correspond to the doppler shift are generated. In other words, the receiver output provides a means for distinguishing scatterers by their velocities relative to the radar antenna. The quantitative results obtained have provided important insights for understanding the interaction of radar waves with rough surfaces.

A major step toward the understanding of sea scattering resulted from observations by D.D. Crombie (1955), R.P. Ingalls and M.L. Stone (1957), and others, which indicated that doppler-shifted signals were caused by scatterers that satisfy the Bragg resonance condition

$$L = (1/2)\,\lambda \sec\theta$$

where L represents the separation of (contributing) scatterers and θ is the incidence angle (see chap. 3). The studies lead to an important concept through the efforts of a number of researchers including J.W. Wright (1966, 1968) and F.G. Bass et al. (1968). In the resulting theoretical formulation, the sea surface consists of large swells on which short gravity and capillary waves are superimposed. From the electromagnetic viewpoint, the smaller waves are treated by previously developed theories for slightly rough surfaces (see sec. 3.12). Effects of the macrostructure (sea and swell) are included by considering changes in scattering caused by the "tilting" (local interference of fig. 4-19) of the small structure (short gravity and capillary waves) by the dielectric background (sea and swell). Resulting theories have elucidated a number of previously unexplained microwave observations, but there are several questions regarding spectral width and polarization sensitivity that are yet to be answered (also see secs. 3.15 and 6.22).

For microwaves, the tilting plane model provides numbers for average σ° (radar cross section per unit area) that are close to those actually observed, but the model does not account for the fact that average σ°_{HH} sometimes exceeds average σ°_{VV}. For low radar frequencies (wave height much less than λ), the Bragg reflection theory predicts an upper limit for σ°_{VV} of -17 dB; this has been validated by Naval Research Laboratory measurements at 10 MHz (Barrick 1972).

Measured spectra are broadened by motion of the radar beam

(Ridenour 1947, p. 658). M.B. Laing (1971) gives a comprehensive analysis of the effects of aircraft motion on spectral width. A major source of broadening is illustrated here. The doppler shift caused by the relative motion of a radar and a target is given by

$$f_D = 2V/\lambda$$

where V is the line-of-sight component of the approach velocity of radar and target. If the angle of the target from the ground track of the radar is χ, the doppler frequency caused by radar motion is $2V_r \cos\chi/\lambda$, where V_r is the radar velocity. A band of frequencies is therefore generated by radar motion because a band of angles $\Delta\chi$ exists (to illuminate an extended area of sea).

Doppler frequencies can be measured with a "coherent" radar and for sea echo this measurement yields a band of frequencies, even if the radar is stationary and the radar beam is so narrow that angle χ is nearly zero. The band of frequencies, the "intrinsic" spectrum, exists for sea echo because the instantaneous velocities (with respect to line-of-sight) of the collection of scatterers that comprise the sea surface form a continuous distribution peaked near some average value.

B.L. Hicks et al. (1960) reported results of measuring the intrinsic spectrum of the sea with a coherent airborne radar operating at 3.2 cm. The measurements yielded the intrinsic spectrum directly, because by making measurements along the ground track the width of the induced spectrum could be made small relative to that of the measured spectrum. The spectra were obtained from recordings that were 15 seconds long which corresponded to 3,750 feet along the sea surface.

The bandwidth of the spectrum was found to be approximately proportional to wind speed. The investigations were made for depression angles between 1° and 10°. The data seemed to show that for some experimental conditions there exists a correlation of clutter bandwidth with depression angle, but not for all angles. When a correlation was observed, the bandwidth decreased with an increase in depression angle. For a depression angle of 4°, it was found that a correlation existed between the bandwidth B, significant wave height $h_{1/3}$ and a period T that corresponds to the maximum of the energy spectrum for the water waves themselves when this spectrum is plotted as a function of frequency. The correlation found in the data was expressed as

$$B = Dh_{1/3}/T$$

where B is in velocity units ($f_{1/2}\lambda/2$) and D is 11 ± 1 (dimensionless). The equation matched the experimental data within about 10 percent for bandwidths in the range of 2 to 5 knots (64 to 160 Hz) and wind speeds in the range of 8 to 19 knots.

For low sea states the spectrum had a Gaussian shape and the half-power width was found to be 2 to 3 knots (65-100 Hz). As the wind increased and the whitecaps became evident, the spectrum broadened and became asymmetric. Asymmetry existed only with the presence of whitecaps. The authors inferred that sea echo for the range of the sea conditions observed is caused by two sets of scatterers. One set produces a symmetrical ("core") spectrum nearly Gaussian in shape with a half-power bandwidth of about 3 knots. The authors conjectured that the core spectrum is produced by small (capillary) surface waves with wavelengths approximately half that of the radar. The other set of scatterers, presumed to be connected with the whitecaps, produces a spectrum that is displaced in the direction of the wind velocity and causes the composite spectrum to be asymmetrical.

It is interesting to note that Hicks et al. (1960) expressed the belief that the motions of the radar scatterers are like the orbital motions of water particles, and that the doppler spectrum width observed was proportional to the ratio $h_{1/3}$. More recently, Bass et al. (1968) have given a theoretical basis for the proportionality of doppler bandwidth to the ratio of wave height to wave period.

G.R. Curry (1965) reported measurements with ground-based coherent radars operating at 425 MHz and 1,320 MHz. The narrow doppler-frequency resolution available (1.0 Hz) permitted echo from sea and rain to be resolved into separate spectral components. The spectra due to rain clouds were centered at frequencies corresponding to the radial velocity of surface winds (12-17 knots) at both frequencies. It was reported that the main spectral component caused by the sea corresponded to about 0.15 times the wind velocity. A smaller component having a slightly higher velocity was attributed to windblown spray. This smaller "spray" doppler component presumably corresponds to the unresolved asymmetrical "hump" reported by Hicks et al. (1960). For both the sea and cloud echoes, the spectral widths corresponded to velocity ranges of about 2 knots.

V.W. Pidgeon (1968) hypothesized that the two peaks reported by Curry, with each circularly polarized radar, was due entirely to sea echo. By extrapolating dual polarization results at *C* band (5.7 GHz), he provided credence to his concept that the separation of peaks reported by Curry depends on a differential doppler shift of vertically and horizontally polarized sea echo.

Pidgeon reported monostatic and bistatic results for depression angles between 0.2° and 8°. He found the mean doppler shift for horizontal polarization to be two to four times as great as the doppler shift for vertical polarization for the same or similar wind and wave conditions. The doppler shift for horizontal polarization was reported to be dependent on wave height and wind speed jointly, but the shift for vertical polarization was apparently dependent on wave height only. For both polarizations, the mean doppler shift appeared to be cosine law dependent on the angle between the wind-wave direction and the radar-beam direction. The doppler bandwidth, however, seemed to be independent of this angle. For both horizontal and vertical polarizations, the doppler bandwidth was reported to be directly proportional to the mean doppler shift when looking into the wind and waves. The doppler bandwidth for horizontal polarization usually exceeded that for vertical polarization, and the ratio of the two was strongly dependent on sea conditions. Bandwidth ratios nearly as large as 2:1 were sometimes observed.

G.R. Valenzuela and M.B. Laing (1970) analyzed doppler shift and doppler bandwidth data obtained with the NRL four-frequency (0.43, 1.2, 4.5, and 8.9 GHz) phase-coherent pulse radar. Effects of spectrum broadening due to airplane motion were calculated and thereby removed from the measured data. The results were for *VV, HH, HV,* and *VH* polarizations for depression angles between 5° and 30°. Sea conditions included 0.3- to 0.6-meter waves with 0.5- to 1-m/sec winds and 8-meter waves with 23- to 24-m/sec winds.

From equation (5.1), doppler bandwidth $f_{1/2}$ (in frequency units) for a set of isotropic scatterers increases linearly with radar frequency F and $f_{1/2}\lambda/2$ (doppler bandwidth in velocity units) is independent of F. Although the long accepted concept that $f_{1/2}$ for the sea increases with F has not been challenged, Valenzuela and Laing (1970) reported the new result that $f_{1/2}\lambda/2$ *decreases* slowly with increases in F. This means that $f_{1/2}$ for the sea increases somewhat slower than the linear increase with F associated with isotropic scatterers. Also, from averages over prevailing wind conditions and depression angles, Valenzuela and Laing reported increases in doppler bandwidth with wave height in accordance with an $h^{1/2}$ dependence instead of the h dependence reported by Hicks et al. (1960).

Valenzuela and Laing (1970) observed polarization sensitivities that were in general agreement with those reported by Pidgeon; that is, the bandwidths and doppler shifts for *HH* polarization exceed those for *VV* polarization. The bandwidths seemed to be greatest when looking upwind, intermediate for crosswind, and least for looking downwind. For *VV* polarization, bandwidth observed was essentially independent of depression angle but for *HH* polarization the bandwidth decreased with increases in depression angle. The differences in doppler shifts for *HH* and *VV* polarizations were found to be wind dependent for wind speeds up to 15 m/sec but the differences tended to be independent of wind speed above 20 m/sec.

Laing (1971) has reported on later studies of doppler spectra using the same radar. The data were for sea conditions from relative calm to 24-meter-per-second winds and 8-meter significant wave heights. Three of the four frequencies—0.43, 1.2, and 4.5 GHz—were used. The bandwidth of the spectra was found to be insensitive to whether or not the wind direction was upwind or downwind, but it was found to be sensitive to wave height. However, the magnitude of differential doppler (the difference in mean frequency between the spectra for *HH* and *VV* polarizations) was found to be sensitive to wind direction, having an upwind/downwind ratio of about two. Laing reported that other observed dependencies on radar frequency and polarization were in agreement with those reported earlier (Valenzuela and Laing 1970).

There have been a number of papers on the use of Bragg scattering to determine magnitudes and directions of surface winds and waves. The interest is in monitoring sea scatter over large areas with radars operating in the 1 to 30 MHz frequency range that permits over-the-horizon propagation via the ionosphere (Barrick 1972; Hasselman 1971; Munk and Nierenberg 1969). A.E. Long and D.B. Trinza (1973) used the relative amplitudes of the first-order approach and recede Bragg-scatter contributions to map wind directions of a storm at sea. Winds of the storm at long range (500 to 1,000 nautical miles) over a large area (10^5 square miles) in the North Atlantic were mapped from the U.S. Naval Research Laboratory facilities at Chesapeake Bay, Maryland. The work has been continued by further measurements with the Chesapeake Bay radar (Ahearn et al. 1974) for environmental conditions that include a relatively small hurricane, a large storm, and a relatively calm ocean. Wave direction and surface roughness determinations were made for ranges of 600 to 2,200 nautical miles. From these results wind speed and direction were inferred.

Over-the-horizon radars ultimately will be used to map sea conditions (with surface winds inferred) over very large areas. The first-order Bragg scattering should provide reliable information on wave direction and the second-order Bragg scattering will be useful for estimating wave height and wind speed (Barrick 1974; Stewart 1974). Much progress has been made on the theoretical interpretation of Bragg scatter, and it seems that surface currents and current (depth) gradients can also be inferred from the Doppler sea-echo records that provide wave direction and height (Barrick et al. 1974).

References

Ahearn, J.L., S.R. Curley, J.M. Headrick, D.B. Trizna, "Tests of Remote Skywave Measurement of Ocean Surface Conditions," *Proceedings of the IEEE*, vol. 62, pp. 681-87, June 1974.

Andrews, G.F. and H.W. Hiser, "Radar Sea State Analyzer," Final Report, Contract N00019-68-C-0393, University of Miami, December 1969.

Barlow, E.J., "Doppler Radar," *Proceedings of the IRE*, vol. 37, pp. 340-55, April 1949.

Barrick, D.E., "First-Order Theory and Analysis of MF/HF/VHF Scatter from the Sea," *IEEE Transactions on Antennas and Propagation*, vol. AP-20, pp. 2-10, January 1972.

Barrick, D.E., "Determination of Significant Waveheight from the Second-Order HF Sea-Echo Doppler Spectrum," Annual Meeting of the International Union of Radio Science, Boulder, Colorado, October 14-17, 1974.

Barrick, D.E., J.M. Headrick, R.W. Bogle, and D.D. Crombie, "Sea Backscatter at HF: Interpretation and Utilization of the Echo," *Proceedings of the IEEE*, vol. 62, pp. 673-80, June 1974.

Bass, F.G., I.M. Fuks, A.I. Kalmykov, I.E. Ostrovsky, and A.D. Rosenberg, "Very High Frequency Radiowave Scattering by a Disturbed Sea Surface, Part II: Scattering from an Actual Sea Surface," *IEEE Transactions on Antennas and Propagation*, vol. AP-16, pp. 560-68, September 1968.

Bigelow, H.B. and W.T. Edmondson, *Wind Waves at Sea, Breakers and Surf*, U.S. Navy Hydrographic Office, H.O. Pub. No. 602, Washington, D.C., 1947.

Bishop, Geffry, "Amplitude Distribution Characteristics of X-Band Radar Sea Clutter and Small Surface Targets," Royal Radar Establishment Memo No. 2348, 1970.

Boring, J.G., E.R. Flynt, M.W. Long, and V.R. Widerquist, "Sea Return Study," Engineering Experiment Station, Georgia Institute of Technology, Final Report, Contract Nobsr-49063, August 1957.

Bowditch, Nathaniel, "American Practical Navigator," U.S. Navy Hydrographic Office, H.O. Publication No. 9, 1966.

Crombie, D.D., "Doppler Spectrum of Sea Echo at 13.56 Mc/s," *Nature*, vol. 175, p. 681, April 1955.

Curry, G.R., "Measurements of UHF and L-Band Radar Clutter in the Central Pacific Ocean," *IEEE Transactions on Military Electronics*, vol. MIL-9, pp. 39-44, January 1965.

Daley, J.C., W.T. Davis, J.R. Duncan, and M.B. Laing, "NRL Terrain Clutter Study, Phase II," Naval Research Laboratory Report 6749, October 21, 1968.

Daley, J.C., W.T. Davis, and N.R. Mills, "Radar Sea Return in High Sea States," Naval Research Laboratory Report 7142, September 25, 1970.

Fishbein, William, S.W. Graveline, and O.E. Rittenback, "Clutter Attenuation Analysis," U.S. Army Electronics Command, Technical Report ECOM-2808, March 1967.

George, S.F., "The Detection of Nonfluctuating Targets in Log-normal Clutter," Naval Research Laboratory Report 6796, 1968.

Guinard, N.W., J.T. Ransone, Jr., M.B. Laing, and L.E. Hearton, "NRL Terrain Clutter Study, Phase I," Naval Research Laboratory Report 6487, May 10, 1967.

Guinard, N.W. and J.C. Daley, "An Experimental Study of a Sea Clutter Model," *Proceedings of the IEEE*, vol. 58, pp. 543-50, April 1970.

Hasselman, K., "Determination of Ocean Wave Spectra from Doppler Radio Return from the Sea Surface," *Nature Physical Science*, vol. 229, pp. 16-17, January 4, 1971.

Hayes, R.D. and M.W. Long, "Study of Polarization Characteristics of Radar Targets," Engineering Experiment Station, Georgia Institute of Technology, Quarterly Report No. 6, Contract DA-36-039-sc-64713, January 1957.

Hayes, R.D., C.H. Currie, and M.W. Long, "An X-Band Polarization Measurements Program," *Record of the Third Annual Radar Symposium*, University of Michigan, February 1957.

Hayes, R.D., J.R. Walsh, D.F. Eagle, H.A. Ecker, M.W. Long, J.G.B. Rivers, and C.W. Stuckey, "Study of Polarization Characteristics of Radar Targets," Engineering Experiment Station, Georgia Institute of Technology, Final Report, Contract DA-36-039-sc-64713, October 1958.

Hayes, R.D. and J.R. Walsh, Jr., "Some Polarization Properties of Targets at X-Band," *Transactions of the 1959 Symposium on Radar Return*, University of New Mexico, May 11-12, 1959.

Hicks, B.L., N. Knable, J.J. Kovaly, G.S. Newell, J.P. Ruina, and C.W. Sherman, "The Spectrum of X-Band Radiation Back-Scattered from the Sea Surface," *Journal of Geophysical Research*, vol. 65, pp. 825-37, March 1960.

Hiser, H.W. and G.F. Andrews, "Relations Between Radar Sea Clutter and Existing Local Weather Conditions," Final Report, Contract N00019-67-C-0479, University of Miami, September 1968.

Ingalls, R.P. and M.L. Stone, "Characteristics of Sea Clutter at HF," *IRE Transactions on Antennas and Propagation*, vol. AP-5, pp. 164-65, January 1957.

Ivey, H.D., M.W. Long, and V.R. Widerquist, "Polarization Properties of Echoes from Vehicles and Trees," *Record of the Second Annual Radar Symposium*, University of Michigan, 1956.

Kerr, D.E. (Ed.), *Propagation of Short Radio Waves*, Massachusetts

Institute of Technology, Radiation Laboratory Series, vol. 13, p. 580, McGraw-Hill Book Company, Inc., New York, New York, 1951.

Kinsman, Blair, *Wind Waves*, Prentice-Hall, Englewood Cliffs, New Jersey, 1965.

Laing, M.B., "The Upwind/Downwind Dependence of the Doppler Spectra of Radar Sea Echo," Naval Research Laboratory Report 7229, January 19, 1971.

Lawson, J.L. and G.E. Uhlenbeck, *Threshold Signals*, Massachusetts Institute of Technology, Radiation Laboratory Series, vol. 24, McGraw-Hill Book Company, Inc., New York, New York, 1950.

Linell, T., "An Experimental Investigation of the Amplitude Distribution of Radar Terrain Return," 6th Conference of the Swedish National Committee on Scientific Radio, March 13, 1963. English translation dated October 1966.

Long, A.E. and D.B. Trizna, "Mapping of North Atlantic Winds by HF Radar Sea Backscatter Interpretation," *IEEE Transactions on Antennas and Propagation*, vol. AP-21, pp. 680-85, September 1973.

Long, M.W., "On a Two-Scatterer Theory of Sea Echo," *IEEE Transactions on Antennas and Propagation*, vol. AP, pp. 667-72, September 1974.

Macdonald, F.C., "Characteristics of Radar Sea Clutter, Part I: Persistent Target-Like Echoes in Sea Clutter," Naval Research Laboratory Report 4902, March 19, 1957.

Munk, W.H. and W.A. Nierenberg, "High Frequency Radar Sea Return and the Phillips Saturation Constant," *Nature*, vol. 224, pp. 1285, 1969.

Myers, G.F., "High-Resolution Radar, Part IV: Sea Clutter Analysis," Naval Research Laboratory Report 5191, October 21, 1958.

Nathanson, F.E., *Radar Design Principles*, McGraw-Hill Book Company, Inc., New York, New York, 1969.

Norton, K.A., L.E. Vogler, W.V. Mansfield, and P.J. Short, "The Probability Distribution of the Amplitude of a Constant Vector Plus a Rayleigh-Distributed Vector," *Proceedings of the IRE*, vol. 43, pp. 1354-61, October 1955.

Pidgeon, V.W., "Doppler Dependence of Radar Sea Return," *Journal of Geophysical Research*, vol. 72, pp. 1333-41, February 15, 1968.

Ridenour, L.N., *Radar System Engineering*, Massachusetts Institute of Technology Radiation Laboratory Series, vol. 1, McGraw-Hill Book Company, Inc., New York, New York, 1947.

Rivers, W.K., "Low-Angle Radar Sea Return at 3-mm Wavelength," Engineering Experiment Station, Georgia Institute of Technology, Final Report, Contract N62269-70-C-0489, November 15, 1970.

Schmidt, K.R., "Statistical Time-Varying and Distribution Properties of High-Resolution Radar Sea Echo," Naval Research Laboratory Report 7150, November 9, 1970.

Stewart, R.H., "Evaluation of HF Measurements of Oceanic Winds," Annual Meeting of the International Union of Radio Science, Boulder, Colorado, October 14-17, 1974.

Trunk, G.V., "Coherent Detection of Non-Fluctuating Targets in Contaminated-Normal Clutter," Naval Research Laboratory Report 6858, March 21, 1969.

Trunk, G.V., "Radar Properties of Non-Rayleigh Sea Clutter," *IEEE Transactions on Aerospace and Electronic Systems*, vol. AES-8, pp. 196-204, March 1972.

Trunk, G.V. and S.F. George, "Detection of Targets in Non-Gaussian Sea Clutter," *IEEE Transactions on Aerospace and Electronic Systems*, vol. AES-6, pp. 620-38, September 1970.

Valenzuela, G.R. and M.B. Laing, "Study of Doppler Spectra of Radar Sea Echo," *Journal of Geophysical Research*, vol. 75, pp. 551-63, January 20, 1970.

Valenzuela, G.R. and M.B. Laing, "On the Statistics of Sea Clutter," Naval Research Laboratory Report 7349, December 30, 1971.

Valenzuela, G.R. and M.B. Laing, "Point-Scatterer Formulation of Terrain Clutter Statistics," Naval Research Laboratory Report 7459, September 27, 1972.

Wong, J.L., I.S. Reed, and Z.A. Kaprielian, "A Model for the Radar Echo from a Random Collection of Rotating Dipole Scatterers," *IEEE Transactions on Aerospace and Electronic Systems*, vol. AES-3, pp. 171-78, March 1967.

Wright, J.W., "Backscattering from Capillary Waves with Application to Sea Clutter," *IEEE Transactions on Antennas and Propagation*, vol. AP-14, pp. 749-54, 1966.

Wright, J.W., "A New Model for Sea Clutter," *IEEE Transactions on Antennas and Propagation*, vol. AP-16, pp. 217-23, March 1968.

6 Average and Median Cross Sections

Introduction

6.1 General Characteristics of σ°

Average radar cross section is the parameter most often used to specify strength of land or sea echo. Another parameter used is median cross section. Since echo fluctuates with random amplitude, echo strength needs to be described by specifying the fraction of time that the echo is between various levels. Although an average or a median value does not uniquely define echo strength, it is very useful to know and to understand the many factors that control average and median radar cross section.

A knowledge of the magnitude of land and sea echo for different wavelengths and various environmental conditions is of importance to the radar system designer and to the theoretician who wishes to better understand scattering mechanisms. The determination of wavelength dependence is exceedingly difficult because the scattering cross section is dependent on many uncontrollable environmental factors, and a measurement requires the absolute calibration of at least two radar systems; then there are at least two sets of system errors involved.

During the early days of radar there were considerable data on wavelength dependence (Davies and Macfarlane 1946; Goldstein 1946; Kerr 1951, pp. 481-587) and the preponderance of data indicated that the smaller the wavelength the larger the value of echo strength for any given incidence angle, polarization, and surface condition. A comprehensive radar measurement for the sea has been difficult because of the variability, unpredictability, and inability to describe the sea surface conditions accurately. However, a comprehensive measurement of radar cross section for land has proved to be no less difficult because of the wide variety of terrain types and prevailing surface conditions (seasons, moisture content, ice and snow cover).

Radar echo or return is caused by the various scattering mechanisms within the resolution cell of a radar. Goldstein introduced the quantity σ°, radar cross section per unit area of sea surface, to provide a normalized parameter that could be used to describe radar cross section of the sea. Using the definition of σ°, the radar cross section σ equals $\sigma^\circ A$, where A is the area of the mean sea surface contained within the radar's cell of resolution. Details regarding the definition of σ° are given in section 2.3. In using the quantity σ° to describe echo intensity, it is tacitly assumed that the echo is caused by a large number of scattering mechanisms that are distributed uniformly throughout the physical area illuminated by the radar. Even though scatterers are not usually uniformly distributed throughout the illuminated area, the quantity σ° is useful because it provides a measure of echo strength that is normalized with respect to the surface area viewed by the radar.

From an electromagnetic point of view, the range of possible incidence angles can be divided into three reasonably distinct regions: near grazing incidence, plateau region, and near vertical incidence. Within each of these regions the dependence of σ° on incidence angle and the dependence of σ° on wavelength can be characterized to some extent. However, the boundaries of the three regions change with wavelength, surface condition, and polarization.

In the near grazing incidence region (see fig. 6-1), σ° increases rapidly with increases in grazing angle and with decreases in transmitted wavelength. For the plateau region, σ° changes slowly with increases in grazing angle and changes slowly with transmitted wavelength; it appears that σ° for transmitting and receiving horizontal polarization is more dependent on wavelength than σ° for transmitting and receiving vertical polarization. The magnitude of σ° increases for increases in surface roughness for small grazing angles and for the plateau region. For near vertical incidence, σ° tends to decrease for increases in surface roughness and the dependence of σ° on wavelength is weak.

6.2 Differences between Average and Median Values

Echo strength from terrain fluctuates over a very wide range, and it is difficult to accurately calibrate a receiver over such a range. This is one of the reasons why average RCS, $\overline{\sigma}$, is difficult to accurately determine. The median RCS, σ_m, that is, the RCS which is exceeded one-half the time, is easier to measure because receiver nonlinearities are less troublesome and fewer calibrations are required.

For an essentially homogeneous echoing area, $\overline{\sigma}$ and σ_m are expected to be equal within a few decibels. G.R. Valenzuela and M.B. Laing (1971, 1972) have studied many of the statistical details of land and sea echo. They concluded that the distributions lie somewhere between the Rayleigh and the lognormal distributions. It still seems, as has been considered the case for many years, that it is a reasonable first approximation for land and sea echo to be assumed Rayleigh in character. Depending on various radar parameters and the terrain, large amplitude peaks occur more often than is

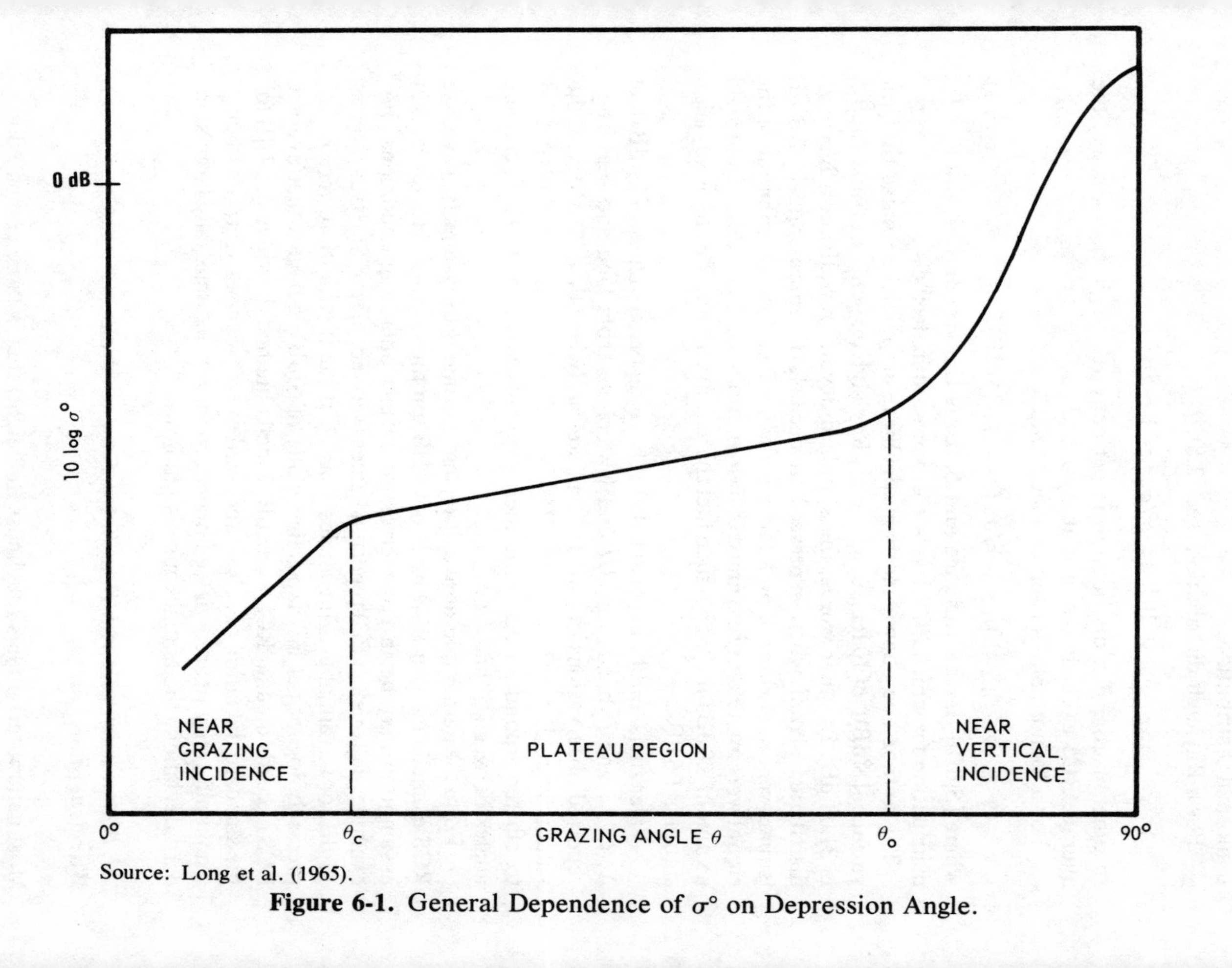

Source: Long et al. (1965).

Figure 6-1. General Dependence of σ° on Depression Angle.

predicted by that distribution but usually less often than is predicted by the lognormal distribution.

For a Rayleigh distribution (sec. 2.8),

$$10 \log_{10} (\overline{\sigma}/\sigma_m) = 1.6 \text{ dB} .$$

In other words, if a distribution is perfectly Rayleigh in character, the average RCS exceeds the median by 1.6 dB.

For a lognormal distribution (sec. 2.8),

$$10 \log_{10} (\overline{\sigma}/\sigma_m) = 0.115[S(\text{dB})]^2 \tag{6.1}$$

where $[S(\text{dB})]^2$ is the variance and $S(\text{dB})$ is the standard deviation of a distribution for which *RCS* levels are expressed in decibels.

The author has shown mathematically in unpublished notes that the variance $[S(\text{dB})]^2$ of $10 \log_{10} \sigma$, where σ is Rayleigh (exponential) distributed, is 31.03 dB^2. In other words, the standard deviation $S(\text{dB})$ of a Rayleigh distributed signal when expressed in decibels is approximately 5.6 dB (square-root of variance). As a further illustration, if the distribution of the signal were mistakenly assumed to be lognormal, equation (6.1) would yield a value of 3.6 dB for $\overline{\sigma}/\sigma_m$ when in fact the actual value of $\overline{\sigma}/\sigma_m$ for a Rayleigh signal is 1.6 dB.

Valenzuela and Laing (1971, 1972) have reported values of $[S(\text{dB})]^2$ at *X, C, L,* and *P* bands for *HH* and *VV* echoes from land and sea. They reported observations of $[S(\text{dB})]^2$ that are rarely less than 20 dB^2, and the values are seldom as large as 60 dB^2. Generally, the values of $[S(\text{dB})]^2$ tend to cluster around 30 dB^2. The measurements were made for incidence angles between 5° and 60°.

From the paragraphs above, it seems reasonable to assume that average RCS exceeds median RCS by about 3 dB, on the average. However, that assumption may be in gross error for certain operating conditions. For example, T. Linell (1963) reported values (see sec. 6.8) of $S(\text{dB})$ as large as 17 dB for terrain at grazing angles near 1°. If the distribution were in fact precisely lognormal in shape, that value of $S(\text{dB})$ indicates that average RCS exceeds median RCS by 33 dB! Linell obtained the value of 17 dB for the standard deviation of echoes received over a scanned sector. Although a nonuniformity in area viewed may account for the unusually large standard deviation, the actual cause is unknown.

6.3 Planar Surfaces with Roughness Comparable to a Wavelength or Less

Most natural terrain has roughness (in height) that is large compared to a wavelength. Examples include mountainous terrain and a typical ocean surface. The peaks and valleys alone comprise a roughness that is large compared to wavelength, but there is a small-scale roughness that is caused by leaves, twigs, ripples, etc. In this section we are concerned with a division of roughness between the very smooth (with respect to radar wavelength) and that comparable to a wavelength. The surfaces that are considered smooth include asphalt and concrete roads and capillary waves on an otherwise smooth ocean; examples of rougher surfaces include terrain with vegetation such as weeds or a rough road.

For a surface very smooth compared to a wavelength, such as an asphalt or concrete road or water with small capillary waves, radar cross section per unit area increases rapidly (as fast as $\sin^4 \theta$) with increased incidence angle—this corresponds to the near grazing incidence region of figure 6.1. For smooth surfaces, the terminology "near grazing incidence" is in fact a misnomer because it extends up to large angles of incidence. For smooth ground covered by grass of a few inches height, radar cross section per unit area varies more slowly with incidence angle (often approximated by a $\sin \theta$ dependence); this corresponds to the plateau region of figure 6-1.

When terrain has a heavy vegetation cover such as grass or weeds, there is little difference between the radar cross sections for horizontal or vertical polarization. For smooth terrain, that is, one with roughness that is small compared to a wavelength, the radar cross section may be much smaller for horizontal polarization than for vertical polarization. For smooth water surfaces, the dependence on incidence angle at small angles varies as $\sin^4 \theta$. For solid surfaces there is observed a $\sin^4 \theta$ dependence for horizontal polarization, and sometimes a variation this rapid is observed for vertical polarization. These observed dependencies are predictable on the basis of theory for planar surfaces with a slight roughness. The variation with incidence angle is affected by the dielectric properties of the surface material, and this is readily apparent from theoretical investigations (sec. 3.11).

The radar cross section depends on transmitter wavelength because cross section depends on surface roughness expressed in terms of wavelength and because complex dielectric constants are also dependent on wavelength. If surface roughness is eliminated from the variables, much of the wavelength dependence is eliminated. However, complex dielectric constant is dependent both on the type of terrain (including water content) and on the wavelength at which it is measured. In general, when radar frequency is increased, radar cross section per unit area is increased.

6.4 Very Small Incidence Angles

For sufficiently small grazing angles, even a very rough surface may appear "smooth" by exhibiting the characteristics depicted by the near grazing incidence region of figure 6-1.

Lord Rayleigh suggested a way of distinguishing between the extremes of a *smooth* and a *rough* surface. Rayleigh's criterion for a surface to be considered smooth is usually given as

$$\Delta h \sin\theta < \lambda/8$$

where Δh is the rms height of surface irregularities, and θ is the angle between incident rays and a plane surface representing the average of the irregularities. For applications involving search radar at sea, $\sin\theta$ might be as small as 10^{-2} or 10^{-3}. Therefore, according to the Rayleigh criterion, the surface might appear smooth for microwaves even through peak-to-trough wave height is very large.

Effects of depression angle dependence for very small angles are particularly apparent when echo strength is investigated as a function of range. The extreme sensitivity of $\sigma°$ to depression angle (as fast as $\sin^4\theta$) of sea echo was underscored in a paper by Martin Katzin (1957), but the same sensitivity is sometimes observed for extended ground areas. For example, radar cross section per unit area for an airport runway or a paved road may be substantially larger than for surrounding vegetation at short range, but much less at greater ranges. For a given range interval, the relative magnitudes will depend on a number of factors, including radar wavelength.

Experience has shown that average echo power, exclusive of propagation effects, from ships and boats exhibit an R^{-8} range dependence at distant ranges and an R^{-4} dependence at short ranges. The R^{-8} range dependence has been explained on the basis of an interference effect between the direct and reflected rays (see sec. 4.12). According to theory and experience, for the R^{-8} effect in range (corresponds to σ proportional to R^{-4}) to occur requires that the radar frequency, angle of incidence, and sea surface be such that the surface is smooth from an electromagnetic point of view.

Suppose that there are a sufficient number of targets present so that they are essentially uniformly distributed throughout the illuminated area A. In other words, assume that the number of targets, spacing, and sizes are such that on the average $\sigma°$ is constant at fixed range (independent of precise pointing of the radar beam) and proportional to R^{-4}. Under these conditions, since illuminated area is proportional to range, effective σ ($\sigma°$ times A) will be proportional to range. Since echo power is proportional to σ, it is then apparent that $\sigma \propto \sigma° \cdot R^{+1} \propto R^{-4} \cdot R^{+1} = R^{-3}$. Similarly, at ranges for which target power varies as R^{-8}, we would expect that σ for the extended group of targets would be proportional to R^{-7}.

With the various assumptions, it seems obvious that for sea targets at long range average echo power can vary as R^{-3}. Therefore, if a rough surface appears smooth (from an electromagnetic standpoint) to the reflected wave (forward scattering in specular direction), $\sigma°$ for an extended

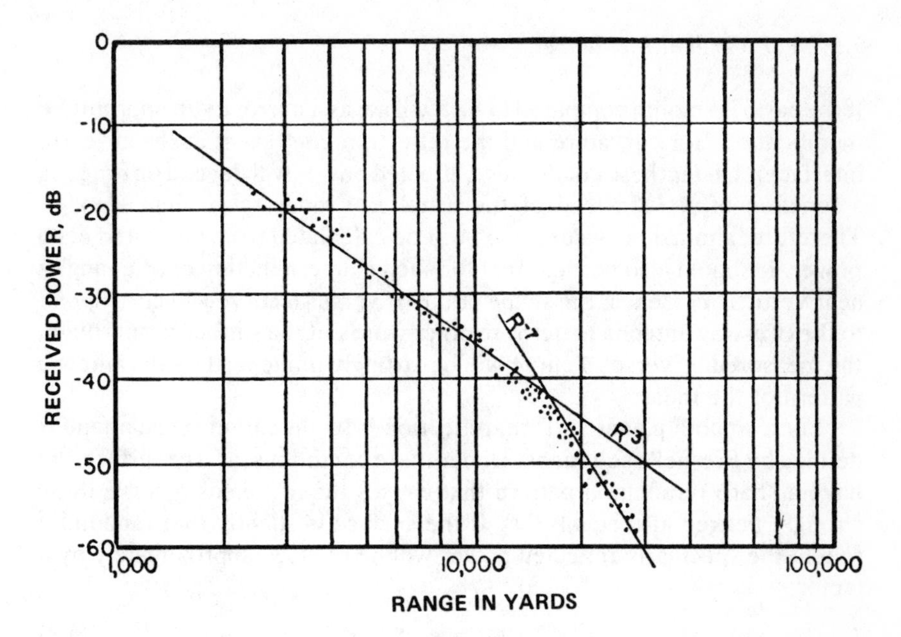

Source: Katzin (1957).

Figure 6-2. Received Sea-Clutter Power Versus Range. λ = 3.2 cm, altitude 1,000 feet. The two lines drawn through the measured points represent R^{-3} and R^{-7} variations, respectively. Figure 6-2 illustrates two regions: $\sigma°$ independent of range (R^{-3} region) and $\sigma°$ proportional to $\sin^{+4}\theta$ (R^{-7} region).

group of targets can vary as R^{-7} at long ranges and as R^{-3} at short ranges —as has frequently been observed for sea clutter (see fig. 6.2). Katzin reasoned that sea clutter is the result of individual, prominent waves that serve the role of targets as in the paragraphs above. These thoughts are given in more precise terms in section 6.6.

The explanation above requires that the sea serve in two roles within the R^{-7} region: to appear as a flat smooth surface from an electromagnetic standpoint and thus cause a strong forwardly reflected wave, and to provide individual targets that are the source of backscattering. Although it is a bit mind-boggling to visualize the interference theory because of effects of shadowing at small incidence angles, there is no question that waves do simultaneously play the two roles; for example, a boat exhibits an R^{-8} range dependence under appropriate conditions and simultaneously sea echo has been observed from all ranges (and directions) between the radar and the boat.

6.5 Near Vertical Incidence

If the earth is smooth compared to λ, it will act as a mirror (with appropriate modifications for curvature and the reflection coefficient at the air-earth interface). Under these conditions, reflected power will depend on range as a function of R^{-2} (instead of the usual R^{-4} radar range dependence). Therefore, anomalous values of σ° will be calculated from measured echo power. It should also be clear that for such a determination of σ° at angles near vertical incidence, the shape of the σ° versus θ curve will correspond to the two-way antenna pattern. In other words, if the surface is not rough, the measured σ° versus θ curve will be strongly influenced by the antenna pattern of the radar.

For a smooth planar surface, reflection is by definition specular and is defined by Fresnel's equations. Under these conditions, a large surface will have a sharp reradiation pattern that causes the σ° versus θ curve to be strongly peaked at θ equals 90°. If the surface is slightly (but randomly) rough, the specularly reflected power will be reduced approximately by a factor

$$\langle \mathcal{R}_s^2 \rangle = e^{-(\Delta\Phi)^2} \tag{6.2}$$

where

$$\Delta\Phi = 4\pi \Delta h \sin\theta/\lambda$$

and Δh is the standard deviation of the surface height variations (see sec. 4.9). The surface roughness also introduces a non-specular reradiation component, but this component is considerably smaller than the specular component if $\Delta\Phi$ is less than unity.

Since the factor $\langle \mathcal{R}_s^2 \rangle$ is down to 0.2 if $\Delta h \sin\theta$ is $\lambda/10$, significant specular reflection does not occur for a rough surface. By this example it is clear, however, that for smooth surfaces σ° can in fact increase with increases in wavelength. For most terrain of interest, the surface will be rough at near vertical incidence angles in the sense that

$$4\pi(\Delta h/\lambda)\sin\theta > 1 \quad \text{or} \quad \Delta h > \lambda/4\pi \tag{6.3}$$

Jacques Renau and J.A. Collinson (1965) have made detailed measurements at optical wavelengths for surfaces rough in the sense of equation (6.3). They expressed the belief that for these conditions σ° is independent of wavelength, increases with rms slope of the surface, and is relatively independent of rms height. From radar experiments, σ° usually is reported to vary inversely with wavelength.

At radar wavelengths, the dielectric properties of the surfaces are also functions of λ; therefore, the reradiation is wavelength dependent because of material as well as shape and roughness. If the physical dimensions of two targets are the same, $\sigma°$ is largest for the target with the largest dielectric constant. The actual scattering problem is even more complicated for terrain (as compared with the sea) because electromagnetic waves can penetrate the terrain and the internal scattering (also wavelength dependent) can contribute to the wavelength dependence of $\sigma°$.

6.6 Classical Interference Effect

The rapid change in signal strength versus range (discussed in sec. 6.4) is attributed to a forward scattering that causes the field at a point h above the mean surface to be reduced. Let F represent the relative electric field strength compared with its free space value. Then from section 4.5, for a small grazing angle θ and microwaves, $|F|^4$ at a point near a smooth surface can be expressed as

$$|F|^4 = |4\pi (h/\lambda) \sin\theta|^4 \tag{6.4}$$

From section 4.9, a surface with roughness of rms height Δh will cause appreciable forward scatter and thereby appear electromagentically smooth if

$$\Delta\phi = |4\pi (\Delta h/\lambda) \sin\theta|$$

is sufficiently small. Therefore, if $\Delta\phi$ is small, equation (6.4) may be valid even though the surface is physically rough.

Echo power is proportional to $|F+^4|$ and therefore forward scattering from terrain can strongly influence the effective cross section of targets. For example, the origin of the $\sin^4\theta$ dependence sometimes observed for target *RCS* at small grazing angles can be accounted for by equation (6.4).

Equation (6.4) can also be used to help explain effects of interference (caused by the terrain) on echo from the terrain. The terrain then serves two roles: as a forward reflector and as a source of backscatter. Now, for specificity, consider sea echo and assume that the backscatter is caused by a relatively few waves which exceed some average height and that these waves collectively have an effective height h_e above an effective plane surface. Then, for sufficiently small values of $\Delta\phi$, the cross section of the waves would be reduced by a factor

$$|F|^4 = |4\pi (h_e/\lambda) \sin\theta|^4 \tag{6.5}$$

From the definition of F, if $|F|^4 = 1$ the sea is electromagnetically rough. Let θ_c be the angle at which equation (6.5) equals unity. Then θ_c, called the critical angle, can be determined from

$$\sin\theta_c = \lambda / 4\pi h_e \tag{6.6}$$

The critical range R_c (e.g., 17,000 yards in figure 6-2) is the radar range for which θ equals θ_c. Thus, when the flat earth approximation is valid and H represents antenna height ($H >> h_e$), $\sin\theta_c$ equals H/R_c.

Values of θ_c obtained from R_c measurements of the sea (sec. 6.14), when used with equation (6.6), yield effective heights h_e that are approximately equal to the rms wave height Δh. Then from equation (6.2), at θ equals θ_c, one finds $\Delta\phi$ equals unity and the surface roughness reduces the power reflected in the specular direction by the factor 0.37. R_c (and therefore θ_c) is at a range (see fig. 6-2) where surface behavior as regards forward scatter is intermediate between smooth and rough. It is therefore satisfying that equation (6.2), which was obtained from theory, yields the factor 0.37 for measured values of R_c.

Radar Cross Section for Land

6.7 Nature of $\sigma°$ for Land

The cross section for land is difficult to document clearly because of the wide variety of terrain types and roughness and the difficulty in describing surface area illuminated by the radar. Even for a given terrain and specified radar system parameters, amplitude distributions (both shapes and amplitudes) are functions of incidence angle, and therefore normally change with range. Even though target areas are often not homogeneous, $\sigma°$ (radar cross section per unit of illuminated surface area) is a useful parameter for describing extended target areas because radar cross section is normalized with respect to illuminated area. For a radar in rough terrain, for example, mountains or woods, incidence angle may be an undefinable parameter. However, for a radar located above the terrain peaks there is a general dependence of $\sigma°$ on depression angle corresponding to that depicted in figure 6-1. In this chapter, for brevity, $\sigma°$ will represent the average value unless it is otherwise specified.

For rough terrain, it is expected that echo strength changes with pulse length (e.g., consider effects of shadowing caused by isolated trees). Therefore, $\sigma°$ will be *dependent* on pulse length. Since echo strength is also dependent on depression angle (range dependent), $\sigma°$ will also vary with range. Therefore, for rough terrain changes in illuminated surface area caused either by changes in pulse length or range will change $\sigma°$.

Radar measurement programs have yielded a wide range of values for $\sigma°$ because of inhomogenities in terrain. Even for terrain that is reasonably homogeneous, wide differences in $\sigma°$ can be caused by the effects of the pattern propagation factor (with associated variability due to ground moisture, surface roughness, and polarization). Because of the wide variability in $\sigma°$, it has not been possible to identify a consistent dependence of $\sigma°$ on transmitter frequency. It is obvious that $\sigma°$ depends on ground roughness, moisture content, dielectric properties of the ground and trees, and the gross size or extent as well as the detailed dimensions of the vegetation — the effects of the properties are all related to wavelength. However, it is clear that, on the average, $\sigma°$ increases slowly with increases in radar frequency.

6.8 *Average $\sigma°$ for Various Terrains*

Some years ago the author was involved in a short-term measurements project on the polarization and statistical characteristics of motor vehicles and trees at 6.3 and 35 GHz (Ivey, Long, and Widerquist 1955). Specialized measurement radars that were not mobile and could not be operated simultaneously were used. Because of logistics, a number of weeks elapsed between measurements at the two frequencies. The ground could be classed as flat to rolling.

Because of the radar height, trees were viewed from broadside. Signal strength obviously was a function of azimuth, elevation, and range. In order to determine $\sigma°$, the illuminated area A was defined as the area (normal to the beam) that the trees (within the range gate) subtended. Average $\sigma°$ was determined by reading a paper-tape recording on which a signal generator calibration was included. Some averaging was obtained with an r-c filter, and the remainder was provided by the data analyst.

Average $\sigma°$ was determined for a number of different pointing directions and weather conditions. Deciduous trees were observed for a number of different polarizations. Amplitude distributions, when graphed on semilog paper, approximated straight lines and therefore were considered to be well represented by Rayleigh statistics.

For that investigation (Ivey, Long, and Widerquist 1955) the averages for $\sigma°_{HH}$ and $\sigma°_{VV}$ of trees were about -10 dB at 35 GHz and -20 to -12 dB at 6.3 GHz. These values were obtained with the radar by viewing trees at broadside (like standing at the edge of a forest). The area of the illuminated surface was defined by the one-way, half-power points of the antenna.

The project at 6.3 and 35 GHz was followed by one for developing and using a 9.3-GHz mobile radar (Hayes, Currie, and Long 1957). The objective was to obtain measurements for as many different types of terrain and environmental conditions as possible. Statistical and polarization measurements were made at various sites under a wide range of weather conditions. Average values of $\sigma°$ for horizontally, vertically, and circularly polarized transmissions and receptions were obtained. The data were investigated for pine trees and deciduous trees as a function of surface moisture and seasons of the year. The radar was mobile and the measurements were made for various terrain types (from essentially flat to mountainous).

Average radar cross section per unit area for trees was found to be about −10 dB, but it varied widely. Figures 6-3 and 6-4 give some of the results reported (Hayes et al. 1958). The symbols $S-1$, $T-2$, etc. represent site 1, target area 2, etc. Each horizontal tick represents one average value measurement.

Usually $\overline{\sigma}_{HH}$ exceeded $\overline{\sigma}_{VV}$, but this was not always so. On the average $\overline{\sigma}_{HH}$ probably exceeded $\overline{\sigma}_{VV}$ by 1 or 2 dB. The wide ranges in measured values of $\overline{\sigma}_{HH}$ relative to $\overline{\sigma}_{VV}$ strongly suggest effects of multipath reflection from the ground, and the fact that the average of the $\overline{\sigma}_{HH}$ measurements tended to be somewhat greater than the $\overline{\sigma}_{VV}$ measurements helps to support this opinion.

Irrespective of polarization, there seemed to be, in general, little difference in cross section for wet trees and dry trees, and no significant trend in the cross section was discerned for different seasons. For all transmissions and receptions, cross sections per unit area of pine trees tended to be slightly lower than those for deciduous trees, but the difference was probably no greater than 3 dB.

For either deciduous or pine trees the ratios of $\overline{\sigma}_{HH}/\overline{\sigma}_{HV}$ and $\overline{\sigma}_{VV}/\overline{\sigma}_{VH}$ were 3 to 10 dB. Since $\overline{\sigma}_{HH}$ was on the average 1 to 2 dB greater than $\overline{\sigma}_{VV}$, $\overline{\sigma}_{HH}/\overline{\sigma}_{HV}$ was on the average 1 to 2 dB greater than $\overline{\sigma}_{VV}/\overline{\sigma}_{VH}$. Data for circular polarization were not, of course, taken simultaneously with those for linear polarization. The results indicated $\overline{\sigma}_{12}/\overline{\sigma}_{11}$ was 2 to 4 dB on the average.

C. R. Grant and B.S. Yaplee (1957) reported results of determining $\sigma°$ for vertical polarization for several terrain types at wavelengths of 3.2 cm, 1.25 cm, and 8.6 mm. Through novel use of a continuous wave doppler system, average $\sigma°$ was measured over a wide range of incidence angles from bridges. To obtain different terrain types, sites were selected in Texas, Louisiana, and Georgia. With the simple systems used, target area was selected through antenna pattern directivity. The values of $\sigma°$ calculated are obviously an average over the antenna beamwidth. According to the authors, the error is negligible at incidence angles (with respect to vertical) greater than 5°, and the value of $\sigma°$ may be as much as 3 dB too low at normal incidence.

Figure 6-5 shows some of the results reported by Grant and Yaplee for $\sigma°$ versus θ. The figure shows results for ground covered with tall weeds and grass in the spring when the grass was green and the ground wet and marshy, and in the fall when the grass and ground were dry. Notice that (1) there are irregularities in the $\sigma°$ versus θ curves, presumably caused by inhomogenieties in the terrain; (2) there is a large and rapid rise in $\sigma°$ near vertical incidence (due, presumably, to specular reflection), amounting to

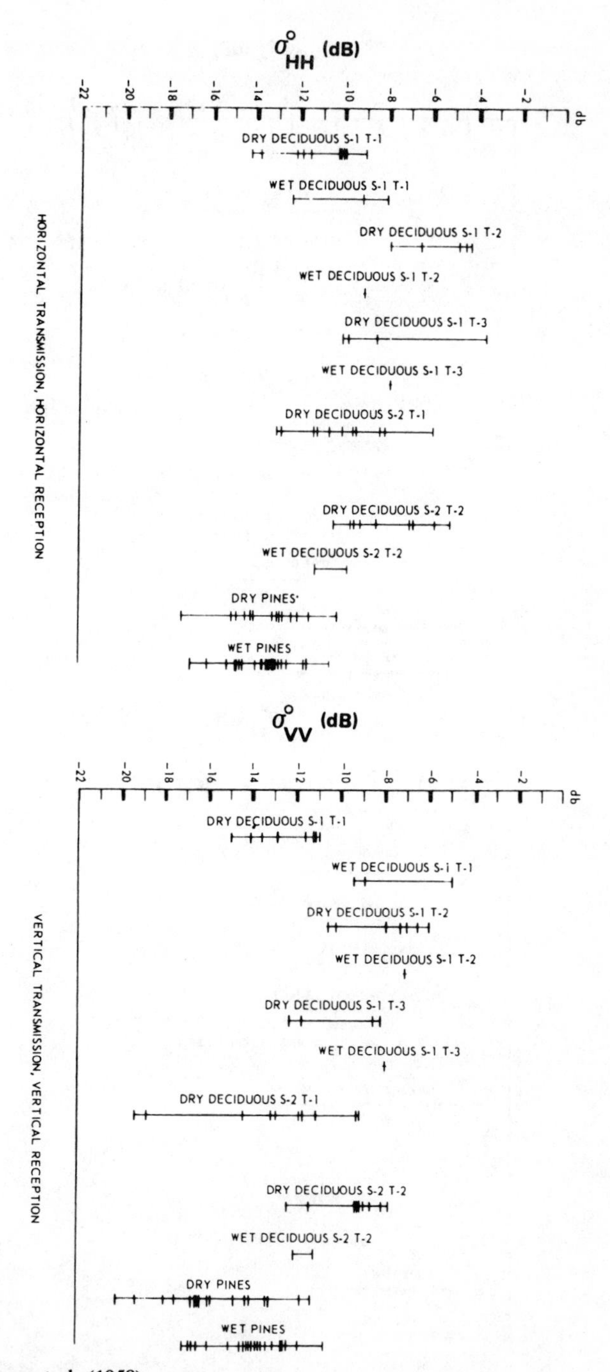

Source: Hayes et al. (1958).

Figure 6-3. Average σ°_{HH} and σ°_{VV} for Trees at *X* Band.

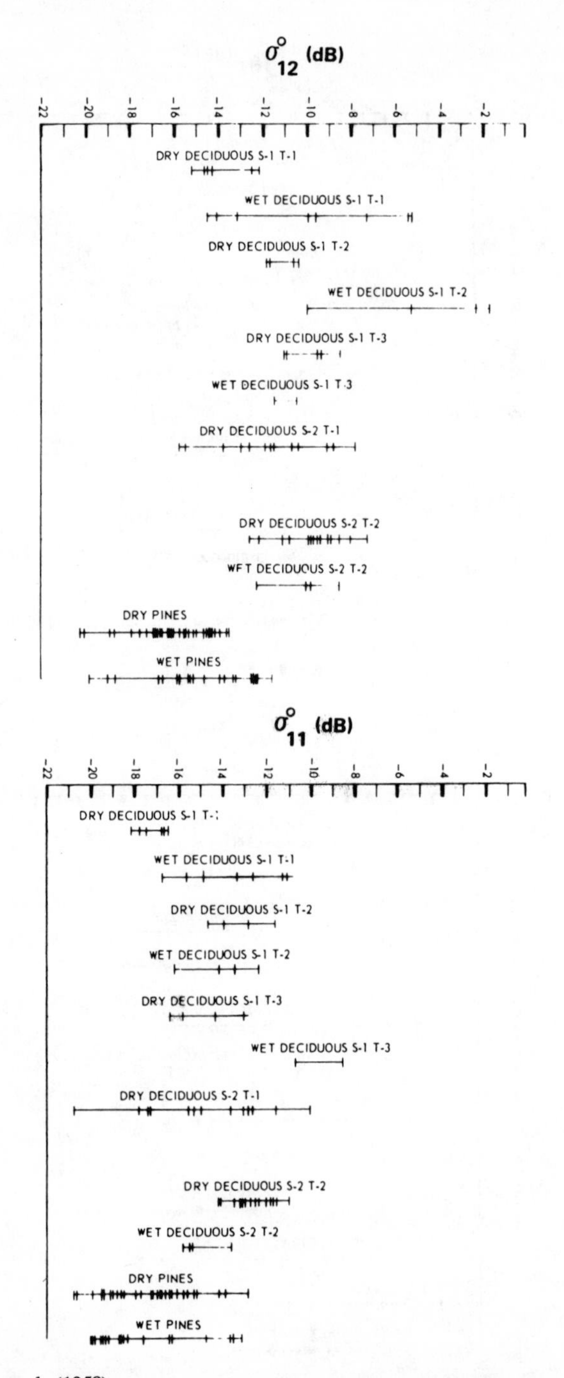

Source: Hayes et el. (1958).

Figure 6-4. Average σ°_{12} and σ°_{11} at X band.

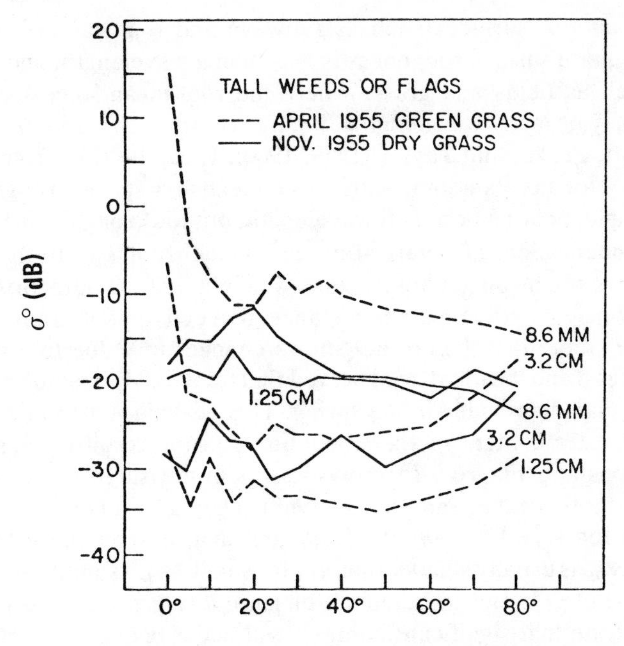

Source: Grant and Yaplee (1957).

Figure 6-5. Comparison of σ°_{VV} for Green Grass and Dry Grass at X, K, and K_a Bands.

15 to 20 dB for 8.6 mm and 3.2 cm when the ground is wet; (3) at 1.25 cm the backscattering by the green grass is much lower than for the dry grass, and there is less evidence of any specular reflection, presumably because this wavelength coincides with a major water absorption band; and (4) σ° increases in the plateau region with increases in moisture at 8.6 mm and 3.2 cm, but decreases with moisture at 1.25 cm.

R. L. Cosgriff, W.H. Peake, and R.C. Taylor (1960) reported an extensive collection of measurements on average σ°. Recall that although σ° is a dimensionless quantity, its determination involves an "absolute" measurement, and the results are reported to be accurate to within 1.0 dB. The cross sections are expressed in terms of the variable γ, where $\sigma^\circ = \gamma \sin\theta$. Cosgriff, Peake, and Taylor made extensive theoretical and experimental studies on five parameters that affect echo magnitude: surface roughness, incidence angle, polarization, complex dielectric constant of the terrain, and radar frequency.

They observed that most terrain can be divided into two distinct categories: *smooth* surfaces such as runways and roadway, where the root-mean-square surface roughness is less than a wavelength; and *rough* surfaces such as fields and grass, where the root-mean-square surface roughness may be many wavelengths. Figures 6-6 through 6-15 were taken from Cosgriff, Peake, and Taylor (1959, 1960). It can be seen from these figures that γ for the "smooth" surfaces depends on surface roughness, incidence angle, polarization, and wavelength; but for "rough" surfaces γ is usually independent of polarization and incidence angle. In figure 6-8 $\overline{z^2}$ and $\rho(r)$ are the mean-square roughness height and the autocorrelation function for height with respect to distance, respectively, of samples measured in the laboratory. Figure 6-9 shows changes in γ due to a natural growth process, and it should be observed that for a fixed vegetation height γ reaches a maximum value in the spring. This peaking is attributed to an increase in water content of the vegetation and the resulting change in dielectric constant. Figure 6-15 shows ranges in measured values of γ_{HH} given by Cosgriff, Peake, and Taylor (1960). They also reported a similar compilation for γ_{VV}. The *ranges* of γ_{VV} are smaller than those for γ_{HH} because (a) γ_{HH} is usually smaller than γ_{VV} for small θ with smooth surfaces, and (b) γ_{HH} and γ_{VV} tend to be equal for large θ (e.g., see fig. 6-7). The figures illustrate that significant contrast will occur between target areas painted on a radar display, and it should be noticed that relative echo strength can depend on incidence angle and radar wavelength.

The extensiveness of surface types reported by Cosgriff, Peake, and Taylor for X, K_u, and K_a bands, as a function of incidence angle, is illustrated by the partial list below:

Roads of various roughnesses

Roads covered with gravel, cinders, and dirt

Effects of snow on road

Effects of rain on roads

Plowed and disked ground

Seasonal changes of grass and crops

Effects of dew and rain on crops

Effects of mowing grass and crops

Another report on terrain characteristics has been published by Ohio State University personnel (Peake and Oliver 1971). The reported values of γ (or σ°) are a few decibels larger than given previously (Cosgriff, Peake, and Taylor 1959, 1960). This helps to emphasize the fact that absolute calibration errors can be nonnegligible even for very carefully designed and instrumented radar experiments. For illustrations of the broad ranges of averages for σ° reported between 9 and 95 GHz, the reader is referred to R.D. Hayes and F.B. Dyer (1973).

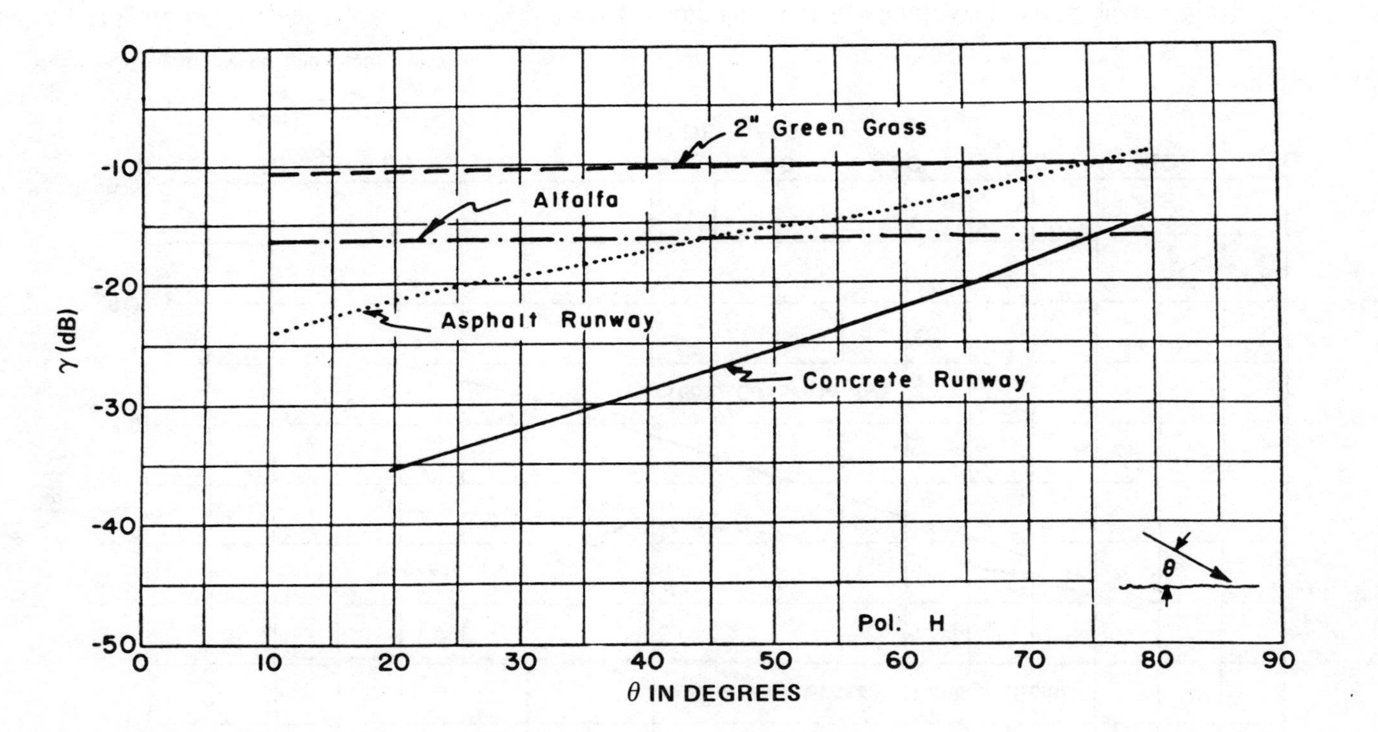

Source: Cosgriff, Peake, and Taylor (1959).

Figure 6-6. γ_{HH} for Paved Runways and Vegetation Versus Incidence Angle θ at K_a Band, Where σ° Equals γ Sin θ.

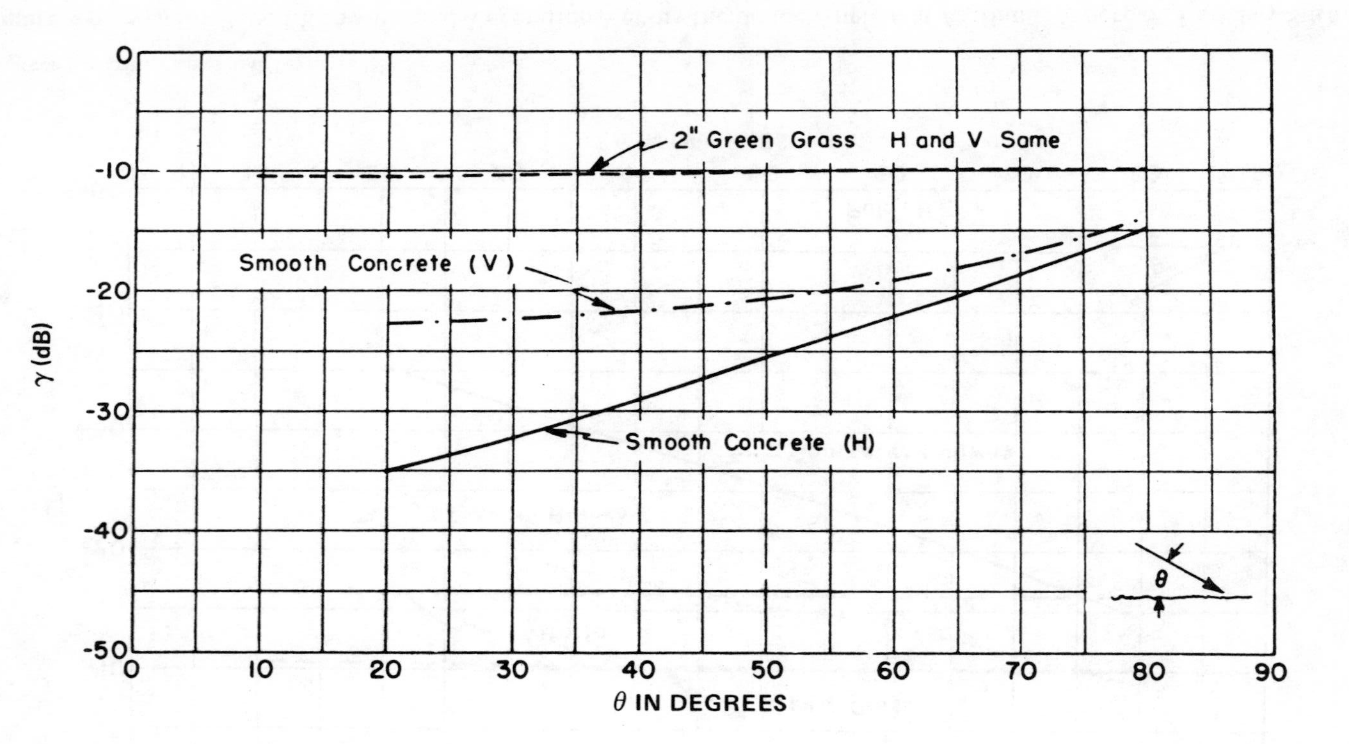

Source: Cosgriff, Peake, and Taylor (1959).

Figure 6-7. γ_{HH} and γ_{VV} for Concrete Runways and Grass at K_a Band, Where σ° Equals $\gamma \operatorname{Sin} \theta$.

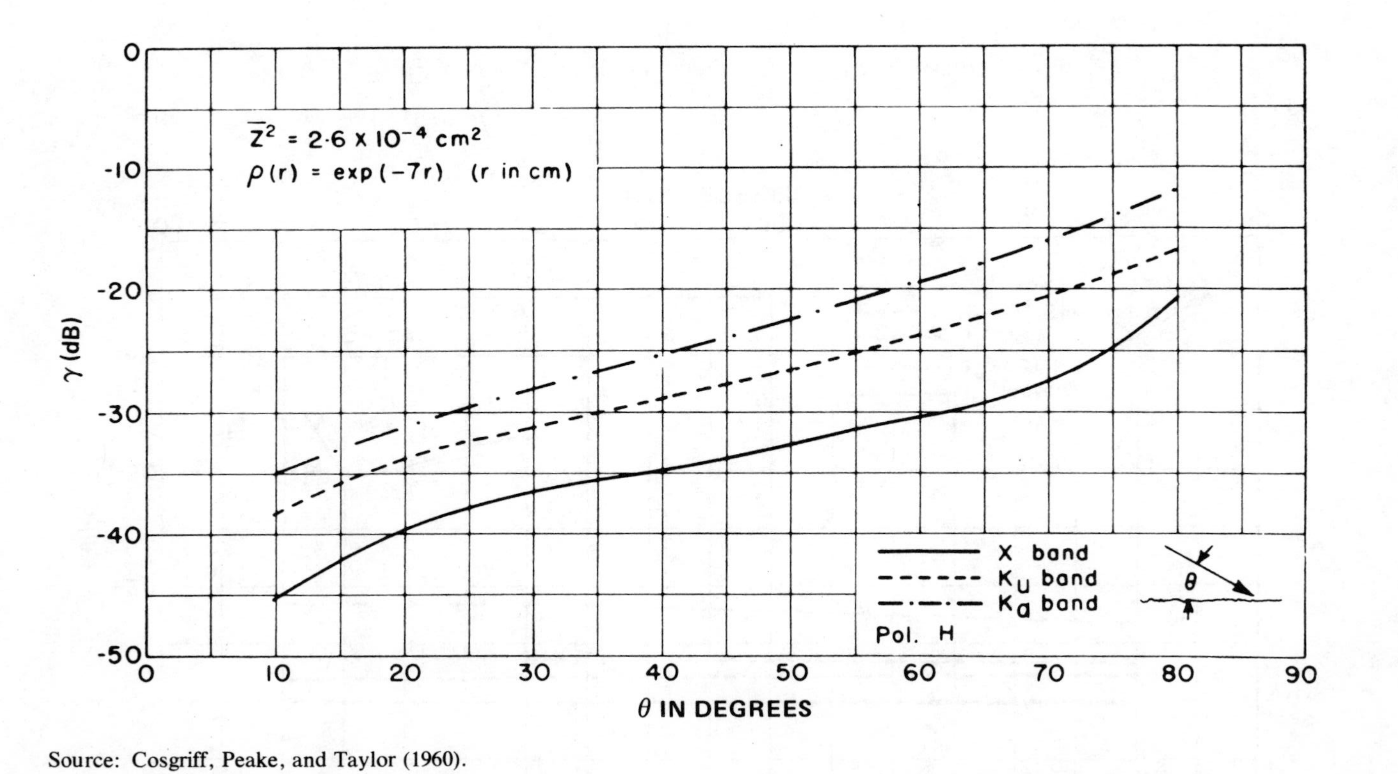

Source: Cosgriff, Peake, and Taylor (1960).

Figure 6-8. γ_{HH} for a Concrete Road at X, K_u, and K_a Bands, Where $\sigma°$ Equals $\gamma \sin\theta$.

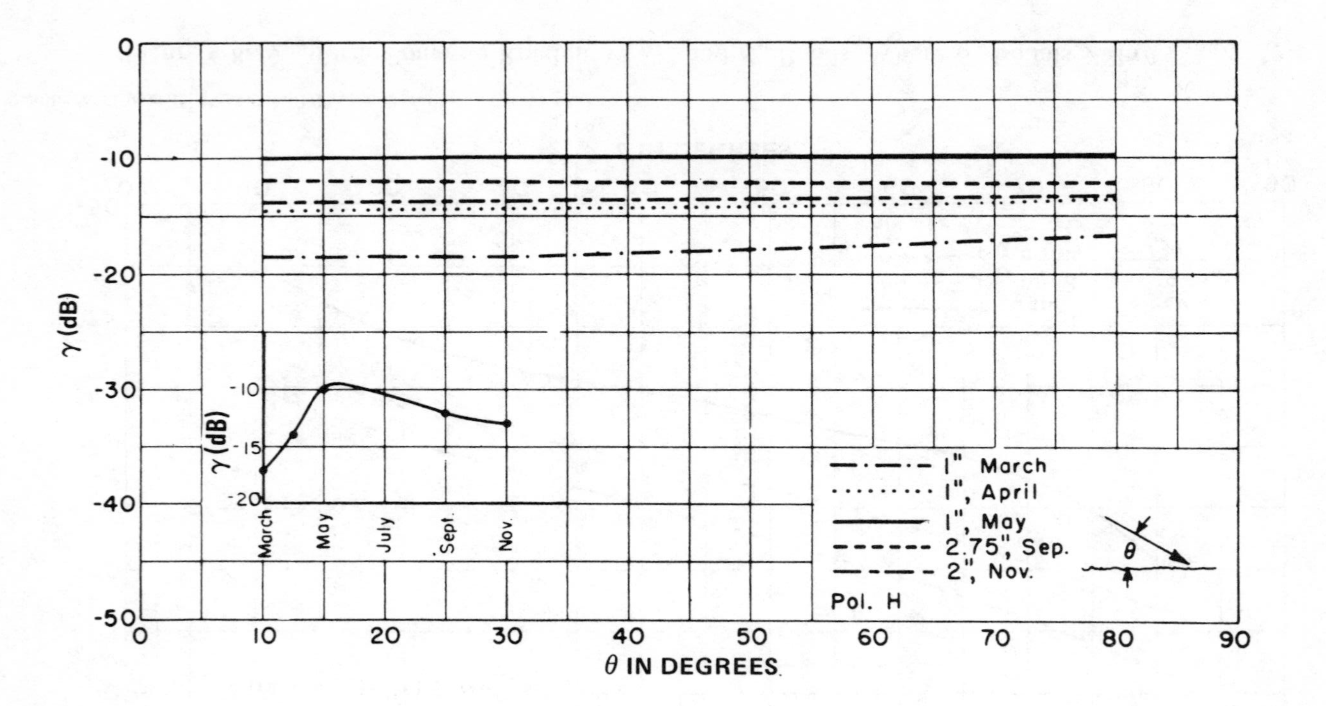

Source: Cosgriff, Peake, and Taylor (1960).

Figure 6-9. Effects of Seasonal Changes on γ_{HH} for Grass at K_a Band.

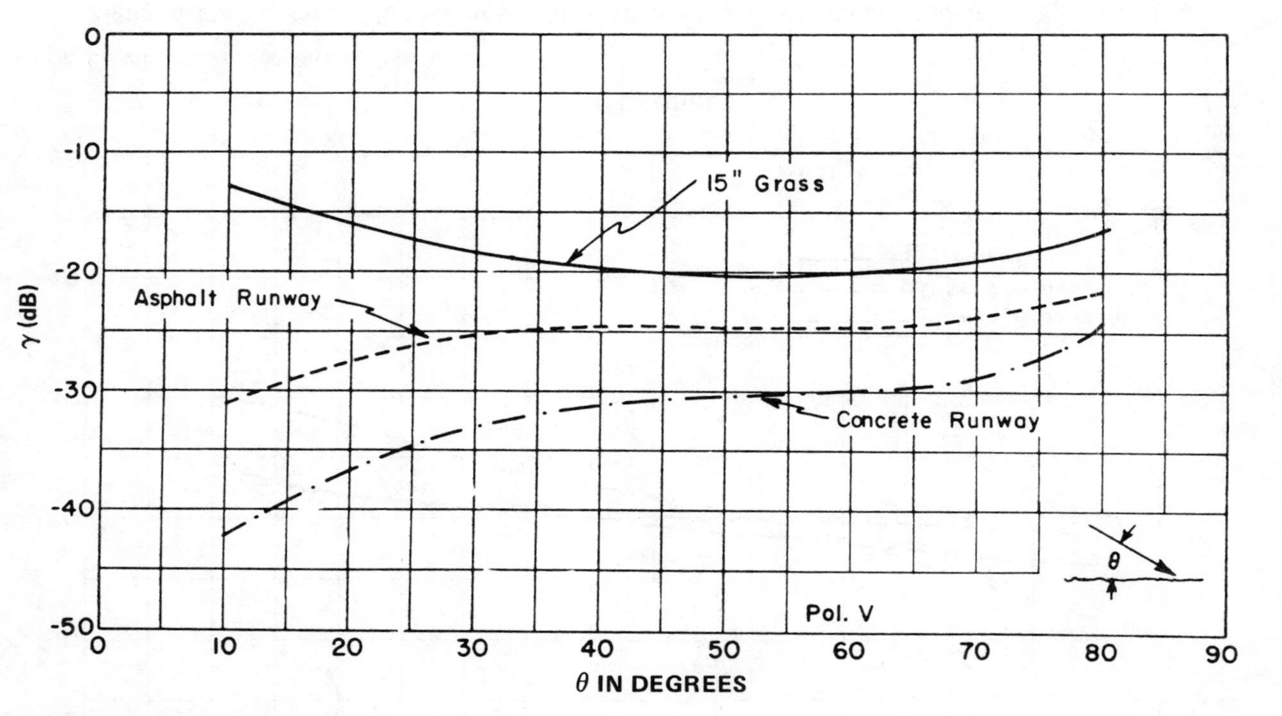

Source: Cosgriff, Peake, and Taylor (1959).

Figure 6-10. γ_{VV} for Grass and Runways at *X* Band, Where σ° Equals $\gamma \sin \theta$.

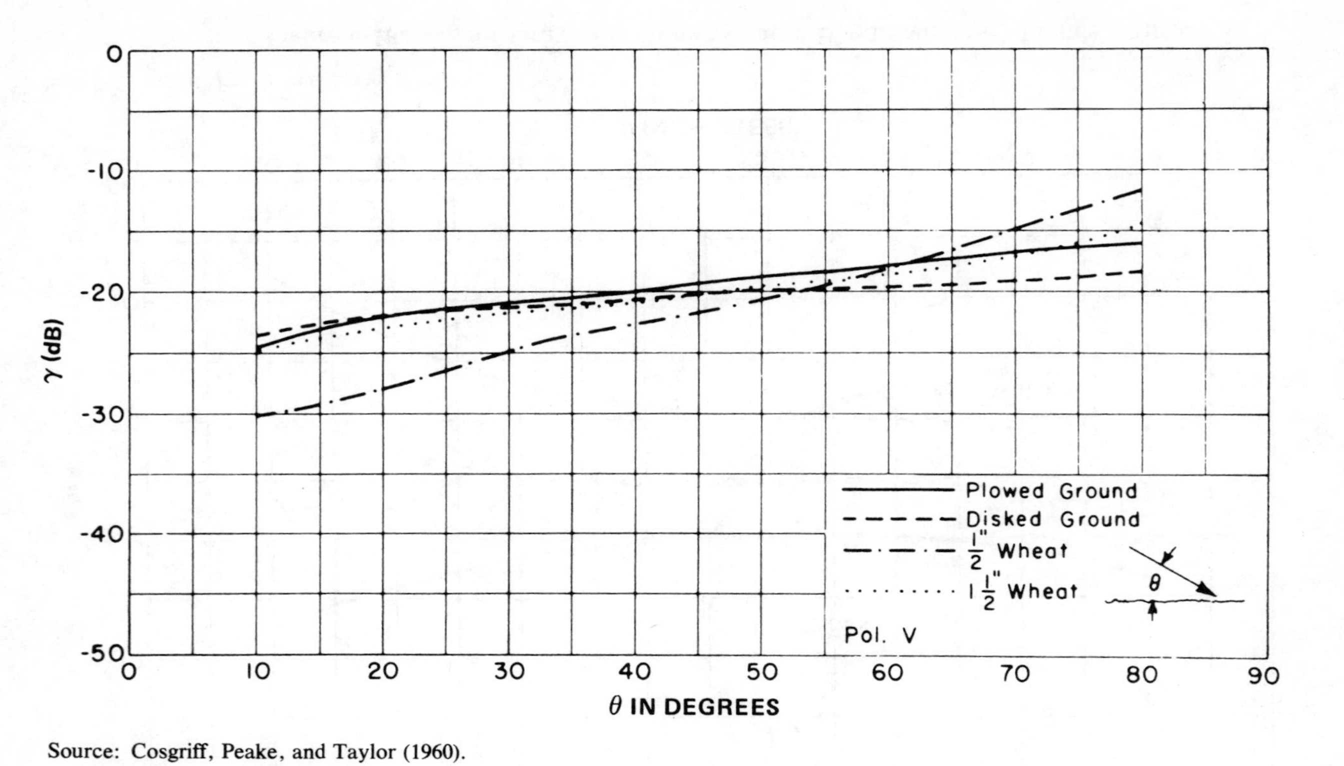

Source: Cosgriff, Peake, and Taylor (1960).

Figure 6-11. Seasonal Changes of γ_{VV} for a Wheat Field at X Band, Where σ° Equals $\gamma \operatorname{Sin} \theta$.

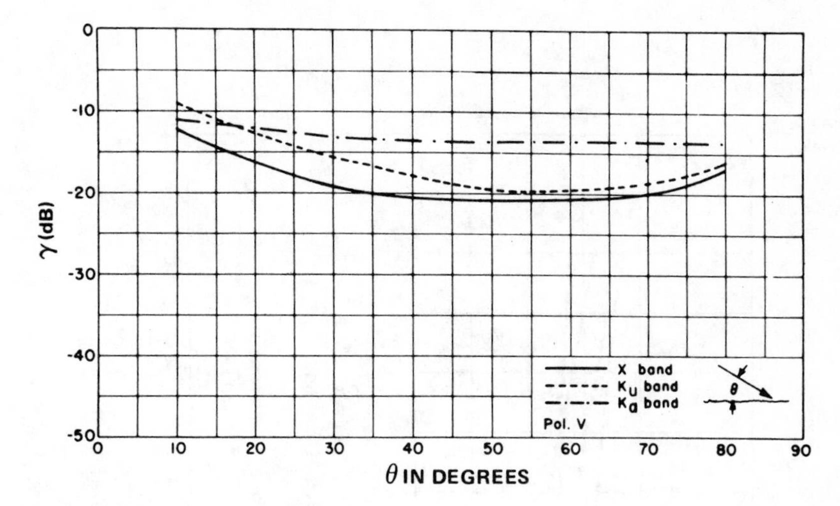

Source: Cosgriff, Peake, and Taylor (1960).

Figure 6-12. γ_{VV} for 15-inch Green Grass in Head at X, K_u and K_a Bands, Where σ° Equals γ Sin θ.

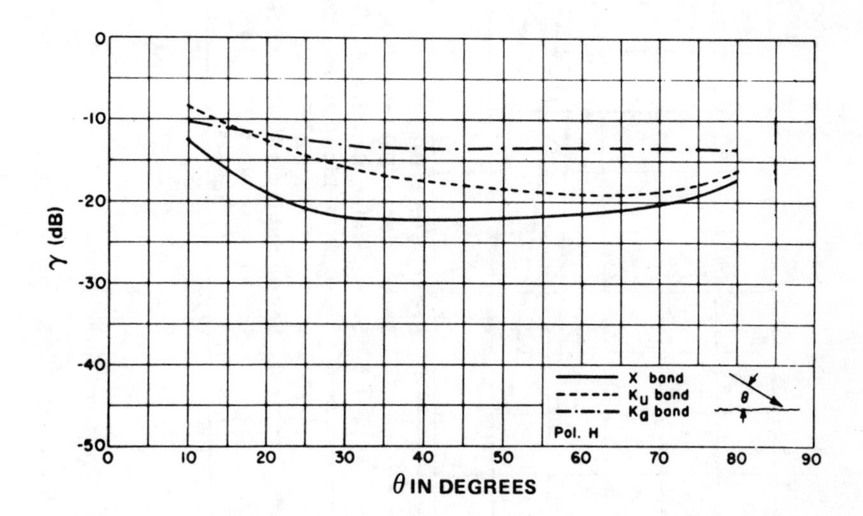

Source: Cosgriff, Peake, and Taylor (1960).

Figure 6-13. γ_{HH} for 15-inch Green Grass in Head at X, K_u, and K_a Bands, Where σ° Equals γ Sin θ.

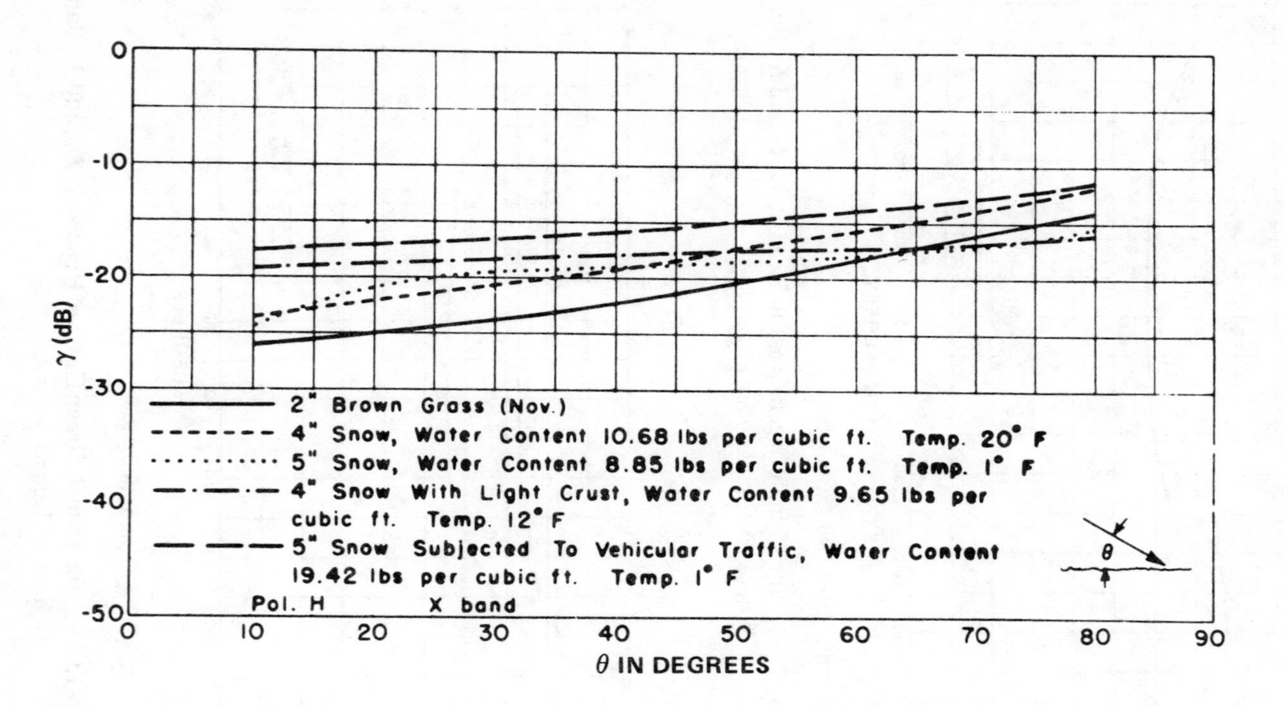

Source: Cosgriff, Peake, and Taylor (1960).

Figure 6-14. Effects of Various Types of Snow Cover on γ_{HH} at X Band, Where $\sigma°$ Equals γ Sin θ.

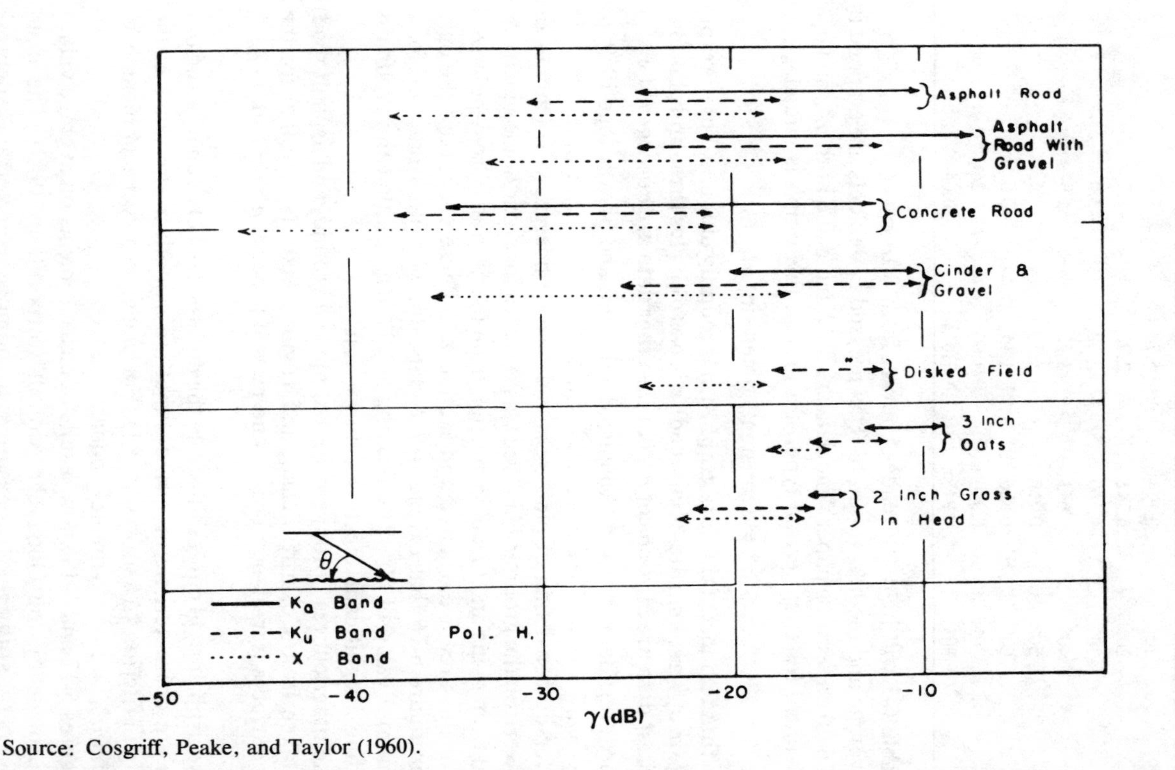

Source: Cosgriff, Peake, and Taylor (1960).

Figure 6-15. Range of γ_{HH} for Various Types of Terrain at X, K_u, and K_a Bands, Where $\sigma°$ Equals γ Sin θ.

Table 6-1
Sin θ Expressed in Decibels

θ	$\sin\theta$	$1/\sin\theta$	$\sin\theta$ (dB)
10°	0.174	5.74	−7.6
20°	0.342	2.92	−4.7
30°	0.500	2.00	−3.0
40°	0.643	1.55	−1.9
50°	0.766	1.30	−1.1
60°	0.866	1.15	−0.6
70°	0.940	1.06	−0.26
80°	0.945	1.01	−0.04

Table 6-1 will help the reader convert from γ to $\sigma°$.

Average and median values of $\sigma°$ expressed in decibels are graphed in figure 6-16 on semilog paper. The data are for X band and both σ°_{HH} and σ°_{VV} are included for each terrain type. If a line on this graph is straight,

$$\sigma° = \text{(constant) times } \sin^n \theta.$$

Notice that for most of the data graphed, n is roughly plus one. For the rapid change in $\sigma°$ versus $\sin\theta$ (asphalt road), n is between plus three and plus four. The forest data are *medians* of $\sigma°$, but the other data are *averages* of $\sigma°$.

The forest data are from (Ament, Macdonald, and Shewbridge 1959, p. 246) X-band airborne measurements made for a heavily wooded area in New Jersey consisting of pine trees and heavy undergrowth. Patches of snow were on the ground. This area was selected for radar homogeneity as judged from maps and visual inspection from the aircraft. Cross-polarized data for this area are discussed in section 7.7. These curves are the only ones in figure 6-16 that extend to θ equals 90°. The other data are for a maximum θ value of 80°. The peaking at 90° suggests that the ground is providing a significant specular contribution.

The other curves (3-foot soy beans, plowed ground, and asphalt road) were taken from Cosgriff, Peake, and Taylor (1960). Notice that for the **rough surfaces (forest, soy beans, and plowed ground) $n \approx +1$ and $\sigma^\circ_{HH} \approx \sigma^\circ_{VV}$.**

The asphalt road curves illustrate the differences between σ°_{HH} and σ°_{VV} which are characteristic of surfaces that are smooth in the sense of the Rayleigh criterion ($\Delta h \sin\theta/\lambda << 1$). For angles near vertical incidence, σ°_{HH} and σ°_{VV} are approximately equal.

Figures 6-17 and 6-18 show results of measurements on average values of σ°_{HH} made by the Goodyear Aircraft Corporation (1959). The main purpose of the study was to measure $\sigma°$ for homogeneous terrain at depres-

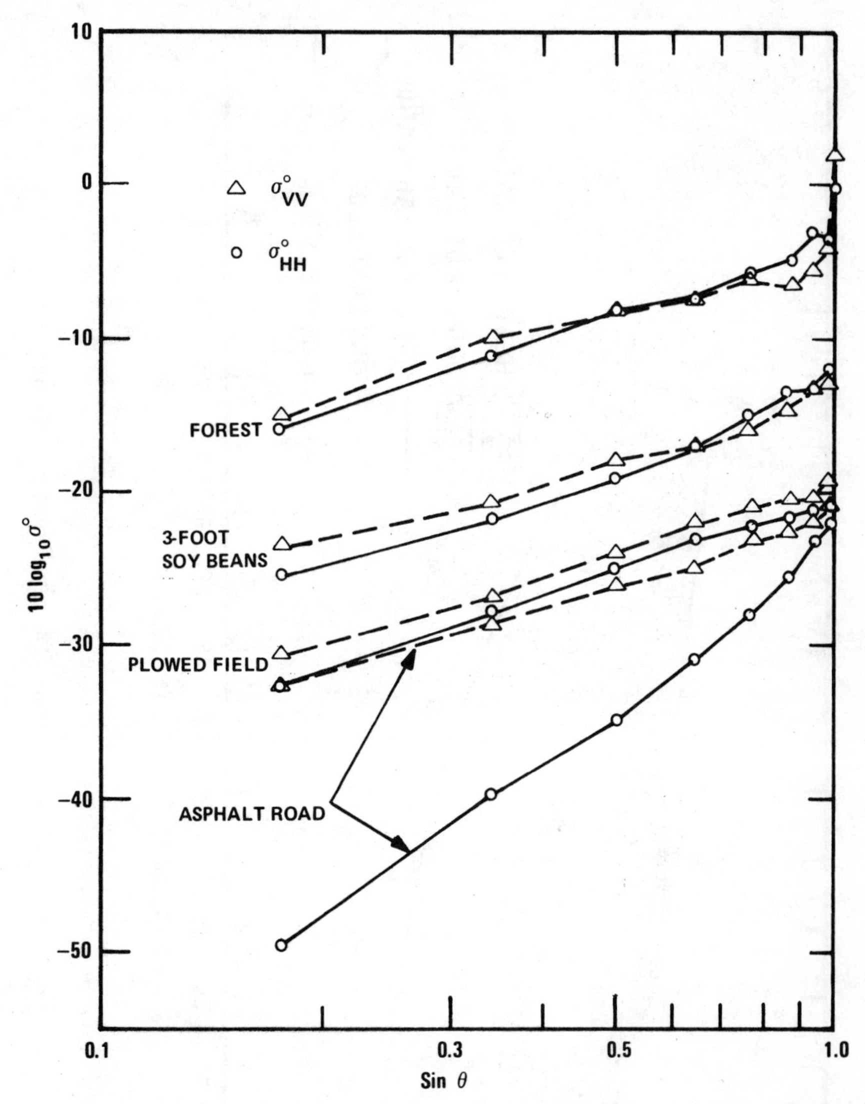

Figure 6-16. σ°_{HH} and σ°_{VV} for a Forest, Soy Beans, a Plowed Field, and an Asphalt Road at *X* Band. The forest data are medians of σ°, but the other data are averages of σ°.

sion angles greater than 10°. However, the Goodyear (GAC) report also included data for angles between 1° and 10° that are graphed in figure 6-18. The airborne radar was operated at *X* band (9,375 MHz).

Figure 6-17 illustrates that reflectivity data can to some extent be separated into bands that depend on gross terrain classification. The GAC

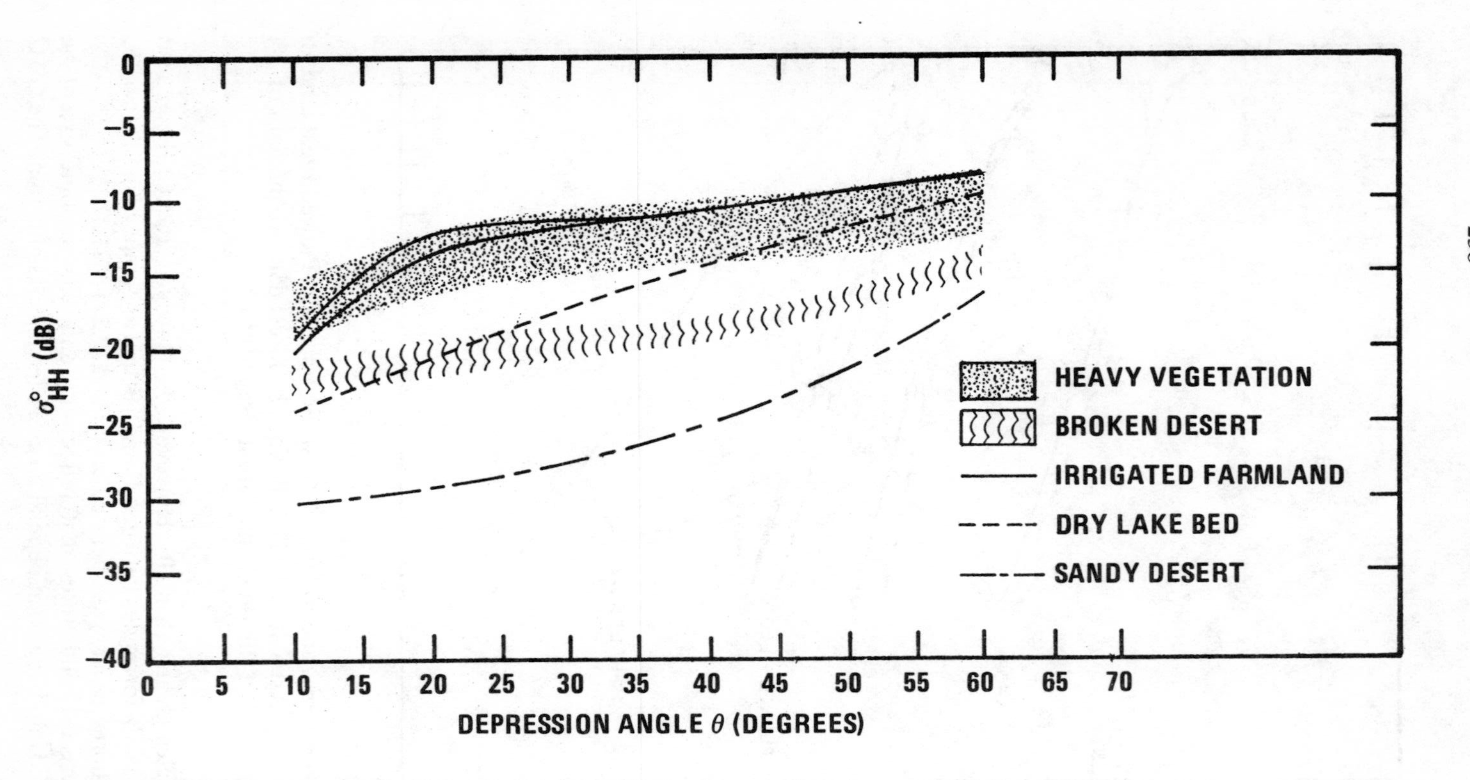

Source: Adapted from Goodyear Aircraft Corporation (1959).

Figure 6-17. σ°_{HH} for Various Terrain Types at *X* Band.

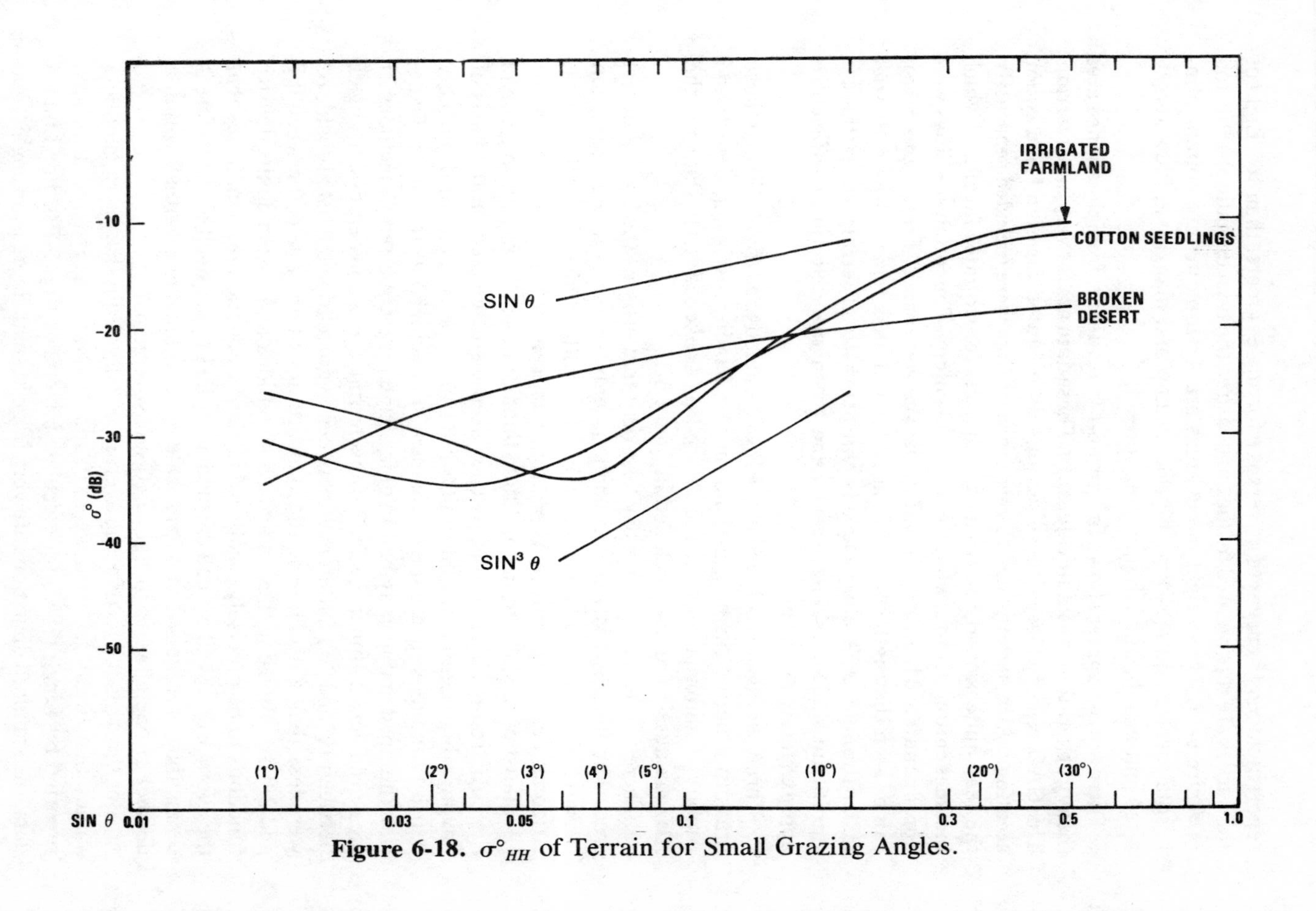

Figure 6-18. σ°_{HH} of Terrain for Small Grazing Angles.

report includes 15 curves of σ° versus θ for angles greater than 10°. Eight of the curves are represented by the shaded area (heavy vegetation) at the top of figure 6-17. These eight target areas are Arizona mature cotton, Minnesota forests and meadow, Florida swamp and mangrove, New Jersey trees and marsh, and Arizona pine forests.

There are two curves shown in figure 6-17 that depart from the shaded area (heavy vegetation) at small values of θ. These data are for very moist terrain. The GAC report classed those terrains as irrigated farmland and cotton seedlings. Although the data for cotton seedlings were included separately, they can quite properly be grouped with those for irrigated farmland which includes cotton, alfalfa, wheat, maize, and onions. The cotton seedlings were approximately 6 inches high and the leaves were small. From figure 6-18 it can be seen that both curves for moist terrain have regions where σ° varies faster than sin θ. A steep slope is characteristic of terrain where there is significant multipath interference and when the reflection coefficient is approximately minus one.

Three curves are included in the lower shaded area (fig. 6-17) and those areas (northern Arizona grassland (dry); Chandler, Arizona, desert; and Amboy, California, desert) are classed as broken desert. Data for the Arizona desert are also included in figure 6-18.

There is one curve (for a dry lake bed) that crosses the broken desert area. The lake bed is flat and contains only sparse, stunted vegetation because of a heavy salt concentration.[a] The bottom curve is for the Yuma, Arizona, desert. It represents σ° from a barren, sandy area.

Statistics of short-term amplitude fluctuations in σ° were also given in the GAC report. The ground patches used were selected on the basis of having a high degree of uniformity in terrain type. Graphs of probability density functions (percentages versus σ° in decibels) were roughly Gaussian (humped and usually bell-like) in shape. The GAC report includes 21 measured density functions for various terrain types and depression angles between 10° and 70°. Differences between average σ° and most probable σ° were less than a decibel and the standard deviations ranged between 0.7 and 2.0 dB. By using these standard deviations and assuming the density functions to be precisely lognormal in shape, equation (6.1) indicates that the average of σ° will exceed the median by less than a decibel. Therefore it seems that average σ°, most probable σ°, and median σ° are all within a decibel of one another for the GAC measurements.

For the irrigated terrain, the average σ° curves dip sharply to minimum values that occur for θ between 1° and 4°. Nulls in σ° have also been reported by Linell (1963) for calculated averages (fig. 6-19). For Linell's data, the standard deviations are very large at small depression angles, and

[a]Microwaves will penetrate dry sand. The steep slope may be caused by reflections from a smooth brine (high water content) surface near the top dry surface (Ulaby, Sobti et al.1974).

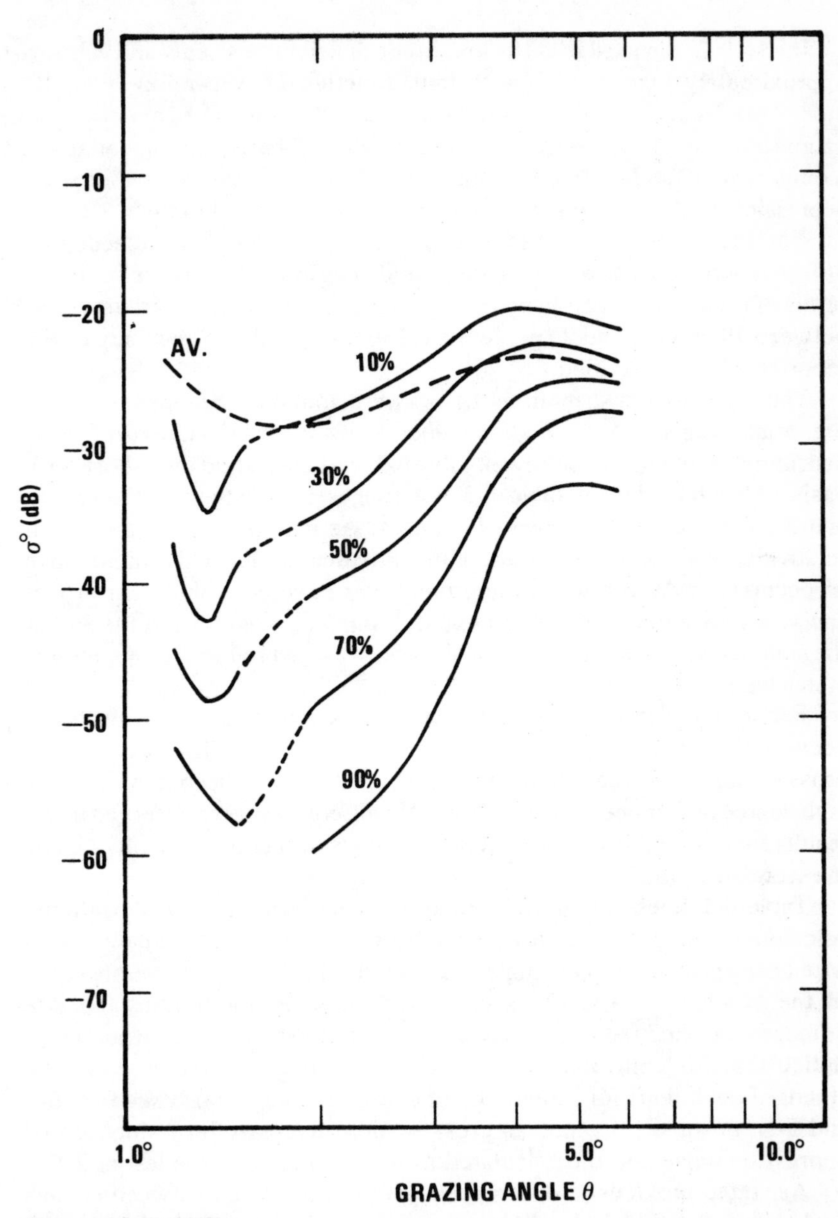

Source: Linell (1963).

Figure 6-19. Various Statistics for σ° of Cultivated Terrain at *X* Band.

the standard deviations always exceeded those of the Goodyear studies. The Goodyear report did not include probability density functions for depression angles less than 10°.

Linell (1963) reported that the amplitude distributions of ground echo are approximately lognormal (see Weibull function in Appendix) when the pulses are from contiguous range cells and the antenna beam scans an azimuth sector. The measurements were made at X band, and the radar was mounted on a 100-foot tower so that clutter could be investigated at small depression angles. Major system parameters are included in table 6-2.

For the statistical studies, the number of pulses that exceeded 12 different percentage levels were measured. These levels were calibrated in terms of radar cross section by the use of a standard target. Various levels between 10 percent and 90 percent were investigated; the accuracy of the reported σ° was specified as 3 dB.

The measured distributions of σ° were found to be approximately lognormal (see sec. 5.2). Median values and standard deviations of pulse distributions given in each row of table 6-3 were measured on a short-term basis. Also included in table 6-3 are ranges to illustrate variations in distributions caused by seasonal changes. As may be seen, there was a relatively strong seasonal variation for cultivated terrain, and a minimum in σ° occurred in March when the ground was covered with 10 cm of wet snow. A maximum in σ° for cultivated terrain occurred toward the end of the summer when a large portion of the area was covered with crops about 1 meter high.

Far less variation of σ° with season was found for trees, probably because the forest consisted mainly of pines and firs. A slight maximum in cross section occurred in November, which may have been due to a very high degree of dampness in the terrain. No differences were observed in the results for horizontal and vertical polarizations, either for the cultivated or the wooded terrain.

Table 6-3 gives measured median σ°, measured standard deviations, calculated average σ° (eq. (6.1)), and depicts changes of these parameters with changes in depression angle and season. The data help illustrate some of the problems in specifying cross section, even for terrains that are homogeneous and relatively easy to describe. To underscore an anomaly, notice that for cultivated terrain (table 6-3) there are increases in the standard deviations for reductions in depression angle (also see figs. 5-6 and 5-7); however, the median cross sections decrease for reductions of depression angle and the calculated average values are smallest at 2.5°.

As stated previously, differences were not observed between echoes for horizontal or vertical polarization. Figure 6-19 shows calculated average values reported by Linell, and measured median levels are indicated by the solid lines. Variations in the ratio of average to median value of σ° versus incidence angle is clearly evident. Notice that in figure 6-19 the angles at which average σ° and median σ° are minimum do not coincide.

L.O. Ericson (1963) has described measurements made on ground clutter during the summers of 1961 and 1962. The objective of the experi-

Table 6-2
System Parameters Used by Linell

Horizontal beamwidth	1.4°
Vertical beamwidth	30°
Range resolution	25 meters
Polarization	changeable, horizontal, or vertical
Calibration	standard target

Table 6-3
Radar Cross Section Per Unit of Illuminated Area $\sigma°$

Depression Angle (Degrees)	*Measured Median (dB)*	*Measured Standard Deviation (dB)*	*Calculated Average (dB)*	*Month*
		Terrain Type: Cultivated		
5	−39	7	−33	March
	−23	6	−19	August
2.5	−51	12	−34	March
	−30	7	−24	August
1.25	−53	16	−24	March
	−38	12	−21	August
		Terrain Type: Forest		
0.7	−42	17	− 9	All but November
	−36	15	−10	November

ments was to determine the average $\sigma°$ as a function of (1) type of terrain, (2) incidence angle, and, if possible, (3) ground conditions. Larger ground areas were illuminated for the flight measurements of Ericson than for the experiments of Linell described above.

In order to cover essentially all types of terrain in middle Sweden, the flights were made over six different courses with a total length of about 500 miles. Depression angles of 1°, 5°, and 30° were used. For each angle the courses were passed in both directions. Major system parameters are given below.

Wavelength	*X* band
Horizontal beamwidth	3.3°
Pulse length and range gate width	1 μ sec
Polarization	horizontal

The ground was illuminated by an antenna mounted in the nose of an aircraft. Depression angle was locked to the horizontal plane by servo-controlled gyro-stabilization of roll and pitch. Antenna azimuth was continuously adjusted during each run so that the direction of propagation was forward, approximately along the ground track of the airplane.

Various geometrical parameters pertinent to interpreting results of the measurements are given in table 6-4.

The signal was stored on magnetic tape and was played back to determine percent of time it exceeded different levels, calibrated in terms of $\sigma°$. Since the distributions of pulses were approximately lognormal, the average value of cross section could be calculated from the median and the standard deviation for each short-term distribution.

Various terrain types were investigated including cultivated swamps, woods, rocks, archipelagoes, small houses surrounded by gardens, small housing districts with buildings up to three floors, and very concentrated higher buildings with four to 12 floors.

For each incidence angle, distributions of the *average* values of $\sigma°$ were plotted for the different kinds of terrain. The results showed no marked differences between the types, but some of this was probably due to the fact that there was averaging over large areas. The numbers in table 6-5 illustrate the wide range of cross sections obtained. The return from built-up areas in the center of Stockholm with buildings four to 12 stories high consistently seemed to yield the largest values of $\sigma°$. For archipelago terrain (small islands mixed with water) and a depression angle of 5°, $\sigma°$ was somewhat smaller than for other classes, ranging between −20 dB and −40 dB. This exception may be dependent on water surface condition, because under calm conditions $\sigma°$ for water (for the depression angles used) is far less than that for terrain. Thus, $\sigma°$ may be strongly dependent on the percentage of illuminated area covered by the islands.

A.R. Edison, R.K. Moore, and B.D. Warner (1960) give terrain data for near-vertical incidence angles at 415 and 3,800 MHz. Terrain investigated included woods, farmland, snow covered farmland, deserts, water, industrial and residential areas, and apartment buildings. For all areas, average $\sigma°$ was found to decrease as the incidence angle was varied away from vertical (normal) incidence. At 415 MHz, the values of $\sigma°$ at vertical incidence was 0.7 for dense woods and 4 for certain city targets. At 3,800 MHz, the values of $\sigma°$ at vertical incidence was 0.8 for woods and 18 for some city targets. For slightly rough water, $\sigma°$ was reported to be 50 at 415 MHz and 200 at 3,800 MHz for near vertical incidence. For an incidence angle 25° from vertical, $\sigma°$ for most farmland and city areas was reported as about 20 dB less than the value at normal incidence. For woods, the decrease in $\sigma°$ at 25° from that of vertical incidence was less than 20 dB, and for water it was usually greater than 20 dB.

Table 6-4
Illuminated Area Versus Geometrical Parameters Used by Ericson (1963)

Depression Angle	*Radar Height*	*Distance to Illuminated Patch*	*Illuminated Patch*		
			Width	*Depth*	*Area*
1°	175 m	10,060 m	580 m	150 m	87,000 m²
5°	875 m	10,060 m	580 m	150 m	87,000 m²
30°	1500 m	3,060 m	176 m	173 m	31,000 m²

Table 6-5
Radar Cross Section Per Unit Area

Depression Angle (Degrees)	*Most Targets (dB)*	*Tall Buildings (dB)*
1	−20 to −40	−20 to −30
5	−15 to −35	−10 to −20
30	−15 to −30	−10 to −20

Measured $\sigma°$ for near vertical incidence can be a sensitive function of antenna beamwidth and range, particularly for a surface with roughness that is small compared to a wavelength. Therefore, the reader is cautioned to carefully assess measurement techniques before using a reported value of $\sigma°$. For example, in contradiction to the results quoted above, it is quite plausible for $\sigma°$ to increase with an increase in λ, because a surface can appear rough at microwave frequencies while appearing smooth for a sufficiently long wavelength.

Figure 6-20 shows Sandia Corporation data (Janza, Moore, and Warner 1959). At near vertical incidence, smooth targets give stronger returns than do rough targets; the converse is true far away from the vertical. This contrast is also suggested by comparing figures 6-11 and 6-16 with figure 6-20. R.K. Moore (1969) cautions that the shape of $\sigma°$ versus θ curves near vertical incidence usually contain deleterious effects of the antenna patterns, because $\sigma°$ usually drops off rapidly as the angle measured with respect to the vertical is increased. For example, most reported $\sigma°$ data for vertical incidence is probably too small because the antenna beam includes a finite range of incidence angles.

6.9 Median $\sigma°$ for Various Terrains

Terrain data cannot be completely understood without knowledge of the

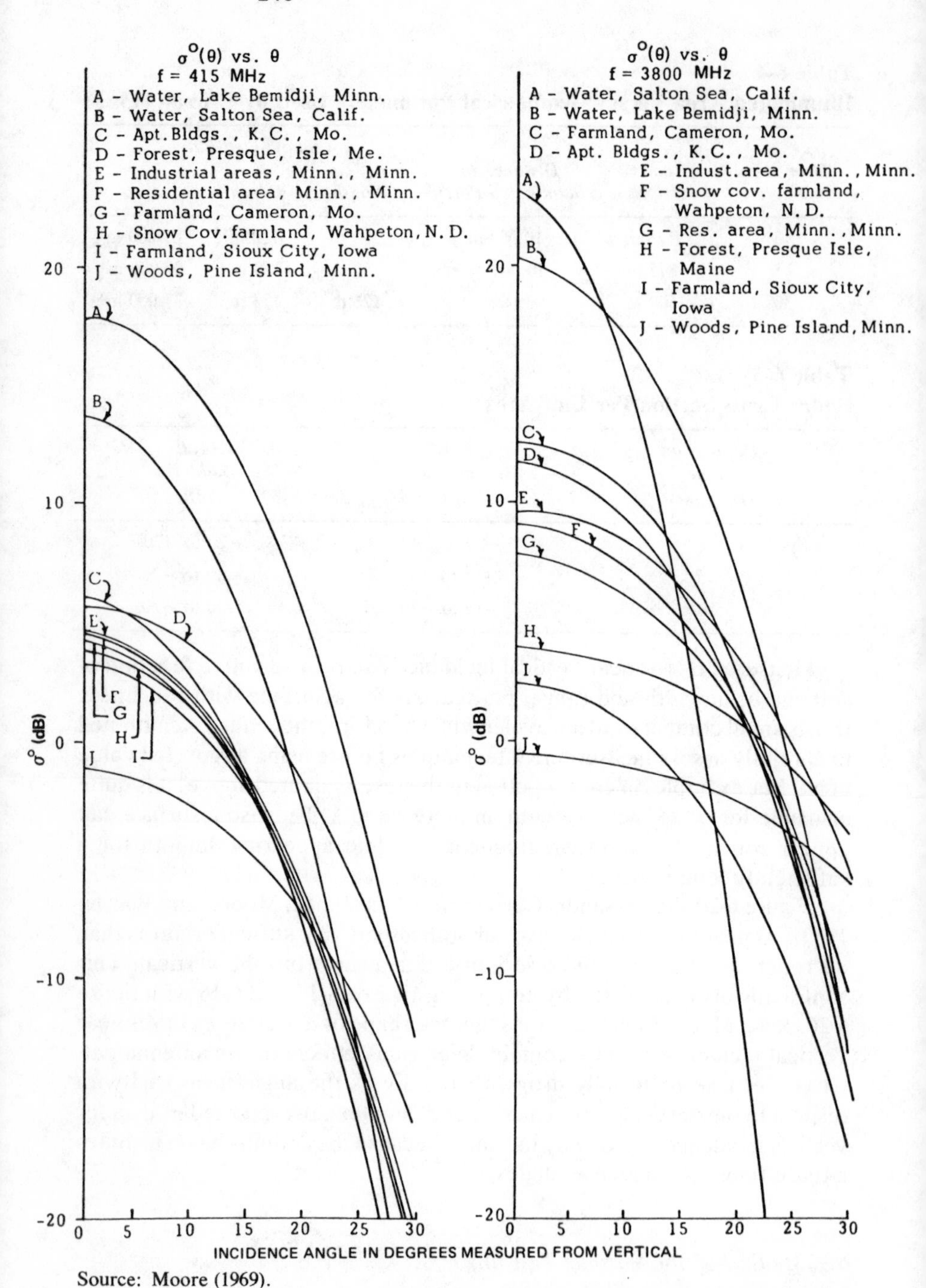

Source: Moore (1969).

Figure 6-20. σ° for Near Vertical Incidence Angles for Various Terrains at 415 MHz and 3,800 MHz.

amplitude distributions. However, a single number to represent intensity is extremely useful for providing insight toward understanding the dependence of echo strength on the many contributing variables. Some more or less obvious advantages for using medians instead of averages are suggested in section 6.2.

Figures 6-21 through 6-26 give some results of Daley, Davis et al. (1968) for depression angles between 5° and 60°. The measurements were accomplished with a 4-frequency radar (*P*, *L*, *C* and *X* bands) capable of transmitting horizontal and vertical polarizations alternately. The parameter reported was the median value of normalized radar cross section $\sigma°$. Cross-polarization ratios for these target areas were also reported by Daley, Davis et al. (1968) and are discussed in section 7.7. As discussed in section 6.2, all values of median $\sigma°$ that are given in figures 6-21 through 6-26 are expected to be 1.6 to 3 dB less than average $\sigma°$.

For rural, desert, and mountainous terrain, Daley, Davis et al. (1968) reported that $\sigma°$ is a slowly increasing function of incidence angle, which may be approximated by the sine function. The sample size used for these surfaces consisted of 10,000 to 20,000 data points, which represents 15 to 30 seconds of flying time (one to two miles of ground track).

Radar cross section changed rapidly with incidence angle for urban data and, in fact, the authors explained observed peaks at 8° as specular reflections from the buildings. According to the authors, large pulse-to-pulse fluctuations present in urban measurements necessitated a further analysis of data in terms of sample size (*SS* in the figures). The figures include results from a short sample (1,000 data points: $SS = 1{,}000$) representing 1.5 seconds of flight or approximately 450 feet along the ground track. Longer samples (26,000 to 37,000 data points) representing larger areas (38 to 54 seconds of flight which equals two to three miles along the ground track) were also used. The short sample-size data were taken primarily from the center of Phoenix, and the longer samples included a lower density of buildings than did the short ones. Accordingly, the curves marked "New Jersey residential" and "Phoenix" have large fluctuations due to inhomogeneities of the terrain, and changes in sample size caused large differences in measured median cross section for those terrains. Measured cross sections for the other terrains reported in figures 6-21 through 6-26 were not sensitive to sample size.

6.10 Dependence on Incidence Angle and Polarization

The main cause for the polarization sensitivity of terrain echo is multipath reflections. For example, smooth earth or water along the path of propagation can cause substantial differences between σ_{HH} and σ_{VV}. The magnitude of reflection from a smooth surface, such as a highway, may also be quite sensitive to polarization.

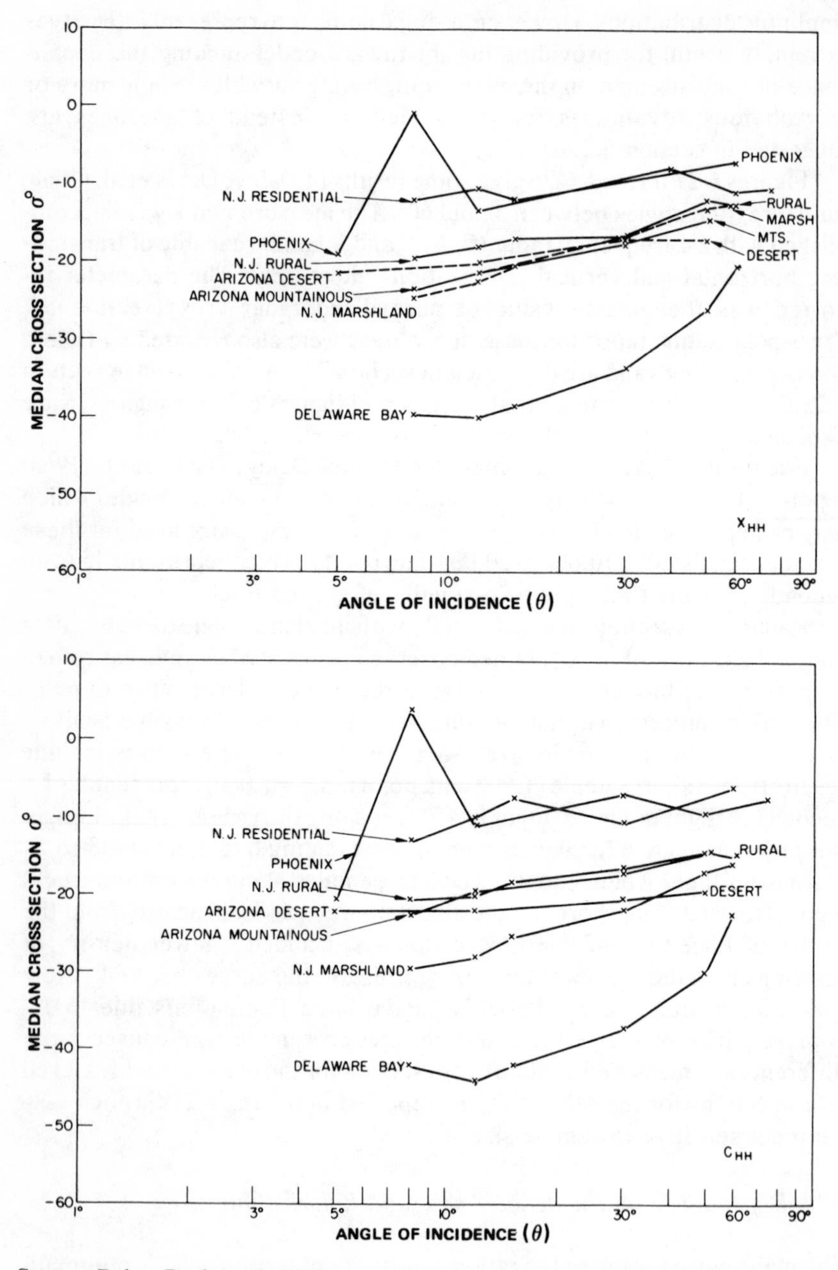

Source: Daley, Davis et al. (1968).

Figure 6-21. Median σ°_{HH} for Various Terrains at X (Upper) and C (Lower) Bands.

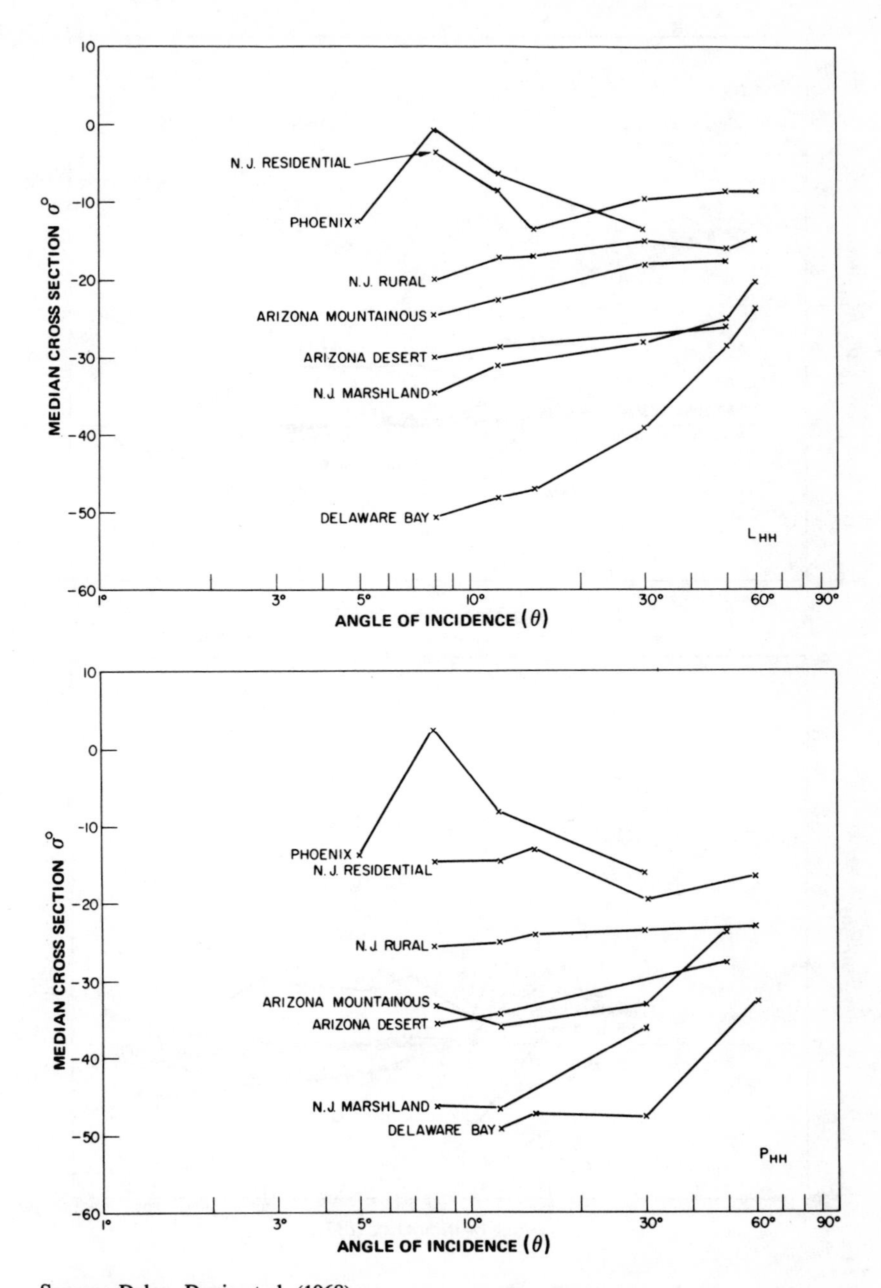

Source: Daley, Davis et al. (1968).

Figure 6-22. Median σ°_{HH} for Various Terrains at L (Upper) and P (Lower) Bands.

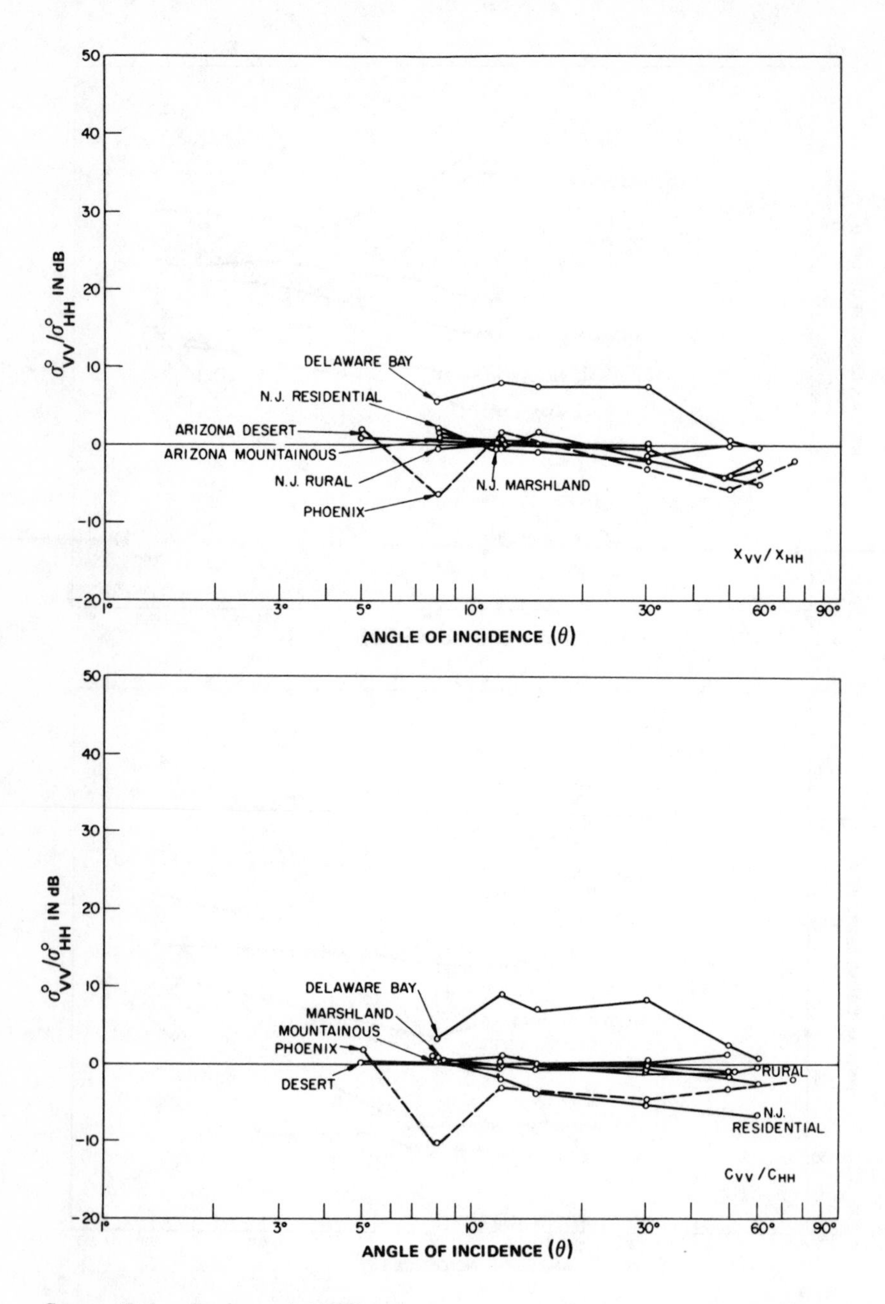

Source: Daley, Davis, et al. (1968).

Figure 6-23. Ratio of Median σ°_{VV} to σ°_{HH} for Various Terrains at X (Upper) and C (Lower) Bands.

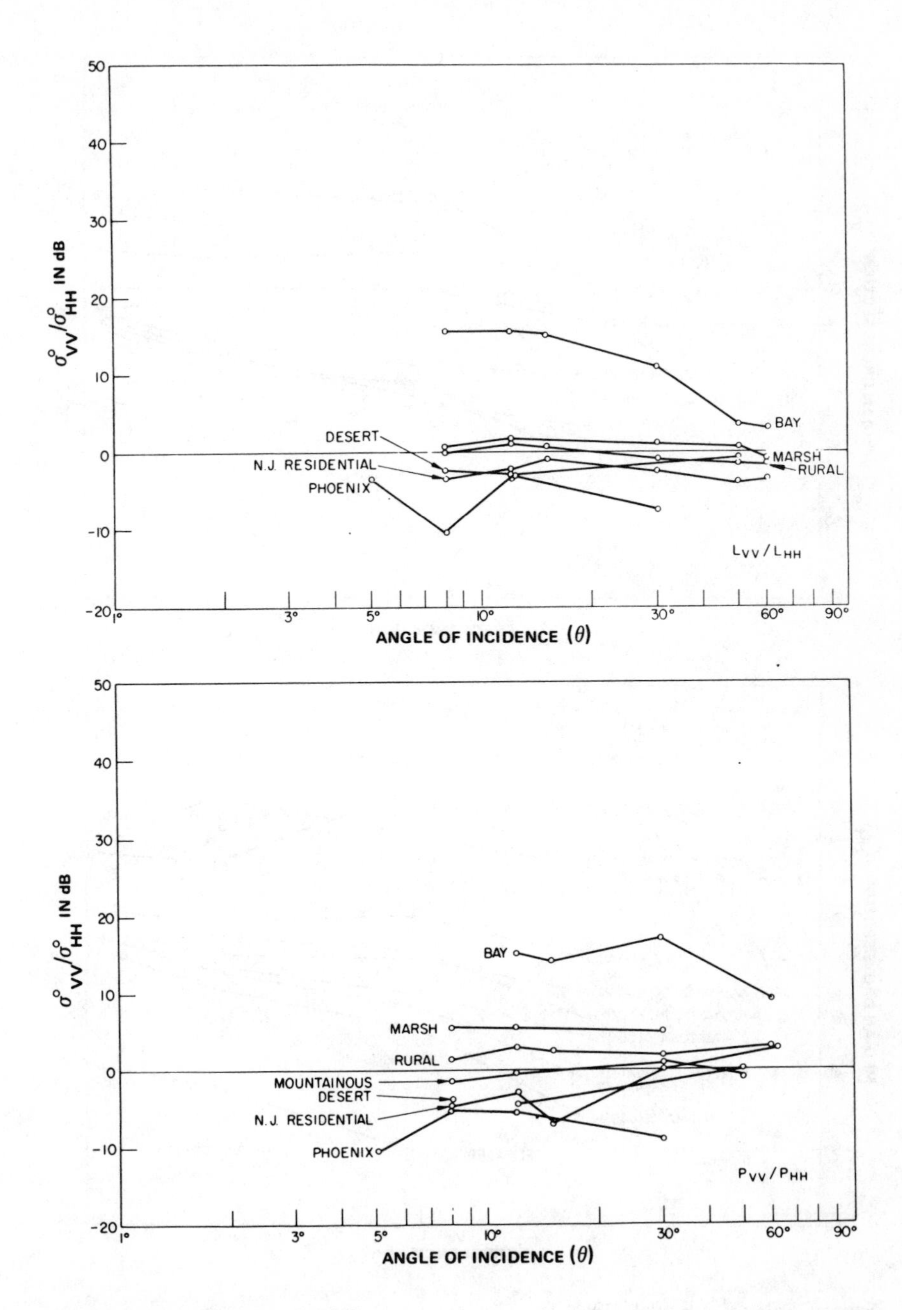

Source: Daley, Davis et al. (1968).

Figure 6-24. Ratio of Median σ°_{VV} to σ°_{HH} for Various Terrains at L (Upper) and P (Lower) Bands.

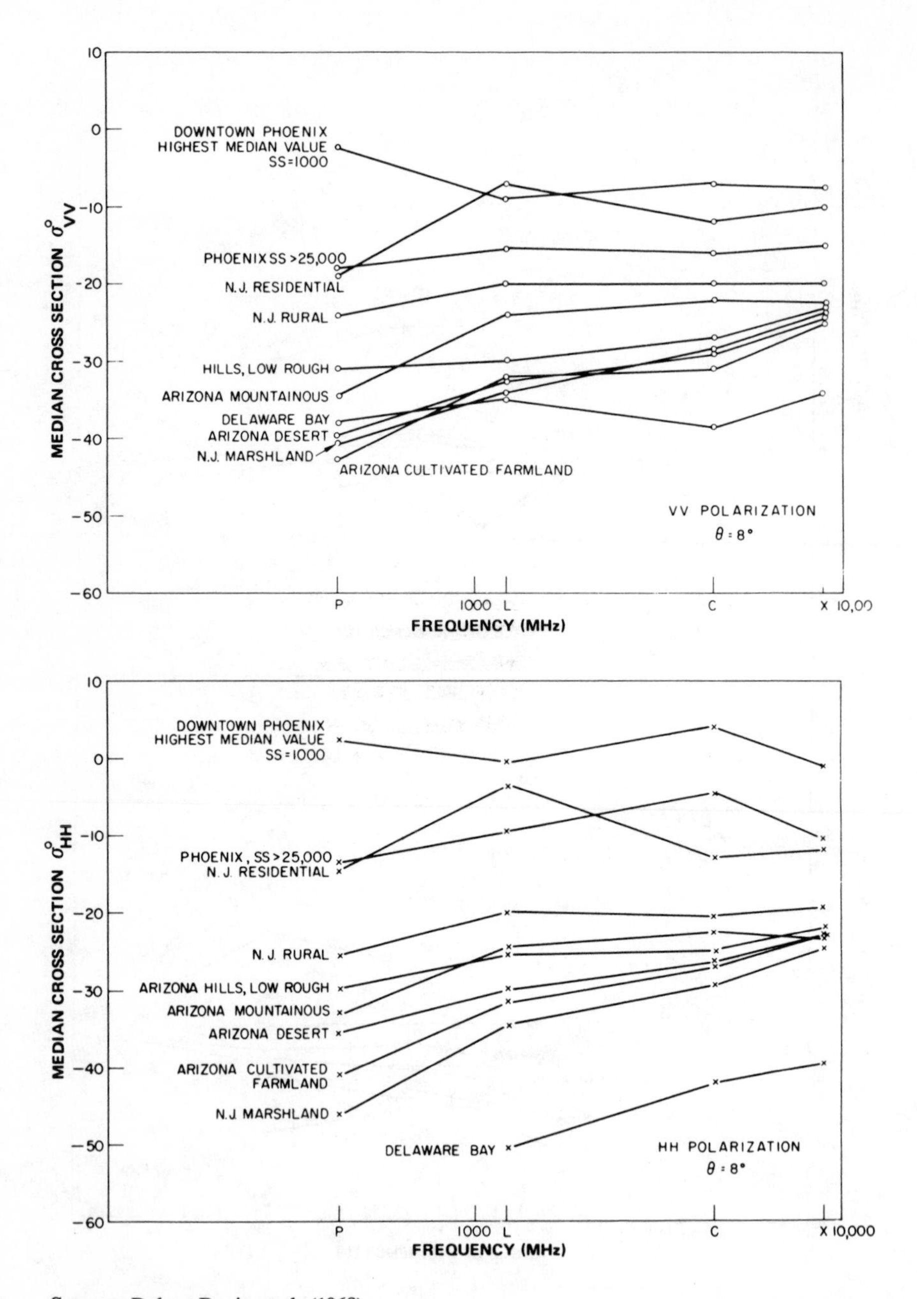

Source: Daley, Davis et al. (1968).

Figure 6-25. Median σ°_{VV} (Upper) and Median σ°_{HH} (Lower) for Various Terrains Versus Wavelength. $\theta = 8°$.

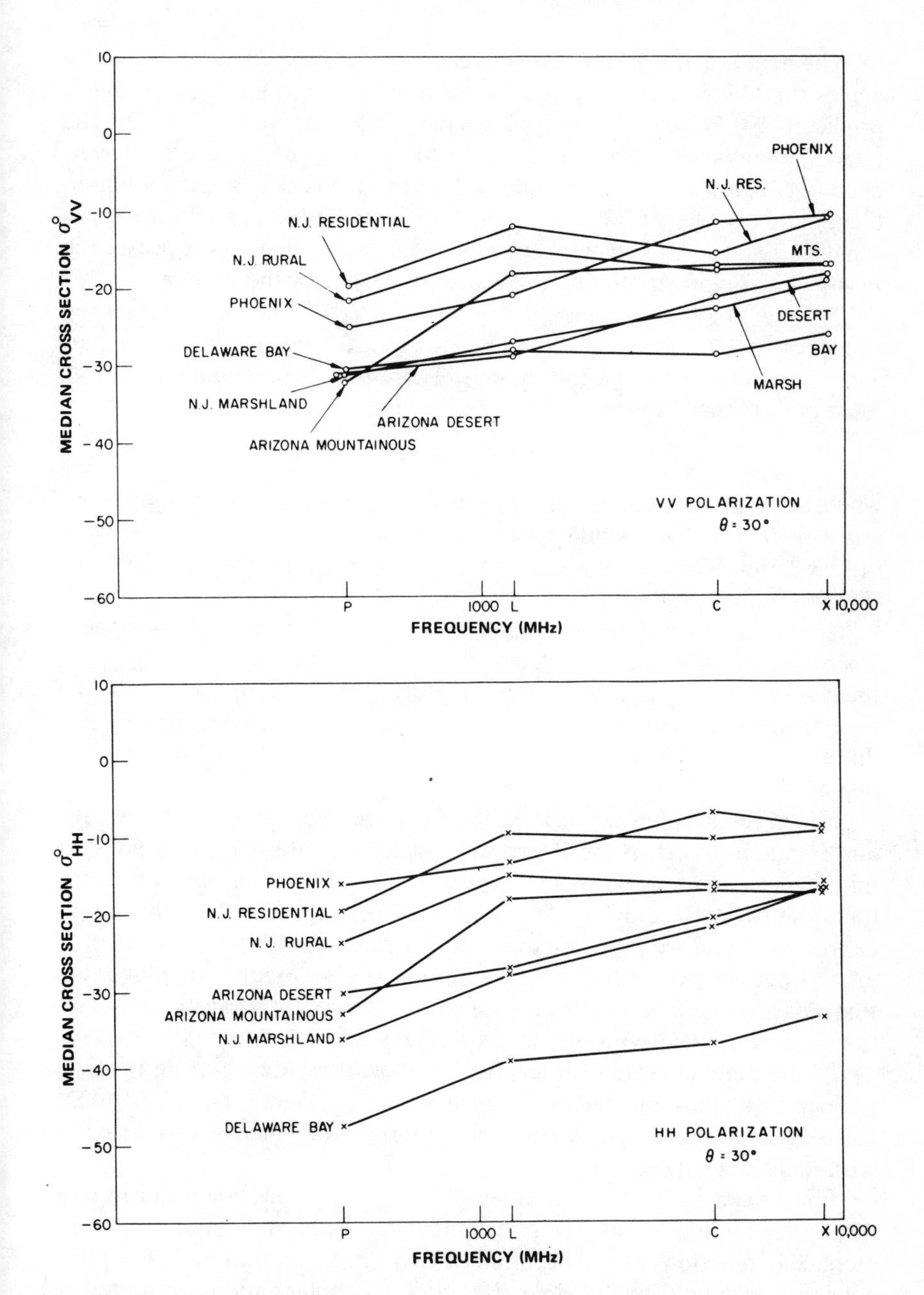

Source: Daley, Davis et al. (1968).

Figure 6-26. Median σ°_{VV} (Upper) and Median σ°_{HH} (Lower) for Various Terrains Versus Wavelength. $\theta = 30°$.

The near grazing or interference region represents a range of incidence angles for which there is a rapid change of σ° with θ. This rapid change is predicted by Peake's theoretical development (see sec. 3.11), but the phenomenon is often accounted for by considering the effects of interference between a direct and a reflected wave (see sec. 6.6). According to theory and experience, σ° may change slowly with θ but can also change as rapidly as $\sin^4 \theta$. The change depends on surface roughness expressed in terms of λ, polarization, and dielectric properties of the surface.

There are few detailed measurements reported for homogeneous terrain at depression angles less than 10°. For the interference effect to be apparent, the surface must appear smooth. Therefore, from equation (6.6), θ_c must be less than is indicated by

$$h_e \sin \theta_c = \lambda / 4\pi$$

where h_e is an equivalent surface roughness. At X band and a depression angle of 30°, a surface would appear smooth if h_e were 1/2 cm or less. With the exception of bodies of water, there are few natural surfaces that satisfy this smoothness criterion.

According to the equation above, roads and runways may appear smooth even for vertical incidence ($\theta = 90°$). Therefore, according to the interference theory σ°_{HH} will change rapidly with θ because phase change for HH polarization is nearly 180° for most surfaces (including water). For those angles (see fig. 4-8) that ρ is close to unity, σ°_{HH} will therefore vary as fast as $\sin^4 \theta$.

For vertical polarization the phase change approaches 180° only for angles much less than the Brewster's angle (a phase change of 90°). At microwaves, Brewster's angle is typically about 20° for earth and 5° to 10° for water (figs. 4-7 and 4-9). Therefore, rapid changes in σ° for changes in depression angle are expected to be less prevalent for VV polarization than for HH polarization at the small depression angles. Figure 6-16 illustrates that measured differences (for the asphalt road data) for these two polarizations are consistent with the discussion above.

In the plateau region the change in σ° with incidence angle is slow, perhaps best approximated by a $\sin \theta$ dependence. Figures 6-16, 6-17, 6-22, and 6-23 include results for *homogeneous* terrain for which σ° varies slowly with incidence angle.

For a radar located so that trees are viewed at broadside or for terrain that is not homogeneous, the σ° versus θ curves are not smooth, slowly increasing functions of θ. The Goodyear Aircraft Corporation (1959) (HH polarization) and Linell (1963) (HH and VV polarizations) reported a minimum in σ° for depression angles between 1° and 5°. This effect may, in fact, be caused by an increase (hump) in σ° caused by nonuniformity of terrain. It seems unlikely that the effect is caused by multipath propagation, because Linell reported identical results for horizontal and vertical polarizations.

Another interesting phenomenon is that of a nearly constant σ° versus θ dependence (a null in the γ versus θ curve) as illustrated in figures 6-12 and 6-13 for vegetation "in head." This suggests that the predominance of echo is from the "heads" or tops; these heads may play the role of an ensemble of isotropic (like spheres) scatterers. Some of the measurements (fig. 6-10) on grass of uniform height also displayed this independence of σ° on θ.

Daley, Davis et al. (1968) reported results of measurements over land, marshes, and the Delaware Bay. The Delaware Bay results help to underscore the differences in σ° between land and smooth sea. Sea results are given in detail in sections 6.13 through 6.21. For the Delaware Bay data, σ°_{VV} exceeds σ°_{HH} by many decibels as is usual for a calm sea; for natural land surfaces, σ°_{HH} is either equal to or slightly greater (as much as 3 dB) than σ°_{VV}; and for residential areas and downtown Phoenix, σ°_{HH} usually exceeds σ°_{VV}, sometimes by as much as 10 dB.

The amount of available data for angles near vertical incidence is limited. The results are especially difficult to interpret for a surface that is smooth with respect to a wavelength, because the backscattering is highly directional. Therefore, the measured values of σ° can be strongly influenced by the beamwidth of the antenna relative to the surface reradiation pattern. Values of σ° for surfaces that are rough compared to a wavelength are less sensitive to antenna beamwidth. Under these conditions it is expected that σ° would be independent of λ were it not for the fact that the dielectric properties of the surfaces are also functions of λ.

6.11 Wavelength Dependence for Land

The totality of experimental results do not yield agreement as to wavelength dependence of σ°. This is partly because the measurements are complicated by the many environmental factors that affect the value of σ°. However, most experimental results indicate that σ° increases inversely with wavelength.

The wavelength dependence of σ° is frequently expressed in terms of λ^{-n} where n is usually between zero and one. In principle the interference effect (sec. 6.6) can cause σ° to vary as fast as λ^{-4} for near grazing incidence angles. Although a change this rapid has been observed for water at small grazing angles, such a fast change has not been reported for terrain.

The highest and lowest operating frequencies used by Cosgriff, Peake, and Taylor (1960) were 35 and 10 GHz. This corresponds to a ratio of 5.4 dB. For vegetation and the plateau region, their reported σ° differed by about 5 dB ($n \approx 1$). For the asphalt parking lot and *HH* polarization (fig.

3-7), measured $\sigma°$ increased 17 dB between X and K_a bands ($n \approx 3$). Figures 6-25 and 6-26 show median $\sigma°$ versus wavelength. Most of the changes in $\sigma°$ between P (428 MHz) and X (9,310 MHz) bands are represented by n between 0 and 1. The most rapid change in $\sigma°$ shown in those figures is for marsh land for which n is nearly 2.

The reader is referred to the discussion in section 6.5 on reflections at near vertical incidence. If the surface is flat and has a smooth "finish" it will act as a mirror (specular reflection). Under these conditions the measured $\sigma°$ will depend not only on the reflection coefficient of the air-surface interface, but also on radar antenna gain and separation distance from the target area. In other words, for smooth flat surfaces a wide range of values of $\sigma°$ can be obtained and these values will not be solely dependent on the target area itself.

For surfaces that are rough compared to wavelength and near vertical incidence, most reports on terrain (see fig. 6-20) and the sea (sec. 6.15) indicate that $\sigma°$ varies inversely with wavelength. For the sea, $\sigma°$ decreases with increases in wind speed for microwaves and is independent of wind speed at 425 MHz. For surfaces that are rough at optical wavelengths, $\sigma°$ is reported to be independent of λ, to increase with rms (root-mean-squared) slope of the surface, and to be relatively independent of rms height (Renau and Collinson 1965).

The magnitude of $\sigma°$ obviously depends on the dielectric properties of the surface, including moisture. The dielectric properties influence the magnitude of the interface reflection between air and the surface and also influence the absorption of the scattering medium. In other words, the scatter problem is made complicated for terrain (as compared to the sea) because electromagnetic waves penetrate the terrain and the internal reradiations can contribute to the echo. Skin depth (distance transversed for power to be 0.135 of its value at the surface) for soil depends on moisture content and wavelength, but for microwaves it is typically in the range of 1 to 10 cm (Ulaby 1974). Therefore, a thin covering material may be penetrated enough that radar scattering can in fact be caused by surfaces not visible to the eye. Clearly the problem of estimating wavelength dependence of $\sigma°$ is highly complicated; for example, moisture content and wavelength both affect *surface* scattering and penetration depth.

6.12 Discussion on $\sigma°$ for Land

Measurements for which the radar beam is fixed in direction often yield values of $\sigma°$ having large differences between successive measurements. Nonhomogenity in terrain will cause fluctuations and inconsistencies in depression angle dependence (see fig. 6-5).

In the case of airborne measurements, the values of σ° versus θ are usually uniform because airborne experiments, by their very nature, have large amounts of averaging. However, two flights over apparently the same type of terrain may at times differ by as much as 10 dB in σ°. The experiments of L.O. Ericson (table 6-3) illustrate this feature. He catalogued σ° averaged over flight segments (1 mile for 5° and 30°, and 2 miles for 1°) and classed each segment according to one of 10 terrain types. Ericson reported that there was not a discernible correlation of average σ° with terrain type over a 500-mile-flight path, and the range of σ° values for each terrain type seems to be approximately equal.

In the Ericson data, tall buildings might be distinguishable because sometimes the averages for σ° are larger than for other classes of terrain. Ericson surmised that the lack of marked differences in the sets of values for the various terrain types was because the averaging was over large areas; but, it might also depend upon various ground conditions because the segments were taken from different runs. The measurements were made during early summer for two years when the trees were in full foliage.

From measurements that are highly repeatable, like those of Cosgriff, Peake, and Taylor (1960), it is known that a slightly different type of surface or type of vegetation can cause a large difference in σ° and in the shape of a σ° versus θ curve.

Different wavelengths are sensitive to different elements on the surface. For example, a surface that might be electromagnetically smooth at one wavelength can be rough at another. At near vertical incidence, the terms horizontal and vertical polarizations are meaningless and for most terrain σ° is independent of the orientation of a linearly polarized wave. However, at vertical incidence, "horizontally" polarized waves will be more strongly reflected by "horizontal" wires and rails than will "vertically" polarized waves. In general, there are differences between the σ° versus θ curves for horizontal and vertical polarizations.

If the shape, size, and roughness of two targets were the same, σ° would be greatest for the target with the largest dielectric constant. This distinction is usually difficult to document for natural targets because identical geometries with differing dielectric properties are rare. However, grass (fixed height) has a greater σ° during summer than fall. Electromagnetically smooth surfaces tend to reflect waves in preferred directions, while rough surfaces cause waves to be reradiated in all directions. A freshly plowed field is obviously rougher before a rain than afterwards. However, the dielectric constant of soil is strongly influenced by moisture content because the dielectric constant of water is large compared with that of dry earth. How much, then, is σ° affected by changes in roughness and how much by changes in dielectric constant? The problem is compounded when considering changes in wavelength because dielectric constant as well as surface roughness are functions of wavelength.

Pieter Hoekstra and Dennis Spanogle (1972) measured σ° for natural snow surfaces at 10 and 35 GHz using grazing angles between 0.3° and 1°. One interesting result is that at 10 GHz σ°_{HH} exceeded σ°_{VV} by about 10 dB. The difference was smaller for wet snow.

A knowledge of soil moisture level is important for various activities including flood forecasting, growing crops, and wildlife management. There have been a number of experimental results that suggest that radar offers potential for estimating soil moisture over large areas within a short time period. For example, observable effects of moisture were mentioned in section 6.8 in the discussions of the work by C.R. Grant and B.S. Yaplee; Cosgriff, Peake, and Taylor; Goodyear Aircraft Corporation; and Linell. More recent investigations directed specifically toward detecting moisture content of terrain have been conducted at the University of Kansas (Ulaby, 1974; Ulaby, 1975; Ulaby, Cihler, and Moore 1974; Waite and MacDonald 1971). The more recent measurements corroborate the early findings that σ° is highly dependent on surface roughness, moisture, microwave frequency and incidence angle. The parameters used by Ulaby (1974) included a 4-8 GHz frequency range; *HH*, *VV*, and *VH* polarizations; and grazing angle from 20° to 90° (nadir). Two plowed bare fields (rms roughness of 2.5 cm and 5.5 cm) were used and moisture content was measured in the top 5 cm of soil. For the surface with 2.5 cm roughness, viewed as "typical" of farming, Ulaby, Cihler, and Moore (1974) report a study of the correlation of σ° with moisture at 4.7, 5.9, and 7.1 GHz. A reasonably high correlation of σ° with calculated mean attenuation coefficient (over a distance of one skin depth below the surface) was reported. The highest correlations obtained in this way for σ° with moisture were for angles within 10° of vertical and for the lowest (4.7 GHz) frequency.

Research has also been under way for a number of years in an effort to develop techniques for mapping soil moisture through vegetation and identifying crop types (Schwarz and Caspall 1968; Ulaby 1975; Ulaby et al. 1972). The early results of the Goodyear Aircraft Corporation on irrigated farmland were mentioned in section 6.8. Ulaby (1975) has reported recent results obtained for corn, milo, soybeans, and alfalfa over the 4-8 GHz frequency range for θ between 20° and 90°. Soil moisture determination was best achieved with the lower frequencies, *HH* polarization and angles near vertical. Ulaby (1975) identifies a number of radar parameters for optimizing crop discrimination, including *VV* polarization and small depression angles.

Clearly it is not possible to precisely describe the quantitative effects of surface material, shape, and roughness on σ°, nor to uniquely relate values of σ° to terrain type or condition. However, figure 6-27 illustrates that

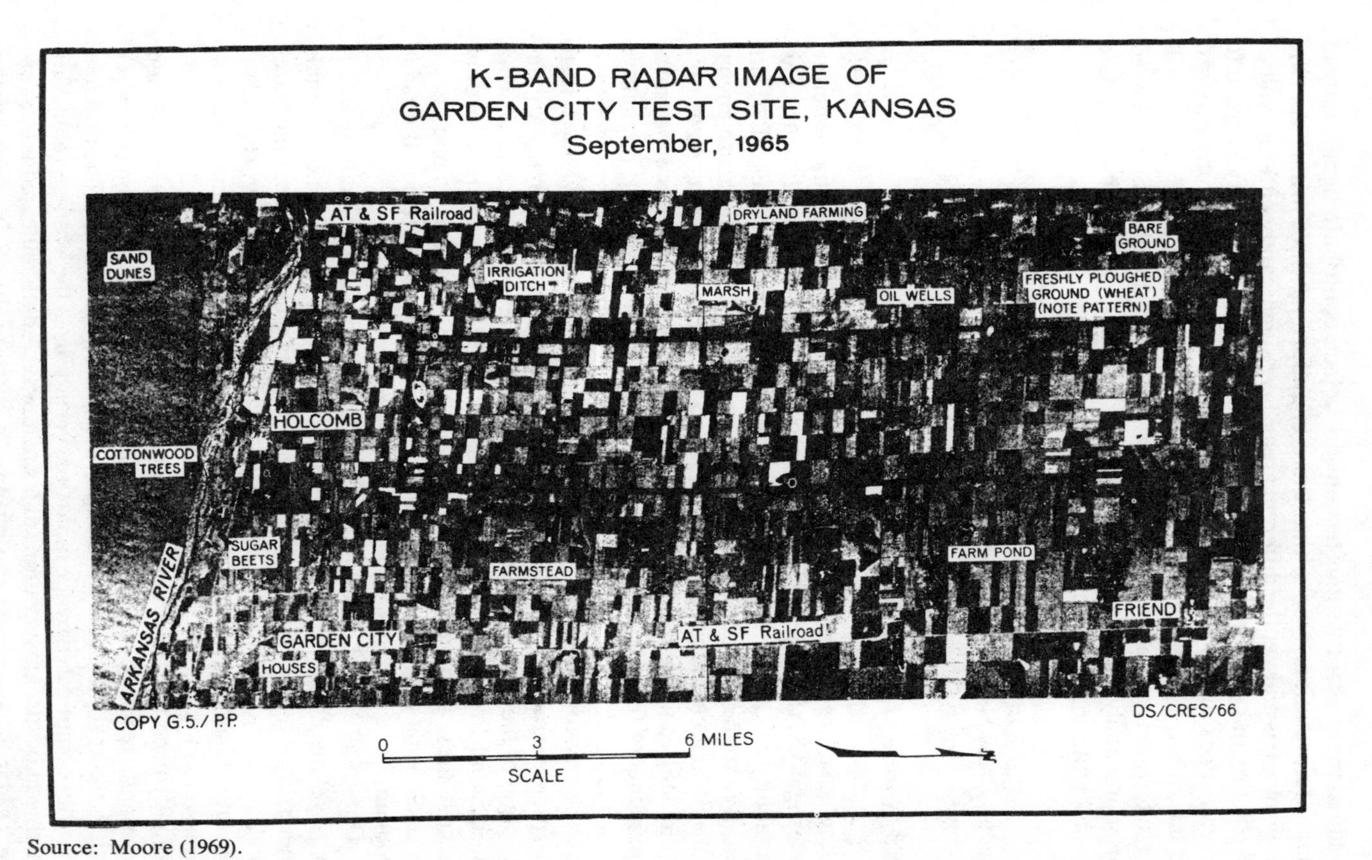

Source: Moore (1969).

Figure 6-27. A Side-looking Radar Display of Garden City Test Site.

graduations in σ° for various terrains do, in fact, contribute significantly to map interpretation. Notice that there is uniformity in σ° within specific types of areas and yet the different areas have significantly different values of σ°. For this site, it has been possible by radar to classify crops into several types or groups (Moore 1969; Skolnik 1970 chap. 25 by Moore). Although there have been several promising studies on classifying according to terrain type and moisture content, the results mainly affirm the need for more analytical and experimental studies supported by adequate ground truth information.

Radar Cross Section for the Sea

6.13 Nature of σ° for the Sea

Although a large quantity of sea-echo data has been reported by numerous investigators, the wide variability of the data makes analysis difficult. This variability is due to a number of reasons. Measured values of σ° are sensitive to the character of the sea surface (whose variations are uncontrollable and can occur quite rapidly) and to variations among measuring equipments and their calibration. In order to understand sea echo, measurements of the sea surface are required in addition to measurements of the sea echo. Problems in making all of these measurements have tended to obscure any correlation which may exist between σ° and the various radar parameters and sea conditions. Usually, even the most rudimentary description of the sea surface conditions (wave height, wave direction, wind speed, wind direction, or a qualitative description) is not reported, not only because these measurements are difficult but also because no one knows precisely what characteristics of the sea surface are important to the radar problem.

Obtaining a valid description of the dependence of radar cross section on wavelength is difficult because one must make measurements simultaneously with at least two radars, while making quantitative sea measurements. The radars must be accurately calibrated; preferably, redundant absolute calibrations should be available for the two or more radars. However, most of the data are not of this type. Since the descriptions ordinally used for the sea are very crude, it is virtually hopeless to obtain meaningful results by comparing data at one wavelength from one observer with those at some other wavelength from another observer. The problem is further complicated by the rapid time variations in the average value of σ° which can occur. Observations at a single wavelength indicate that "average" radar cross section can change as much as 10 dB in a one-minute interval. Therefore, the measurement errors and uncertainties tend to obscure the weak functional relationships that exist between σ° and λ for most incidence angles.

In spite of the fact that few of the existing data are of the desirable quality, the data do indicate trends concerning the dependence of σ° on wavelength. This dependence cannot be stated as a simple functional relationship between σ° and λ because the dependence varies with both the sea state and the angle of arrival. For example, the dependence on angle of arrival is quite noticeable for the small grazing angles below the critical angle (θ_c in fig. 6-1) where the interference effect makes σ° a strong function of wavelength. At the larger incidence angles near vertical incidence, the value of σ° is highly dependent on sea state and actually, for microwaves, decreases as the roughness of the sea increases. Reviews of sea echo are included in D.E. Kerr (1951), D.J. Povejsil, R.S. Raven, and Peter Waterman (1961, chap. 4), Petr Beckmann and André Spizzichino (1963), and M.I. Skolnik (1970).

Various opinions have been expressed regarding the dependence of σ° on θ in the plateau region (which for microwave extends at least from 5° to 50°), but this dependence of σ° on grazing angle is "slight". It is reasonably well established that for the plateau region the slope of σ° versus θ depends on sea roughness, but σ° tends to be independent of θ for rough sea conditions. Therefore, the plateau region is particularly attractive for the study of wavelength dependence because the effect of θ, one of the many variables, is at a minimum.

6.14 Range Dependence at Small Incidence Angles

Average echo power for a homogeneous, rough sea has been reported to vary as R^{-3} and R^{-7} at close and long ranges, respectively. Theoretically, the R^{-3} dependence is expected for a collection of scatterers that fill an antenna beam; the increase in range dependence to R^{-7} has been attributed to an interference effect between direct and reflected electromagnetic waves.

Figure 6-2 typifies data reported by Martin Katzin (1957) on range dependence as a function of incidence angle (range for a given antenna height) for wavelengths of 3.2, 9.1, and 24 cm. The data are for horizontal polarization and six altitudes from 200 to 10,000 feet. Katzin observed two distinct zones where the range of the common boundary area depends on λ: (1) short range for which echo power varies as R^{-3}, and (2) long range for which the power varies as R^{-7}. Illuminated area A is linearly proportional to range (constant beamwidth); therefore, σ° for the short- and long-range zones is constant and varies as R^{-4}, respectively.

Katzin reported (Povejsil, Raven, and Waterman 1961) that limited experimental data suggest the relationship

$$R_c \approx 2Hh_{1/10}/\lambda \tag{6.7}$$

where $h_{1/10}$ is the peak-to-trough height exceeded by one-tenth of the waves (sec. 2.11). His paper included data for which the R^{-3} zone extended to ranges as short as $R_c/4$. Since $\sin\theta$ equals H/R, it may be seen from equation (6.7) that $h_{1/10}\sin\theta$ equals 2λ at a range $R_c/4$; therefore, according to the Rayleigh roughness criterion, the surface is very rough at the short ranges.

There have been numerous unpublished studies on the dependence of σ° on range, but to the author's knowledge V. Müller (1966) is the only person other than Katzin who has systematically examined R_c versus wave height. Müller analyzed substantial data on average value of sea echo power versus range taken from a lightship in the North Sea 30 km off the northwest coast of Germany. The antenna was mounted at a height of 23 meters, polarization was horizontal, transmitter wavelength was 3.2 cm, and measurements were made over a 300- to 7,000-meter range. Müller found R_c to be independent of antenna direction; and the "slopes" for the R^{-3} region were independent of wave height or wind, while the "slopes" obtained for the R^{-7} region were reported to have irregular variations as a function of wave height and wind. Measured range dependence is seldom precisely R^{-3} or R^{-7} (see fig. 6-28 (b)).

Müller determined wave height from stereophotographs. For each wave-height determination, 10 pairs of photographs were taken. From these pictures peak-to-trough height was obtained for those waves that were very high. With this procedure, Müller got 15 to 20 peak-to-trough values and the mean of these very large heights was used to represent wave height. Through personal communications, Müller expressed the opinion that his wave-height determinations are closer to $h_{1/10}$ than to $h_{1/3}$ (see sec. 2.11).

Müller reported measurement results on the dependence of R_c on wave height for peak-to-trough heights of 1/3 to 3 meters, and from his data it is estimated that R_c equals 1,700 $h_{1/10}$, where $h_{1/10}$ represents peak-to-trough wave height determined from stereophotographs. Therefore, by using the parameters of equation (6.7) with Müller's data,

$$R_c \approx 2.5\ (Hh_{1/10}/\lambda) \tag{6.8}$$

This result for R_c is in surprisingly close agreement with equation (6.7) when considering the variability of sea echo, differences in measurement techniques, and difficulties in determining wave height.

Equation (6.6) can be used to express R_c in terms of an effective height h_e of the scatterers above an effective plane surface—the relatively few waves that exceed some average height play the role of targets. Therefore, since $\sin \theta_c$ is H/R_c (flat earth approximation with $H >> h_e$),

$$R_c = 4\pi \; (Hh_e/\lambda) \tag{6.9}$$

Equation (6.9) can be used with equations (6.7) and (6.8) to determine the "effective" height of sea scatterers above the "effective" reflecting plane of the sea. An important assumption used to derive equation (6.9) is that the reflection coefficient for forward scattering from the sea is minus one. This assumption is valid for horizontal polarization, but is invalid for vertical polarization at frequencies below the microwave range.

Significant wave height $h_{1/3}$ is used more often than $h_{1/10}$ to describe wave height. The standard deviation of surface height versus time (fixed position) and rms wave height provide other convenient parameters for specifying wave height. It has been estimated (sec. 2.11) that $h_{1/3}$ and $h_{1/10}$ are, respectively, four and five times the standard deviation of wave height or alternatively, four and five times the rms wave height. Therefore, by comparing the experimentally obtained results (eqs. (6.7) and (6.8)) with the theoretically determined value for R_c (eq. 6.9), we find that the effective wave height h_e is approximately equal to the standard deviation of wave height or, alternatively, the rms wave height. Although the scattering model used represents an oversimplification of the actual sea scattering process, the apparent relationship of transition range to wave height is useful for providing limiting performance criteria for the radar designer.

From experimental results it is known that the ratio $\sigma^\circ_{VV}/\sigma^\circ_{HH}$ depends on depression angle, wavelength, and sea state, but it is generally recognized that σ°_{VV} usually exceeds σ°_{HH}. The ratio decreases as the sea becomes rougher, and it has been observed (for small depression angles) that σ°_{HH} sometimes exceeds σ°_{VV} by a few decibels. Because of differences in the forward scattering reflection coefficients of the sea for *HH* and *VV* polarizations, it has sometimes been erroneously argued that the "classical" interference effect described here will account for differences between σ°_{VV} and σ°_{HH} in the plateau (R^{-3}) region. As already discussed, the effective reflecting plane representing the sea appears rough in the R^{-3} zone and, therefore, has negligible effect on σ°. In other words, the assertion here is that σ°_{VV} and σ°_{HH} would be equal for microwaves if the sea scatterers, per se, were insensitive to polarization.

Clearly distinguishable R^{-3} and R^{-7} zones are not always observed. In fact, E.R. Flynt et al. (1967) reported that on many days range dependence intermediate between R^{-3} and R^{-7} was observed, and frequently there was no distinct transition between two different regions of range dependence. However Flynt et al. did report that on several occasions sea echo varied with range nearly as R^{-3} at short range and as R^{-7} at longer ranges; two such curves are given in figure 6-28(a) for different sea conditions. The upper

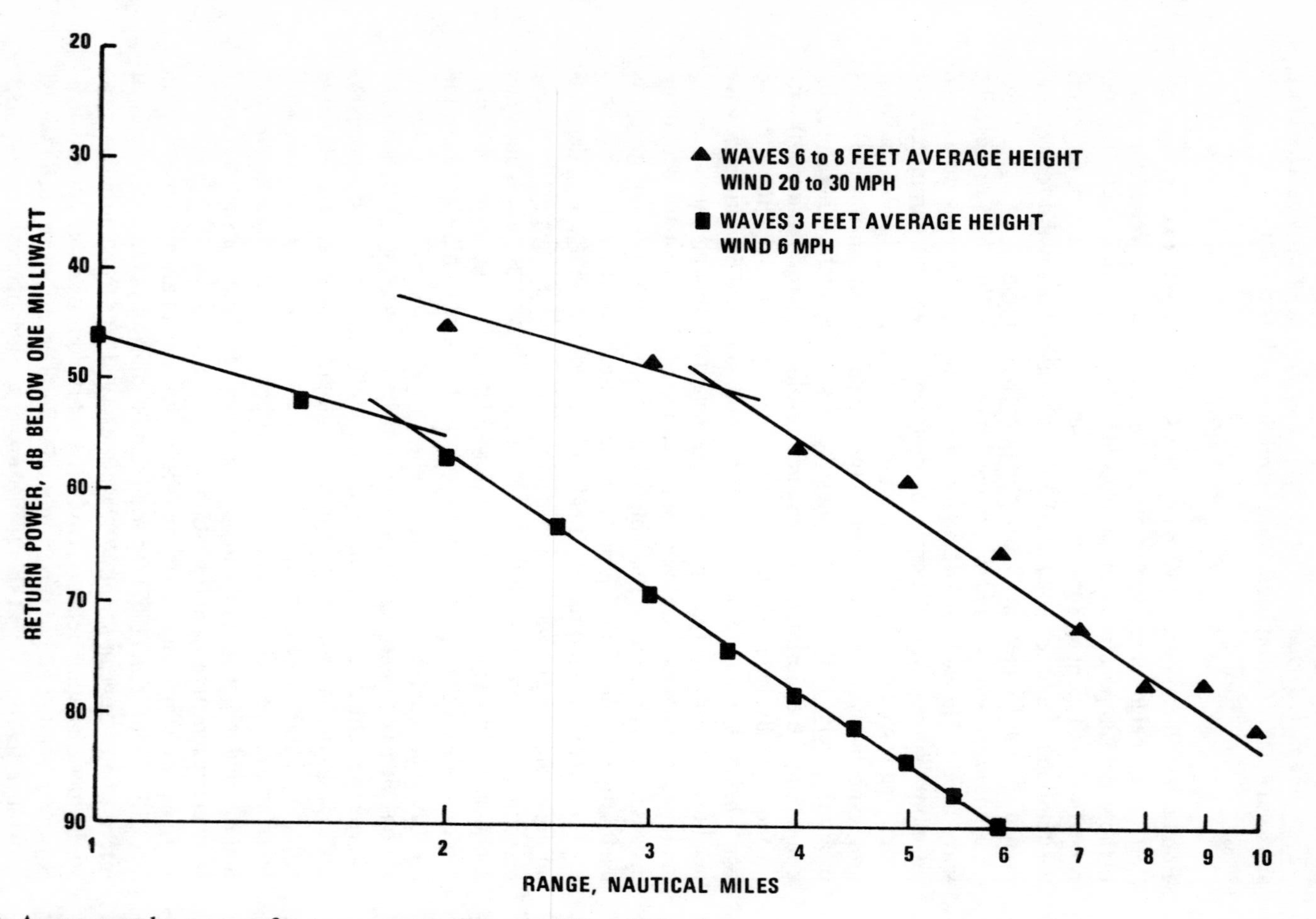

(a) Average echo power for two sea conditions. *HH* polarization.

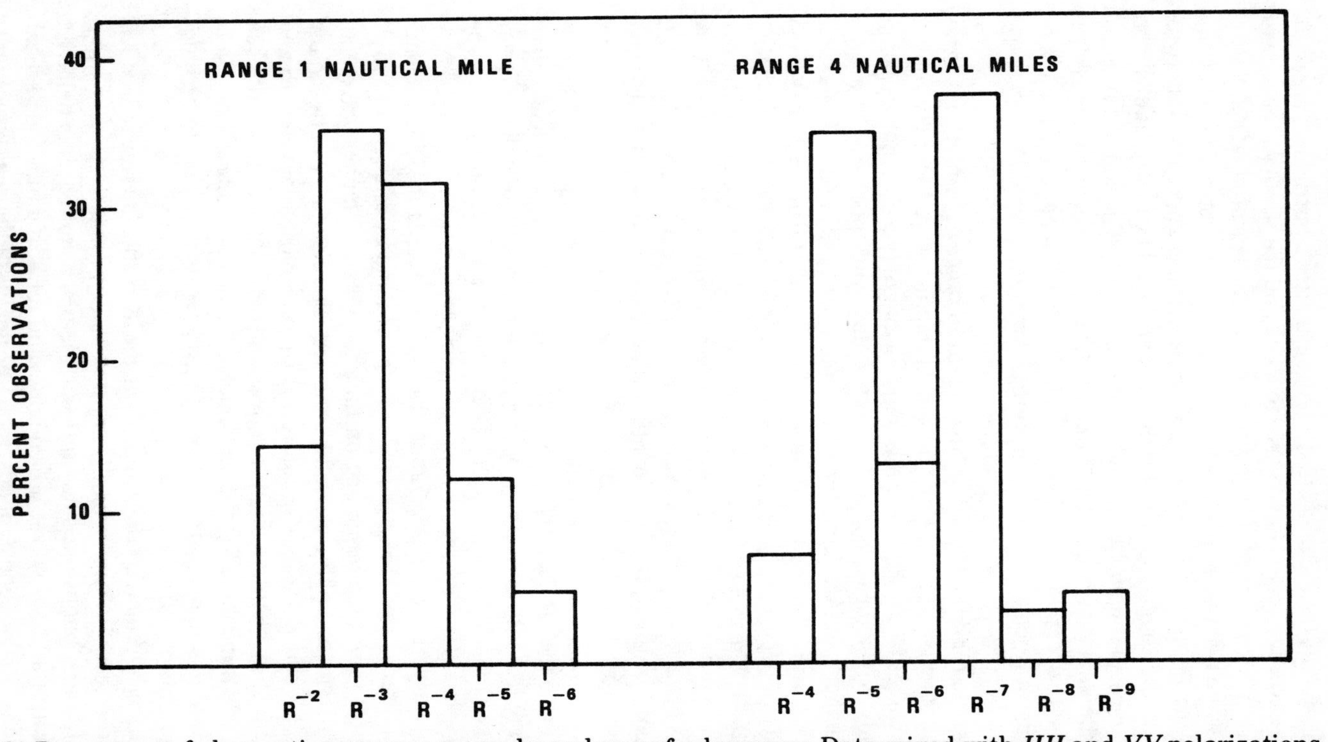

(b) Percentage of observations versus range dependence of echo power. Data mixed with *HH* and *VV* polarizations.

Sources: (a) Flynt et al. (1967); (b) Dyer and Currie (1974).

Figure 6-28. Range Dependence of Sea Echo. *X* band, antenna height 80 feet.

points (fig. 6-28(a)) were measured a few days after a hurricane passed the radar site; average wave height was 6 to 8 feet and the wind was gusty with speeds from 20 to 30 mph. The lower points were measured on a day when average wave height was 3 feet and wind speed was 6 mph. The transition between the R^{-3} and R^{-7} regions occurred at a range approximately twice as great on the rougher day as on the calmer day. The observations were based on measurements being made twice daily over a period of 5 months, at half-mile intervals beginning at one nautical mile and ending where sea echo strength dropped to receiver noise level. The radar operated at X band on the coast at Boca Raton, Florida, with an antenna elevated 80 feet above the sea.

One cause for deviations from the idealized range dependence is atmospheric ducting (sec. 2.12). For example, if ducting were to totally confine the $e-m$ waves to a region near the water so that only azimuthal spreading of the fields exists, the classic R^{-3} and R^{-7} zones would become R^{-1} and R^{-5} zones. Further measurements have been made at the site and with the same general operating parameters used by Flynt et al. Figure 6-28(b) shows probability of occurrences of range slope reported by F.B. Dyer and N.C. Currie (1974). The measurements were made on an average of four times a day over a 300-day period. Therefore, the graph represents a composite of 1,200 runs of various sea and weather conditions. The data are for both HH and VV polarizations; no significant differences were observed for range dependence of HH and VV polarizations. At the 1-mile range, the histogram clusters near an R^{-3} dependence. However, the histogram is distinctly bimodal in shape (R^{-5} and R^{-7}) for the 4-mile range. Dyer and Currie attribute the R^{-5} data to ducting. Ducting should produce a smaller effect at short ranges than at longer ranges. Therefore, the assertion by Dyer and Currie that the R^{-5} data result from ducting seems correct.

In summary, the R^{-3} and R^{-7} zones can be accounted for by interference between a direct "ray" from the radar to the sea and by an indirect "ray" reflecting off (and then impinging on the relatively few higher waves) an equivalent reflecting plane caused by the sea. The boundary between the idealized R^{-3} and R^{-7} zones is controlled by transmitted wavelength and wave height. For microwaves, the reflection coefficient for a smooth sea will be nearly minus one for vertical and for horizontal polarization if θ is much less than θ_c. Therefore, for microwaves, R_c is expected to be independent of polarization.

By comparing equation (6.6) for R_c with the available microwave data, it seems that effective wave height is approximately equal to the rms wave height. Although published data on this subject are limited, it is believed that the results at microwaves are equally valid for horizontal and vertical polarizations. Radar echo depends upon fine details of the sea surface and, therefore, it depends upon wind speed and wave height at a given range. In spite of this, experimental results reveal a reasonably consistent dependency of transition range on wave height and transmitter wavelength. Clearly, ducting can strongly influence the range dependence of sea echo.

6.15 Dependence on Incidence Angle

Experience has shown that the dependence of σ° on incidence angle is of the basic form shown in figure 6-1. There are three general domains of variation of σ° as a function of θ, and the magnitude of the variation in these domains is a function of polarization and environmental conditions. The case of small grazing angles was studied extensively during World War II (Davies and Macfarlane 1946; Kerr 1951). Beckmann and Spizzichino (1963) have provided a review of the small grazing angle and near vertical incidence angle domains. Other general references include M.W. Long et al. (1965), J.C. Daley, J. T. Ransone et al. (1968), M.I. Skolnik (1970), and N.W. Guinard and J.C. Daley (1970).

Because of the interdependence of range and θ, ducting can obviously also influence the measured θ dependence of σ°. Figure 6-29 shows vastly different shapes in σ° versus θ that were obtained by Dyer and Currie (1974) while acquiring the data of figure 6-28(b). The data were obtained with a fixed antenna height and therefore the actual measurement was of σ° versus range. One curve was obtained in the morning (ducting) and the other in the afternoon (nonducting) with measured wave height and wind speed essentially equal. Figures 6-28 and 6-29 help to illustrate that scattering data taken over water must be interpreted with caution.

1. Near Grazing Incidence. The rapid change of slope of the $10\log_{10}\sigma^\circ$ versus $\log\theta$ curves for incidence angles less than the "critical angle" (θ_c in fig. 6-1) can be accounted for on the basis of the interference pattern formed by the difference of path length between the direct ray to the scattering element and the ray reflected from the surface of the sea (see sec. 6.6).

Figure 6-2 shows X-band results of Martin Katzin (1957) that illustrate the two regions: σ° independent of range (R^{-3} region) and σ° proportional to $\sin^4\theta$ (R^{-7} region). Since $\sin\theta$ equals H/R (antenna height divided by range), $\sin\theta$ is often approximately θ for the region of significant multipath interference (see sec. 6.4). For the conditions existing in figure 6-2, the wave height $h_{1/10}$ is about 2 feet; therefore, θ_c is about one degree.

The interference effect can also be used to explain the strong dependence of σ° on wavelength for grazing angles less than the critical angle (see sec. 6.6). Much of the reported microwave data (see sec. 6.16 for longer wavelengths) indicate that σ° varies about as θ^4 in this region. Such behavior agrees with the predictions of the interference theory, provided the

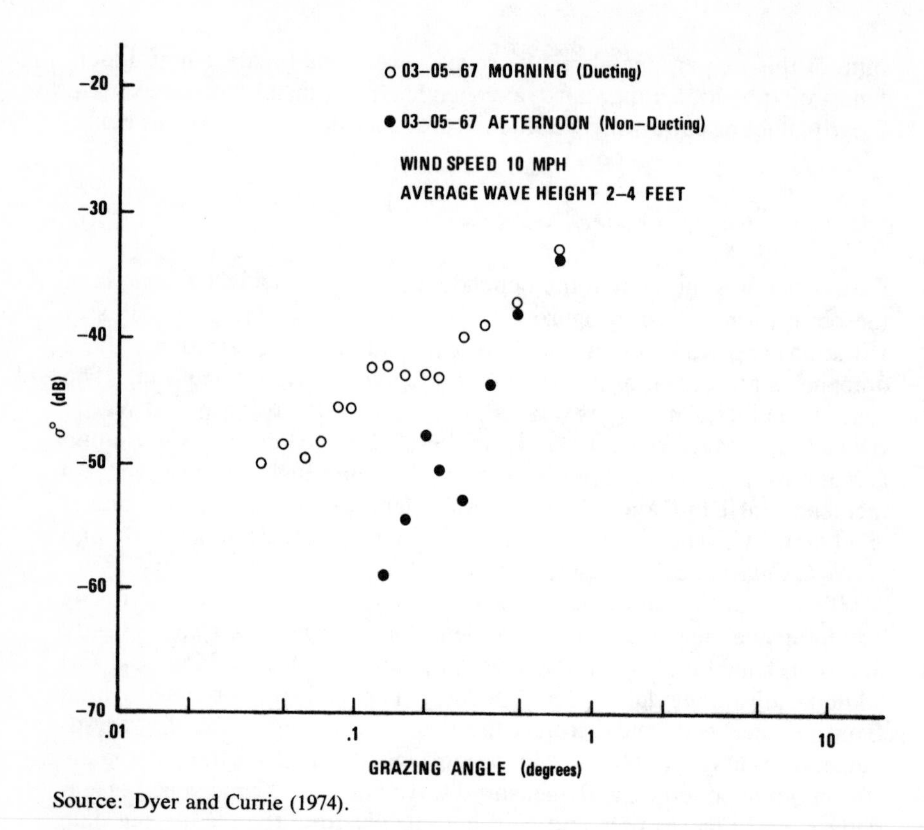

Source: Dyer and Currie (1974).

Figure 6-29. Comparisons of σ°_{HH} Made Under Ducting and Nonducting Conditions. Adapted from Dyer and Currie (1974).

radar cross section of the individual scatterers for a constant illumination amplitude is independent of θ. Most of these data also indicate that for grazing angles below the critical angle, σ° increases rapidly as λ decreases. In some cases σ° has been observed to vary as rapidly as λ^{-4}. Since σ° is expected to be proportional to $|F|^4$ (sec. 4.4), the interference theory predicts a λ^{-4} dependency, provided the radar cross section of the individual scatterers for a constant illumination amplitude is independent of λ.

2. Plateau Region. Various opinions (Davies and Macfarlane 1946; Guinard and Daley 1970; Katz 1963; Kerr 1951) have been expressed regarding the interpretation of measurements on the angular dependence of σ° in the plateau region, but it is generally concluded that the dependence of σ° on grazing angle is "slight". It has been reasonably well established that the slope of σ° versus θ depends on sea roughness, and that σ° tends to be independent of sea state for rougher sea conditions. A discussion on correlating experimental results with theoretical models is included in section 6.21.

Figures 6-30 and 6-31 show a collection of experimental results obtained for σ°_{VV} versus θ for frequencies from 1.2 GHz to 35 GHz and for wind speeds of 10 knots and greater. Note that when these data are considered as a whole, there is good agreement in the slopes of 10 log σ°_{VV} versus θ obtained by individual observers. In addition, the differences between the median of all data presented and the absolute values obtained by various investigators are not substantially greater than the differences obtained on several runs by one observer. Thus, it is apparent that the slope of σ°_{VV} versus θ is independent of wavelength for many purposes; also these data suggest no strong correlation of σ°_{VV} with wavelength for any specific value of θ within the plateau region. The medians of the data shown in figures 6-30 and 6-31 are essentially straight lines between 5° and 50° and have slopes of roughly 1/4 dB per degree. Thus, even though the variation of σ° versus θ in the plateau region is small, there is still a significant increase of approximately 10 dB when the grazing angle is increased from 5° to 45°.

Figure 6-32 shows a collection of results obtained for σ°_{HH} versus θ at wind speeds of 10 knots and greater. Fewer σ°_{HH} data than σ°_{VV} data were available; the σ°_{HH} curves are included for completeness and should not be used to compare σ°_{HH} with σ°_{VV} of figures 6-30 and 6-31. Relative measurements provide a much more accurate comparison of σ°_{VV} with σ°_{HH} since most calibration errors do not affect the ratio $\sigma^\circ_{VV}/\sigma^\circ_{HH}$, whereas figures 6-30, 6-31, and 6-32 show absolute data taken at different times by different experimenters. Results from studies of the ratio $\sigma^\circ_{VV}/\sigma^\circ_{HH}$ are included in section 6.17.

N.W. Guinard and J.C. Daley (1970) have reported measurements on σ° taken simultaneously at *X*, *C*, *L*, and *P* bands for a wide range of sea conditions. The values reported are in terms of median value of σ°, and the difference between a median and an average value is expected to be 3 dB or less (see sec. 6.2). Figures 6-33 through 6-36 show some of the experimental results obtained at *X* and *P* bands for which there is roughly a 20:1 ratio in radar wavelength. Notice that the results are in general agreement with the data reported earlier and given in figures 6-30 through 6-32. The solid lines are from theoretical calculations, and these results are discussed further in following sections.

3. Near Vertical Incidence. A substantial amount of work has been done at microwave frequencies to measure σ° at large incidence angles. Pertinent references include Petr Beckmann and André Spizzichino (1963), E.W. Cowan (1946), G.S.R. Maclusky and H. Davies (1945), J.C. Wiltse, S.P. Schlesinger, and C.M. Johnson (1957), F.C. Macdonald (1956), C.R. Grant and B.G. Yaplee (1957), J.P. Campbell (1959), A.R. Edison, R.K. Moore,

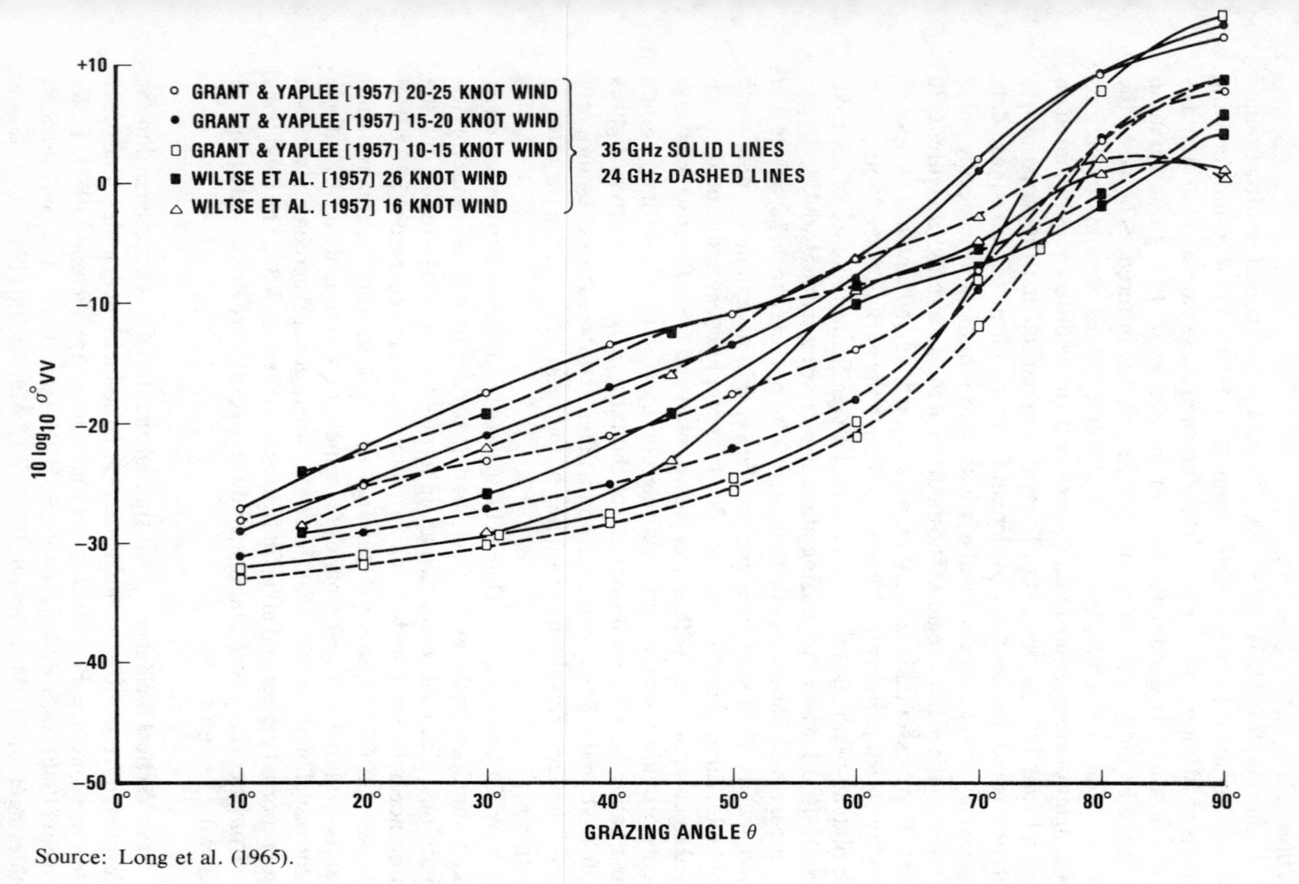

Source: Long et al. (1965).

Figure 6-30. Measured Values of σ°_{VV} at K and K_a Bands.

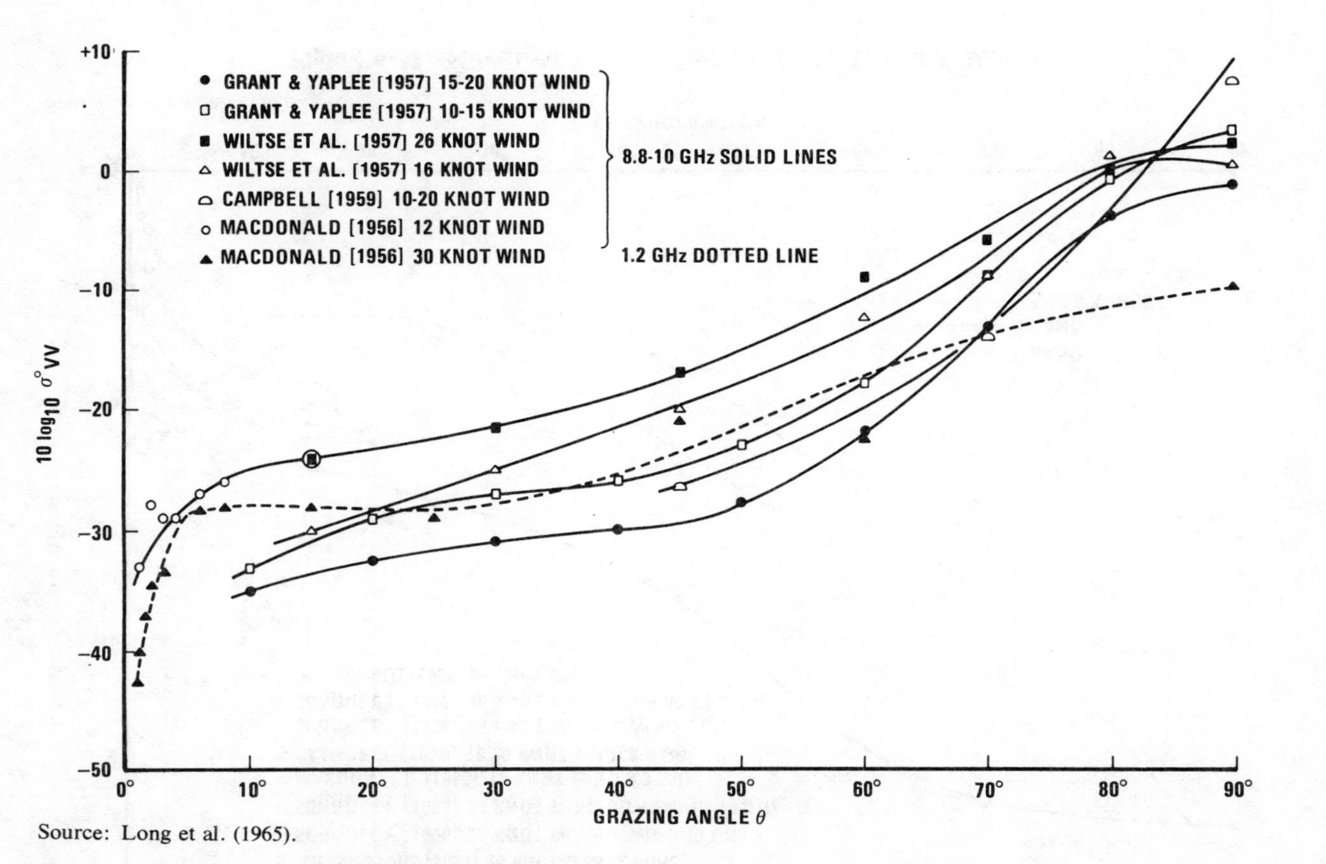

Source: Long et al. (1965).

Figure 6-31. Measured Values of σ°_{VV} at L and X Bands.

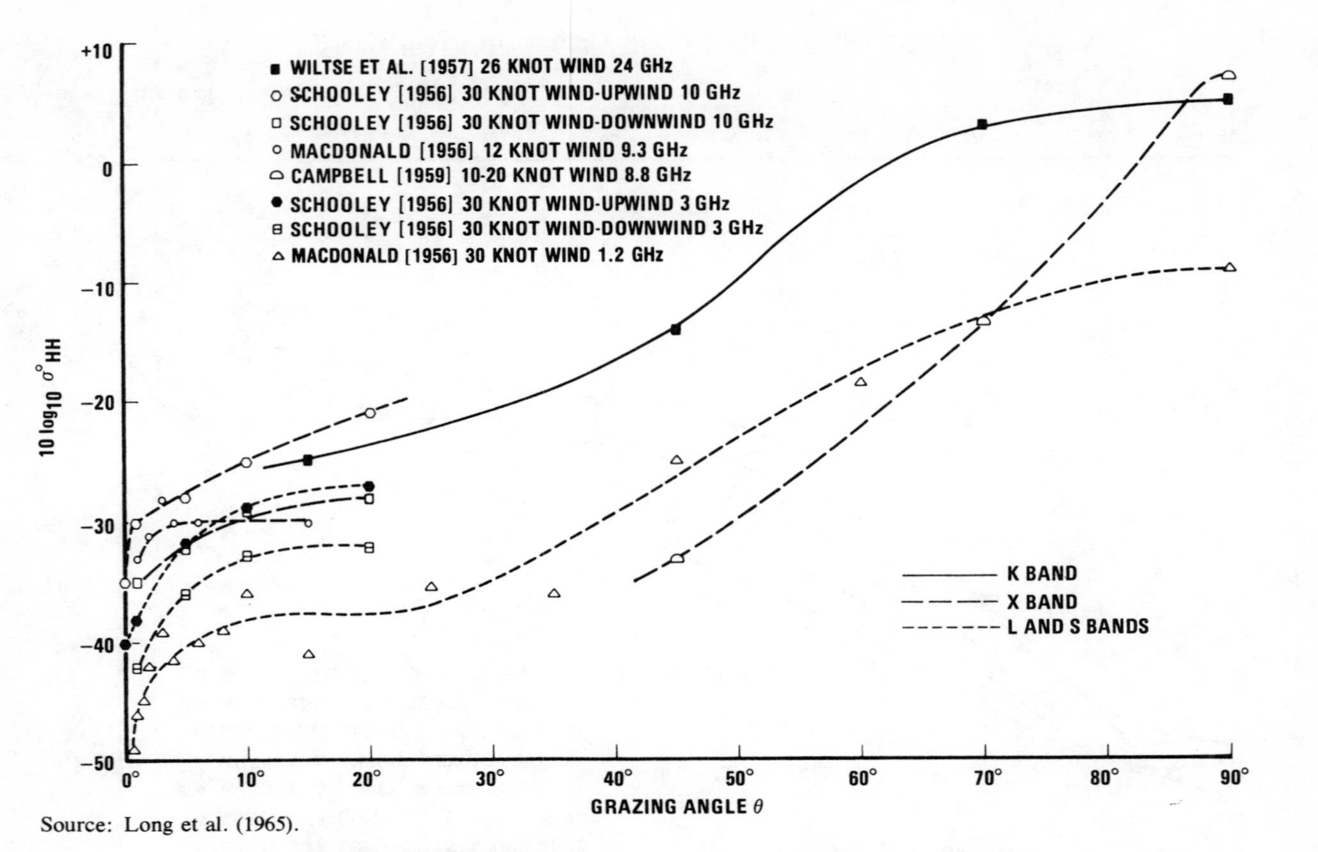

Source: Long et al. (1965).

Figure 6-32. Measured Values of σ°_{HH} at L, S, X, and K Bands.

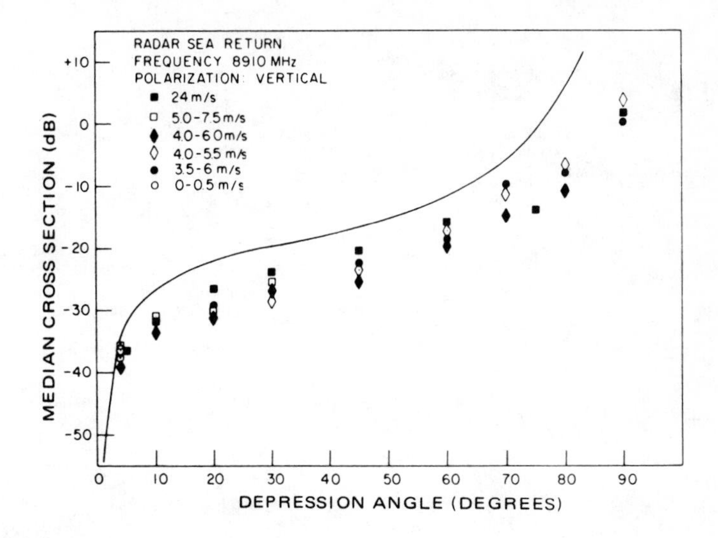

Source: Guinard and Daley (1970).

Figure 6-33. The Variation of Median σ°_{VV} with Grazing Angle and Wind Speed, *X* Band.

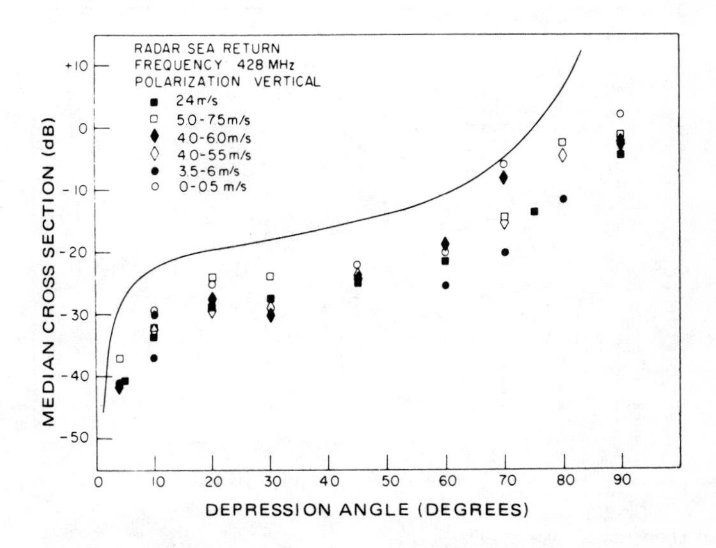

Source: Guinard and Daley (1970).

Figure 6-34. The Variation of Median σ°_{VV} with Grazing Angle and Wind Speed, *P* Band.

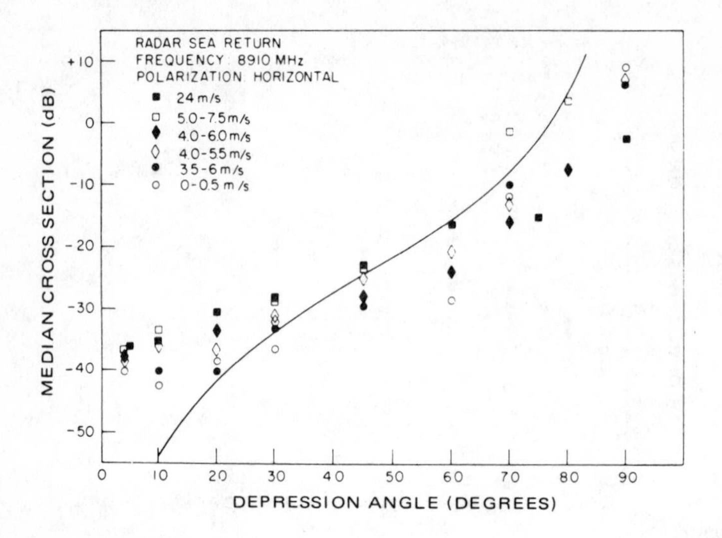

Source: Guinard and Daley (1970).

Figure 6-35. The Variation of Median σ°_{HH} with Grazing Angle and Wind Speed, *X* Band.

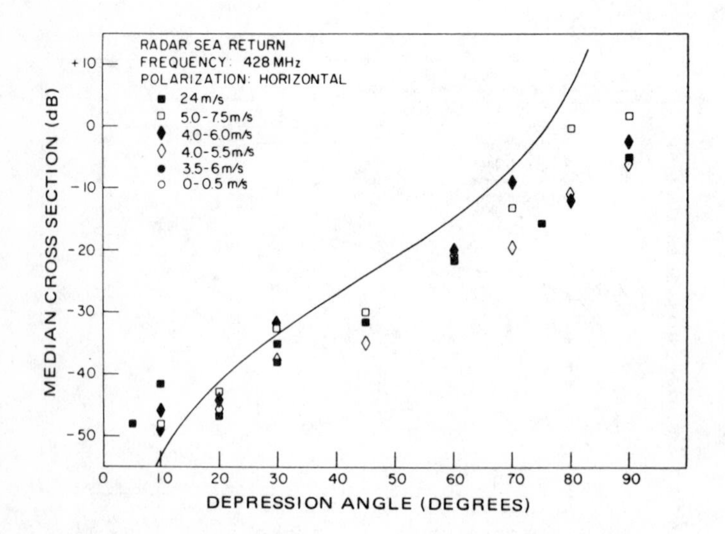

Source: Guinard and Daley (1970).

Figure 6-36. The Variation of Median σ°_{HH} with Grazing Angle and Wind Speed, *P* Band.

and B.D. Warner (1960), C.S. Williams, Jr., C.H. Bidwell, and D.M. Bragg (1960), and J.C. Daley, J.T. Ransone, Jr., and W.T. Davis (1973). Section 6.18 includes values of σ° as a function of wind speed that were reported by Daley, Ransone, and Davis (1973) for angles near vertical incidence.

For θ greater than θ_0 (see fig. 6.1) and for frequencies in the microwave region:

1. σ° increases with radar frequency increases between 428 MHz and X band.
2. At microwaves, σ° decreases with wind speed (i.e., surface roughness) increases.
3. σ° seems to be independent of wind speed at 428 MHz.
4. σ° is slightly larger for the upwind direction than for other directions.

6.16 Incidence Angle Dependence at Low Frequencies

For microwaves, the only strong frequency dependence is associated with the interference effect which determines the location of θ_c and makes σ° a strong function of λ for angles less than θ_c. For a given sea state, θ_c is proportional to the wavelength. For horizontal polarization, the curve of σ° versus θ for low frequencies is expected to be similar to that for microwaves, except the critical angle will be larger. The region of θ^4 dependence is thus elongated and the plateau region shortened. The curves for horizontal polarization thus might appear as in figure 6-37. For example, suppose transmitter wavelength is 136 cm (220 MHz) and h_e is 1 foot (corresponds to a peak-to-trough wave height of 4 or 5 feet), then from equation (6.6) θ_c is 21°. It is quite possible that at frequencies of a few megahertz θ_c may increase so that the plateau region disappears entirely, giving a curve like that marked *HF* in figure 6-37.

For vertical polarization, σ° would not be expected to vary as θ^4 because the reflection coefficient does not approximate minus one over 0 to 14°. The graph of σ° versus θ is expected to resemble that for horizontal polarization in some respects and to differ in other respects. They will be alike because the graph for vertical polarization will also consist of three similar regions. It is expected that they will differ in three important respects: (1) σ°_{VV} will be larger than σ°_{HH} for most grazing angles, (2) the critical angle will not be proportional to wavelength, and (3) the functional dependence of σ° on θ near grazing will not be proportional to θ^4. In order to present the basis for these statements, it is necessary that some additional consideration be given to the interference phenomenon inherent in all three statements.

As discussed in section 6.6, the θ^4 dependence of σ° at small grazing

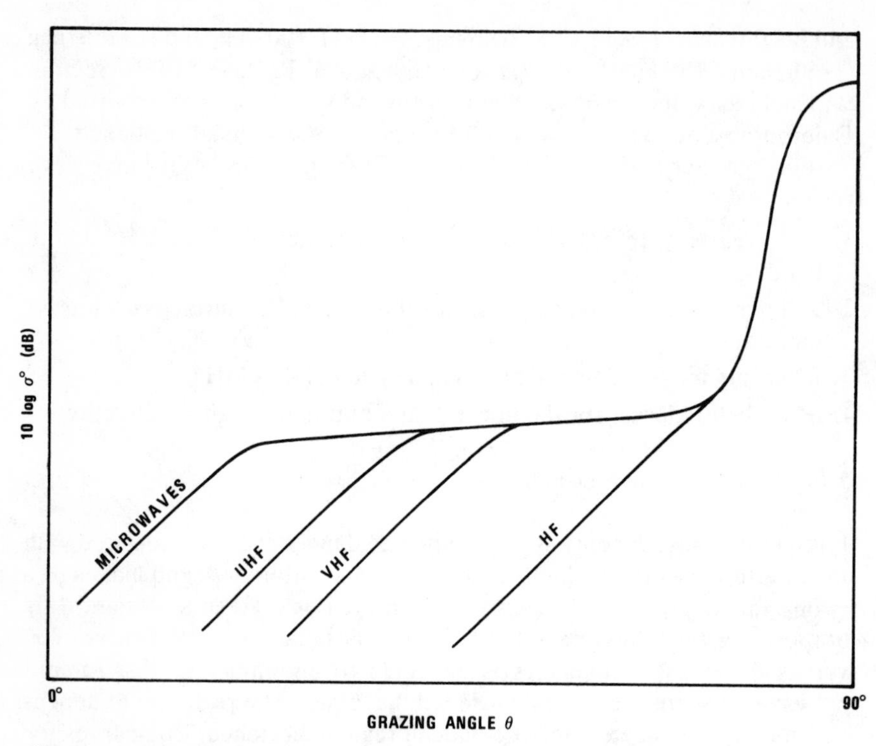

Source: Adapted from Long et al. (1965).

Figure 6-37. Assumed Shape of $\sigma°$ Versus θ Curves for Horizontal Polarization.

angles for horizontal polarization is due to interference. For microwave frequencies, this statement is approximately true for vertical as well as for horizontal polarization, although there are insufficient low angle data for vertical polarization to permit drawing a definite conclusion. However, for lower frequencies, the statement is strictly true only for horizontal polarization.

For horizontal polarization, the reflection coefficient for all angles of incidence has a magnitude of essentially unity and a phase shift of 180°. Therefore, the critical angle will increase approximately linearly with λ in accordance with equation (6.6). For vertical polarization, the situation is quite different. Both the magnitude ρ and phase ϕ of the reflection coefficient vary with grazing angle and with frequency. The angle for which ϕ is 90° (and ρ a minimum) is known as Brewster's angle. At grazing angles less than Brewster's angle, the phase shift exceeds 90°, while at higher grazing angles the phase shift is less than 90°. Thus, destructive interference can occur near the surface only when the grazing angle is below Brewster's angle; in order for the interference to be appreciable, it must be well below Brewster's angle. We would, therefore, expect to find the knee of the σ° curve (critical angle) connecting the plateau region with the rapid drop near grazing to be well below Brewster's angle. However, this has little bearing on the σ° plot for microwave frequencies. For instance, at X band Brewster's angle is about 8° (fig. 4-6), whereas θ_c is usually below 1°. At 1° the reflection coefficient does not differ greatly from unity and the phase shift is very nearly 180°. Therefore, as the grazing angle is decreased, we expect destructive interference to begin occurring for both vertical and horizontal polarizations at almost the same (critical) angle.

At lower frequencies, the situation is quite different. For example, at 1,000 MHz Brewster's angle and θ_c for horizontal polarization may each be about 5°. As the frequency is lowered further, θ_c for horizontal polarization increases to still larger angles and Brewster's angle decreases.

In summary, for low frequencies and for vertical polarization, as the grazing angle decreases destructive interference will not be expected until the grazing angle becomes very small (of the order of 1° at 30 MHz). Therefore, it is expected that the plateau region of the σ°_{VV} curve will extend to much smaller angles than it will for σ°_{HH}. Furthermore, this "critical angle" for low frequency and for vertical polarization will be largely independent of the surface irregularities. It is determined almost entirely by the frequency. Also, below this critical angle the decrease of σ° with θ may reach a rate much greater than θ^4, since decreasing θ "sharpens" the interference pattern in addition to moving the lower lobe higher in space. As the grazing angle approaches zero, σ°_{VV} should approach σ°_{HH} since the vertical reflection coefficient approaches unity with a phase shift of 180°. We thus expect the general shape of the σ° versus θ curves for the lower frequencies to be as shown in figure 6-38. Unfortunately, the available data are insufficient to validate the assumptions used here.

R.P. Ingalls and M.L. Stone (1956) studied sea echo for vertical polarization at 18 and 24 MHz. The antenna was about 50 feet above the water, propagation was by ground wave, and sea echo was measurable out to ranges of 100 miles. Although the object of their study was not to measure σ°, they reported that at 18 MHz the effective value of σ° is of the order of 5×10^{-4}. Expressed in decibels, this is -33 dB, which agrees quite well with σ° reported for the microwave plateau region even though the ratio of wavelengths is roughly 1,000.

W.S. Ament et al. (1958) studied sea return for low grazing angles at 220 MHz with horizontal polarization. The blimp-mounted radar transmitted 5 microsecond pulses at a pulse repetition frequency of 300 per second; the return for different depression angles was separated by range gating. The maximum grazing angle for which data were recorded was about 14°.

The data are shown in three graphs, each displaying the data recorded

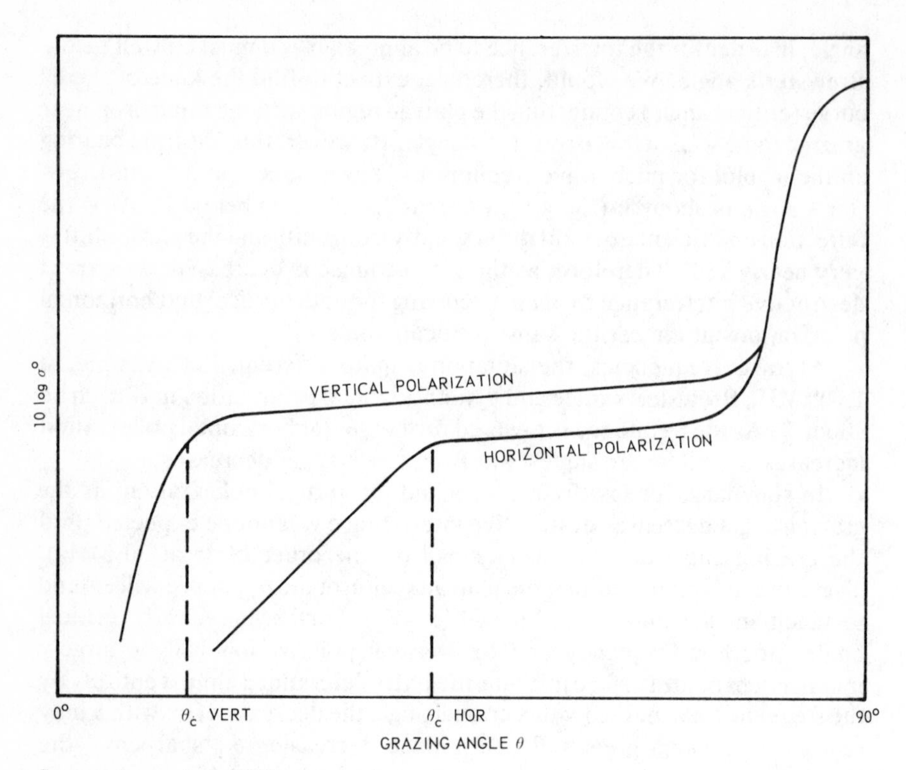

Source: Long et al. (1965).

Figure 6-38. Appearance of $\sigma°$ Versus θ Curves for UHF.

during one day. The graphs show received power as a function of grazing angle, along with a line representing the -80 dB level for $\sigma°$. From these graphs the values of $\sigma°$ can be determined. They cluster very tightly around a θ^4 variation line and reach a maximum value of about -38 dB at 14° grazing. This result is in agreement with expectations since for 220 MHz and for horizontal polarization it is expected that the critical angle will be large (of the order of 20° to 30° for moderate seas). Therefore, the 0° to 14° grazing angles for which data were recorded all lie in the near-grazing angle region, and a θ^4 variation would be expected.

Although the graphs also indicate whether each data point represents an upwind, downwind, or crosswind measurement, no definite dependency of $\sigma°$ on wind direction is discernible. Note that this result differs from that obtained for microwave data, but it does agree with recent observations at *P* and *L* bands reported by J.C. Daley, J.T. Ransone, Jr., and J.A. Burkett (1971).

The discussion given here on extrapolating microwave data was made some years ago (Long et al. 1965) to obtain estimates of σ° for lower frequencies. Recently both theory and experiment (Barrick 1972) have shown that average σ°_{VV} is about -17 dB for frequencies of a few megahertz. This result is approximately independent of θ within 30° of grazing for all but calm seas (Barrick et al. 1974). On the other hand,[b] σ°_{HH} is quite sensitive to θ; near grazing σ°_{HH} is at least 20 dB lower than σ°_{VV}.

For very practical reasons, often it will be difficult to relate theory to experience at low frequencies for horizontal polarization because of a contamination effect. The contamination is introduced by the requirement that the electric field be in both the plane of the exciting dipole and orthogonal to the direction of propagation. This problem does not arise for vertical polarization; but for a horizontally polarized dipole suspended above the sea surface, the radiation striking the surface is truly horizontally polarized only for the azimuths perpendicular to the dipole. For radiation at other azimuths, the electric vector at the sea surface will be slanted and can be broken into horizontal and vertical components. The amount of contamination is a function of both the depression (or grazing) angle and the azimuth angle measured with respect to the main beam axis. For small azimuth angles such as 5° or 10° off the axis, the contamination is negligible for all depression angles; consequently, this is a minor problem in microwave measurements where narrow-beam antennas are used.

For low frequencies, however, the effect can present a serious problem. Low frequency antennas are usually broad-beamed, particularly for airborne antennas which are subject to size and weight limitations. For typical low-frequency antenna patterns, radiation and detection of the vertical component may produce power levels close to those caused by the horizontal component. This effect can cause errors in the measurement of σ°_{HH}, particularly if σ°_{VV} is much greater than σ°_{HH}.

6.17 Dependence of σ° on Polarization

Changes in polarization have different effects on σ° for different surface conditions or for different wavelengths. In a calm sea, σ°_{VV} is usually larger than σ°_{HH}—occasionally by more than 20 dB (Cowan 1946; Kerr 1951; Wiltse, Schlesinger, and Johnson 1957). For rougher seas the two σ°'s are more nearly equal, and under certain circumstances σ° for horizontal polarization has been observed to exceed σ° for vertical polarization.

At incidence angles near the vertical, the terms "horizontal" and "vertical" polarization lose their meanings since when $\theta = 90°$ the electric field vectors in both cases lie in the horizontal plane. As one would expect, the

[b]See footnote, p. 675, of D.E. Barrick et al. (1974).

differences between σ° for "horizontal" and "vertical" polarization vanish as θ approaches 90°.

Herbert Goldstein (Kerr 1951, pp. 495-99) states that observed values of the $\sigma^\circ_{VV}/\sigma^\circ_{HH}$ ratio varied between −8 dB and + 22 dB in a series of MIT Radiation Laboratory observations at 9.2 cm and 3.2 cm for θ between 0.65° and 1.35°. These measurements were considered quite accurate because many of the system errors cancelled when the ratios were taken. There is definite correlation of this ratio with sea state. Large ratio values were found only when the sea was calm, and as the sea became rougher the ratio steadily decreased to about unity in very rough seas at 9.2 cm. However, for 3.2 cm and very rough seas σ°_{HH} was reported to be larger than σ°_{VV}.

It is generally recognized that the following statements hold for the $\sigma^\circ_{VV}/\sigma^\circ_{HH}$ ratio in the plateau region: (1) the ratio increases with an increase in wavelength, (2) the ratio decreases with an increase in sea roughness, and (3) σ°_{HH} can exceed σ°_{VV} for heavy seas and small depression angles. It should not be erroneously assumed from the above that under very rough conditions σ°_{VV} and σ°_{HH} are about equal for all incidence angles. For example, F.C. Macdonald (1963) reported that at *X* band σ° for vertical polarization may be as much as 15 dB greater than for horizontal in calm seas for angles in the plateau region. For wind speeds around 20 knots, the ratio decreases to 3 or 4 dB. At *L* band, the ratios vary from about 25 dB in a calm sea to 11 or 12 dB in a 20-knot wind (Macdonald 1963) for grazing angles of approximately 5° to 30°.

Tables 6-6 through 6-9 give σ° in terms of median values reported by J.C. Daley, W.T. Davis, and N.R. Mills (1970) for significant wave heights of 3 and 16.4 feet taken on February 17 and February 20, 1969, respectively. For completeness, σ° for the cross-polarized echoes are included in tables 6-7 and 6-9. Interdependence of circular- and linear-polarized echo components are discussed in chapter 7.

Graphs of measurements on $\sigma^\circ_{VV}/\sigma^\circ_{HH}$ reported by Daley, Davis, and Mills (1970) are included in figure 6.39. Wind speed and wave height for the various curves are shown in table 6-10.

From figure 6-39 it is clear that the $\sigma^\circ_{VV}/\sigma^\circ_{HH}$ ratio increases with increases in wavelength. By comparing the data for February 17 with that of February 20, there is some decrease in $\sigma^\circ_{VV}/\sigma^\circ_{HH}$ apparent for increases in sea roughness but at a depression angle of 20° the ratio is large even for rough seas.

The fact that at small grazing angles the polarization dependence of σ° changes with sea state or with wave height suggests that this dependency is caused by differences in an interference effect for the two polarizations. The classical interference effect (sec. 6.6) may account for this provided the incidence angle is less than the value of θ_c indicated by equation (6.6). However, the interference effect does not account for the fact that at larger

Table 6-6
Median Values of the Normalized Radar Cross Section $\sigma°$ for Directly Polarized Signals Measured at X, C, L, and P Bands[a]

Feb. 17, 1969: Wind Velocity 5 knots; Wave Height 3 ft.

Depression Angle (degrees)	*Wind Direction*[b]	$\sigma°$(dB)							
		X_{VV}	X_{HH}	C_{VV}	C_{HH}	L_{VV}	L_{HH}	P_{VV}	P_{HH}
5	U	−48.5	−46.0	−44.5	−50.5	−44.5	—	−45.5	—
	D	−48.5	−49.5	—	—	−43.5	—	−44.0	—
	C	−47.5	−47.5	−49.5	−53.0	−44.5	—	−45.0	—
10	U	−46.5	−46.5	−42.5	−49.5	−37.0	−54.5	−38.0	−55.0
	D	−46.0	−54.5	−43.5	−58.5	−36.0	−54.0	−37.5	−54.0
	C	−43.0	−50.5	−40.5	−53.5	−37.5	−55.0	−36.5	−54.5
20	U	−45.0	−48.5	−40.0	−53.0	−34.0	−47.0	−30.5	−52.0
	D	−42.5	−48.5	−38.0	−53.0	−31.5	−46.0	−29.5	−51.5
	C	−38.0	−44.5	−35.5	−46.5	−31.5	−46.5	−31.0	−50.5
30	U	−41.0	−46.5	−38.0	−47.5	−31.0	−41.5	−30.5	−41.0
	D	−39.5	−46.5	−36.0	−47.5	−29.0	−40.5	−29.5	−40.0
	C	−35.5	−43.5	−33.5	−44.0	−30.0	−40.0	−28.5	−40.0
45	U	−36.0	−38.5	−31.0	−38.5	−27.0	−33.5	−26.5	−33.5
	D	−34.5	−38.5	−30.5	−38.0	−26.0	−32.5	−25.5	−33.5
	C	−33.5	−37.0	−29.0	−36.5	−27.0	−34.0	−26.0	−31.5
60	U	−26.0	−26.5	−22.5	−26.0	−20.0	−21.5	−22.0	−25.5
	D	−25.5	−25.0	−22.0	−26.0	−19.5	−21.0	−21.5	−24.5
	C	−24.5	−24.5	−21.5	−25.0	−20.0	−21.0	−20.5	−23.5
75	—	−12.0	−12.0	−8.0	−12.0	—	—	−14.5	−16.0
90	—	+1.5	+1.5	+6.0	+3.5	—	—	+5.0	+4.0

[a]From J.C. Daley, W.T. Davis, and N.E. Mills (1970).
[b]U = upwind, D = downwind, and C = crosswind.

grazing angles (in the plateau region) $\sigma°$ is usually also larger for vertical polarization than for horizontal. Moreover, for moderate and rough seas σ°_{HH} is sometimes greater than σ°_{VV} for small grazing angles (Katz and Spetner 1960; Kerr 1951, p. 512; Long 1965 and 1967; Rivers 1970). Notice that σ°_{HH} exceeds σ°_{VV} for a few cases in figure 6-39.

A model to explain polarization differences has been proposed (Long 1965, 1974) that consists of two mechanisms: (1) a wind-dependent fine structure of the sea (ripples) interspersed with (2) smooth reflecting surfaces (facets). With this model, the backscattering from the ripples is larger for VV than for HH because of a local interference caused by reflections from the smooth surfaces adjacent to the ripples. Furthermore, this model accounts for σ°_{HH} sometimes exceeding σ°_{VV} because the reflection coefficient at a smooth air-water interface (facet) is larger for horizontal than it is for vertical polarization (see fig. 4-4).

Table 6-7
Median Values of the Normalized Radar Cross Section $\sigma°$ for Cross-Polarized Signals Measured at X, C, L, and P Bands[a]

Feb. 17, 1969: Wind Velocity 5 knots; Wave Height 3 ft.

Depression Angle (degrees)	*Wind Direction*[b]	$\sigma°$(dB)							
		X_{VH}	X_{HV}	C_{VH}	C_{HV}	L_{VH}	L_{HV}	P_{VH}	P_{HV}
5	U	−49.5	−53.0	−51.0	—	—	—	−52.0	—
	D	−50.0	—	—	—	−51.5	—	−52.0	—
	C	−50.0	—	−51.0	—	−51.5	—	−52.0	—
10	U	−53.5	−57.5	−53.5	—	−51.0	−52.5	−50.5	−56.0
	D	−55.5	−60.5	−56.5	—	−50.0	−50.5	−50.5	−56.0
	C	−53.5	−57.0	−53.5	−55.5	−51.5	−52.5	−50.5	−56.5
20	U	−51.5	−58.0	−52.0	−51.0	−46.0	−46.0	−46.0	−48.5
	D	−49.5	−55.5	−52.0	−50.5	−44.5	−45.5	−45.0	−49.5
	C	−47.0	−54.0	−48.5	−46.0	−45.5	−46.5	−44.5	−48.0
30	U	−51.5	−54.5	−52.0	−49.5	−44.0	−44.5	−43.0	−46.5
	D	−50.5	−53.5	−51.0	−49.0	−42.0	−42.5	−43.0	−46.0
	C	−48.5	−50.5	−48.0	−46.5	−42.5	−43.0	−42.5	−45.5
45	U	−50.0	−51.0	—	—	−41.0	−40.5	−41.5	−43.5
	D	−48.0	−50.5	—	—	−39.5	−38.5	−42.0	−44.0
	C	−48.5	−49.5	—	—	−41.0	−41.5	−41.0	−43.0
60	U	—	—	—	—	—	—	−38.5	−41.0
	D	—	—	—	—	—	—	−38.0	−41.0
	C	—	—	—	—	—	—	−37.5	−40.0

[a]From Daley, Davis, and Mills (1970).
[b]U = upwind, D = downwind, and C = crosswind.

6.18 Dependence of $\sigma°$ on the Wind and Sea

It is difficult to obtain accurate quantitative information concerning the influence of the sea on $\sigma°$ because the sea's parameters are hard to measure. Some of the sea surface characteristics that are known to influence various features of radar echo include the period and shape of the waves, the wave height, wind ripples, and the presence or absence of whitecaps and spray. Nonetheless, most data indicate that the average and median values of $\sigma°$ increase as the sea becomes rougher for all grazing angles except those near vertical incidence.

For small grazing angles H. Davies and G.G. Macfarlane (1946) observed a rapid increase in $\sigma°$ as wave height increased, until a sort of "saturation" set in. At X band the saturation height is about 2 or 3 feet and it increases as the radar wavelength increases. Such behavior is consistent with the interference theory.

Table 6-8
Median Values of the Normalized Radar Cross Section $\sigma°$ for Directly Polarized Signals Measured at X, C, L, and P Bands[a]

Feb. 20, 1969: Wind Velocity 29 knots; Wave Height 16.4 ft.

Depression Angle (degrees)	*Wind Direction*[b]	$\sigma°$(dB)							
		X_{VV}	X_{HH}	C_{VV}	C_{HH}	L_{VV}	L_{HH}	P_{VV}	P_{HH}
5	U	−35.5	−37.5	−34.5	−39.5	−40.5	−45.0	−41.0	−47.0
	D	−38.5	−43.0	−40.0	−44.0	−43.0	—	−43.0	—
	C	−37.0	−40.5	−39.5	−41.5	−44.0	−48.5	−42.5	—
10	U	−31.5	−35.0	−28.5	−38.5	−36.0	−48.0	−36.0	−46.5
	D	−33.5	−42.5	−31.5	−43.5	−34.5	−51.0	−36.0	−49.0
	C	−34.5	−39.5	−33.5	−41.5	−35.5	−48.5	−36.0	−49.0
20	U	−27.0	−33.5	−25.0	−35.0	−31.0	−41.5	−30.0	−47.5
	D	−29.5	−38.5	−28.0	−40.0	−29.0	−42.5	−28.5	−46.5
	C	−30.0	−38.5	−30.0	−40.0	−30.5	−43.5	−30.5	−45.5
30	U	−25.5	−30.5	−22.0	−32.5	−28.0	−37.5	−28.5	−37.0
	D	−27.0	−35.5	−23.5	−36.5	−26.0	−37.5	−28.0	−36.5
	C	−29.0	−35.5	−25.5	−35.5	−26.5	−36.5	−28.0	−37.5
45	U	−20.5	−24.0	−19.0	−27.0	−27.0	−33.0	−25.5	−31.0
	D	−23.0	−28.5	−21.5	−31.0	−25.0	−32.0	−24.5	−30.5
	C	−25.5	−29.0	−23.5	−30.0	−27.0	−34.5	−25.5	−32.5
60	U	−17.5	−18.5	−14.5	−19.0	−20.5	−22.0	−23.0	−25.5
	D	−18.5	−20.0	−16.0	−21.0	−19.0	−21.5	−22.0	−24.5
	C	−19.5	−20.5	−17.5	−21.0	−22.0	−24.0	−21.0	−24.0
75	—	−9.5	−10.0	−7.0	−9.0	—	—	−13.0	−15.0
90	—	−1.0	−1.0	+2.0	+1.0	—	—	+0.5	−2.0

[a]From Daley, Davis, and Mills (1970).
[b]U = upwind, D = downwind, and C = crosswind.

A sudden increase in average $\sigma°$ (as much as 10 dB in a one-minute interval for a 2-foot average wave height sea) has been observed to occur simultaneously with an abrupt increase in wind speed (fig. 5-24). As discussed in the next section (sec. 6.19), empirical formulas have been obtained from experimental 6.3 GHz data that indicate that for small grazing angles average $\sigma°$ varies as the cube of wind speed for horizontal polarization and as the square of wind speed for vertical polarization. A.H. Schooley (1956) reported "$\sigma°$ is very approximately proportional to the local wind velocity squared" for horizontal polarization. There also seems to be a wind-speed saturation effect similar to that with wave height. For example, $\sigma°$ is less sensitive to a 5-knot change in wind speed at 25 knots than at 10 knots.

The early literature on radar also contained other references to $\sigma°$ increasing with increasing wind speed (Kerr 1951, p. 512; Wiltse,

Table 6-9
Median Values of the Normalized Radar Cross Section σ° for Cross-Polarized Signals Measured at X, C, L, and P Bands[a]

Feb 20, 1969: Wind Velocity 29 knots; Wave Height 16.4 ft.

Depression Angle (degrees)	*Wind Direction*[b]	σ°(dB)							
		X_{VH}	X_{HV}	C_{VH}	C_{HV}	L_{VH}	L_{HV}	P_{VH}	P_{HV}
5	U	−44.0	−47.5	−42.0	—	−47.0	—	−47.0	—
	D	−46.0	−51.0	−43.0	—	−48.5	—	−48.0	—
	C	−44.5	−47.5	−43.5	—	—	—	−47.5	—
10	U	−42.5	−43.5	−41.0	−43.0	−48.5	−50.0	−45.5	−51.0
	D	−46.0	−46.5	−44.5	−46.0	−48.0	−48.5	−46.0	−51.0
	C	−45.5	−47.0	−44.0	−47.0	−47.5	−48.5	−46.5	−51.5
20	U	−39.0	−40.0	−38.0	−39.0	−42.5	−45.0	−42.0	−46.0
	D	−42.0	−43.5	−40.5	−40.5	−42.0	−41.5	−42.0	−45.0
	C	−43.0	−43.5	−41.0	−43.0	−43.5	−45.0	−42.0	−45.0
30	U	−38.0	−39.0	−36.5	−36.0	−41.0	−41.5	−40.0	−43.5
	D	−40.0	−41.0	−38.0	−38.0	−39.5	−39.0	−40.5	−44.0
	C	−40.0	−40.5	−38.0	−38.5	−39.0	−39.5	−40.5	−43.5
45	U	−36.0	−35.0	−34.5	−33.5	−41.0	−40.5	−40.0	−42.5
	D	−38.0	−37.5	−36.0	−36.0	−38.5	−38.0	−38.5	−41.5
	C	−37.5	−37.0	−36.5	−37.0	−41.5	−42.0	−40.0	−43.0
60	U	−35.0	−33.5	—	—	—	—	−38.5	−40.5
	D	−35.0	−34.0	—	—	—	—	−37.5	−39.5
	C	−34.5	−33.5	—	—	—	—	−37.0	−39.5

[a]From Daley, Davis, and Mills (1970).
[b]U = upwind, D = downwind, and C = crosswind.

Table 6-10
Wind Speed and Wave Height for Measurement Reported by Daley, Davis, and Mills (1970)

Date (1969)	*Wind Speed (knots)*	*Wave Height (feet)*
Feb. 6	40	15
Feb. 10	30-33	11.5-13.1
Feb. 11	46-48	21.3
Feb. 13	35-39	23
Feb. 14	37-40	23-26
Feb. 17	5	3
Feb. 18	22	9.8
Feb. 20	29	16.4

Schlesinger, and Johnson 1957). According to C.R. Grant and B.S. Yaplee (1957), for vertical polarization and depression angles less than about 70°,an increase in wind speed from 5 to 25knots can increase σ° by more than 20 dB in the 15 to 35 GHz frequency range.

There have been two recent reports published on the dependence of the median value of σ° on wind speed (Daley, Ransone, and Burkett 1971, 1973). Both investigations used the NRL 4-FR radar that operates at *X*, *C*, *L*, and *P* bands, and the data were collected for wind speeds from essentially 0 to 50 knots. Figures 6-40 and 6-41 are from J.C. Daley, J.T. Ransone, Jr., and J.A. Burkett (1971) and show *HH* and *VV* polarization data taken at *X* and *C* bands for a depression angle of 10°. There was virtually no change in σ° for wind speeds above 20 knots. That report also includes data for an incidence angle of 30° for which it seems σ° reaches a "saturation" level at somewhat smaller windspeeds than at 10°. The *C*-band data show essentially the same wind dependence as the *X*-band data. The *L*- and *P*-band data included in Daley, Ransone, and Burkett (1971) indicate there is much less dependence of σ° on wind speed at the longer wavelengths, and this is consistent with 220-MHz data reported by W.S. Ament et al. (1958).

The data points in figure 6-42 were reported by Geoffrey Bishop(1970) for *X* band at a grazing angle of 1°. Also included as dashed and solid lines are the curves for *X* band included in figure 6-40 (a) and (b).

R.W. Newton and J.W. Rouse, Jr. (1972) have analyzed NASA data taken at 13.3 GHz to determine the capability of airborne and satellite radar for remote sensing of local wind speed. The data were for depression angles between 55° and 90°, the wind speeds were between 5 and 50 knots, and the polarization was vertical. A very apparent observation was that average σ°_{VV} for the measurements continue to increase with wind speed; that is, the results at 13.3 GHz showed the absence of the saturation effect that is illustrated by the NRL data in figures 6-40 and 6-41 for *X* and *C* bands.

Radar cross section for vertical incidence is the subject of J.C. Daley, J.T. Ransone, Jr., and W.T. Davis (1973). For vertical incidence (nadir), there is no difference between *VV* and *HH* polarizations for an idealized, uniformly rough sea. However, data were reported for polarizations identified as *HH* and *VV* and, as expected, there seems to be little significant difference in the results for the two polarizations. Table 6-11 indicates general dependences of σ° on frequency and wind speed.

Virtually all investigators using microwaves agree that σ° is greatest when the antenna is pointing directly into the wind. The increase over the downwind direction is frequently as great as 5 dB, and sometimes as much as 10 dB. Looking across the wind, σ° seems to be about the same as for the downwind.

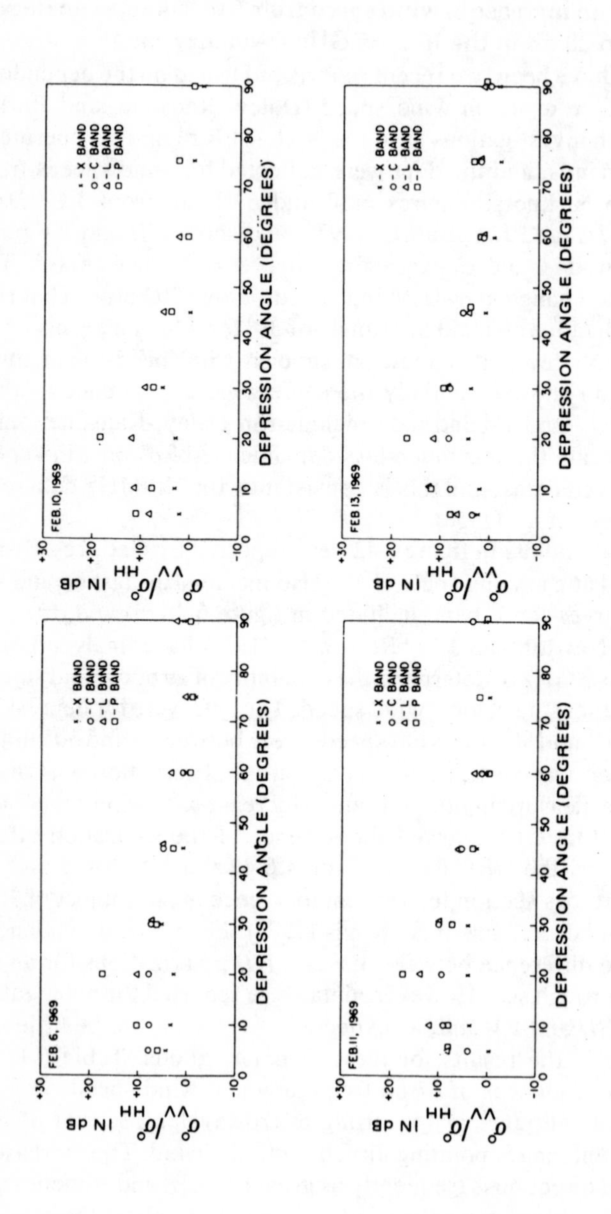

FEB 6, 1969
FEB 10, 1969
FEB 11, 1969
FEB 13, 1969
x - X BAND
o - C BAND
Δ - L BAND
□ - P BAND
σ°VV/σ°HH IN dB
+30
+20
+10
0
-10
0
10
20
30
40
50
60
70
80
90
DEPRESSION ANGLE (DEGREES)

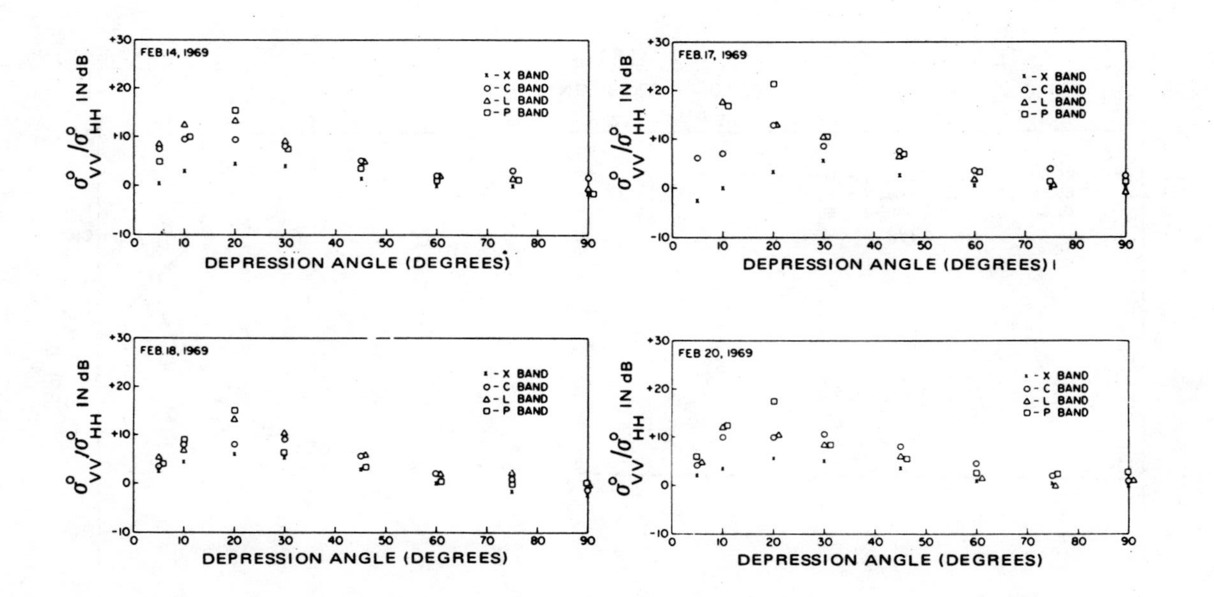

Source: Daley, Davis, and Mills (1970).

Figure 6-39. Ratio of σ°_{VV} to σ°_{HH} Versus Depression Angle.

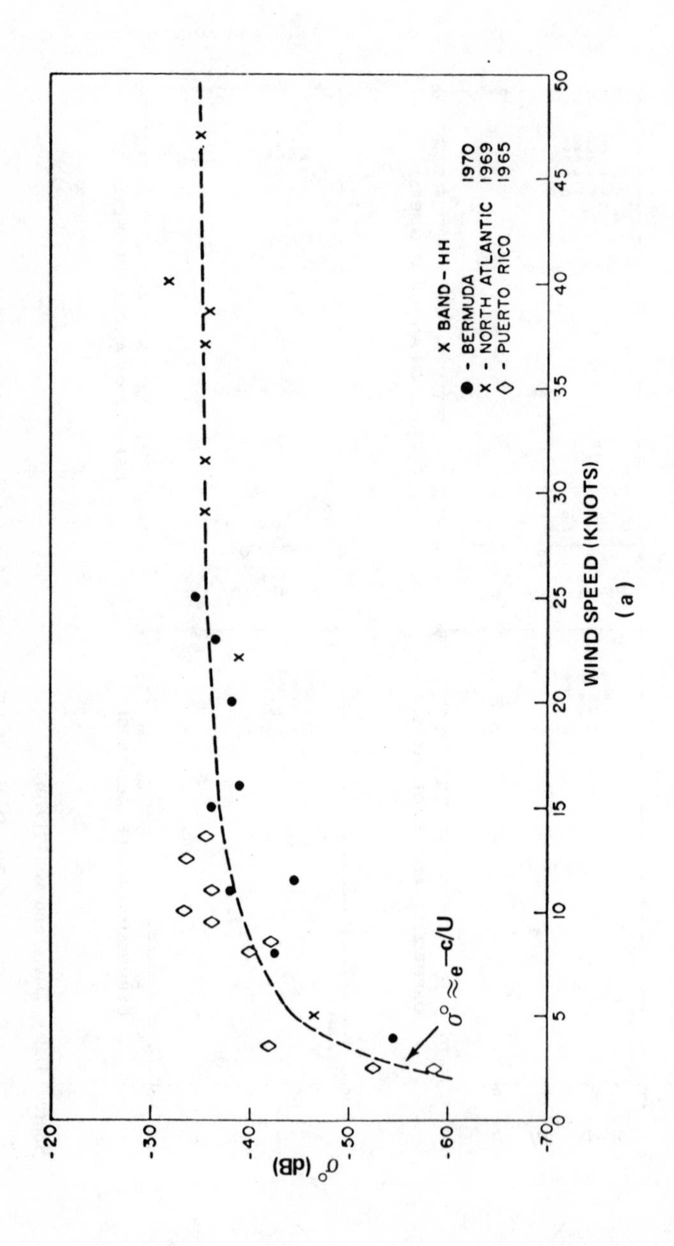

σ^o (dB)
-20
-30
-40
-50
-60
-70
0
5
10
15
20
25
30
35
40
45
50
WIND SPEED (KNOTS)
(a)
$\sigma^o \approx e^{-c/U}$
X BAND - HH
● - BERMUDA 1970
X - NORTH ATLANTIC 1969
◇ - PUERTO RICO 1965

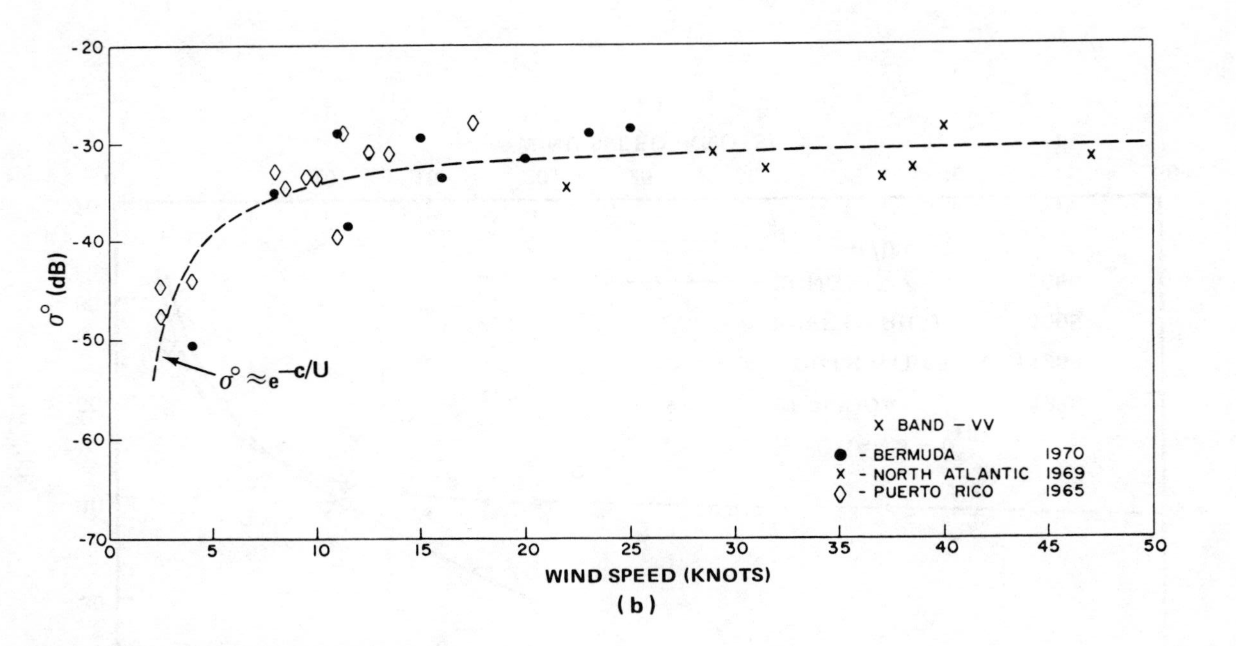

Source: Daley, Ransone, and Burkett (1971).

Figure 6-40. Median σ°_{VV} of the Sea Versus Wind Speed (*X* Band, Upwind, 10° Depression Angle).

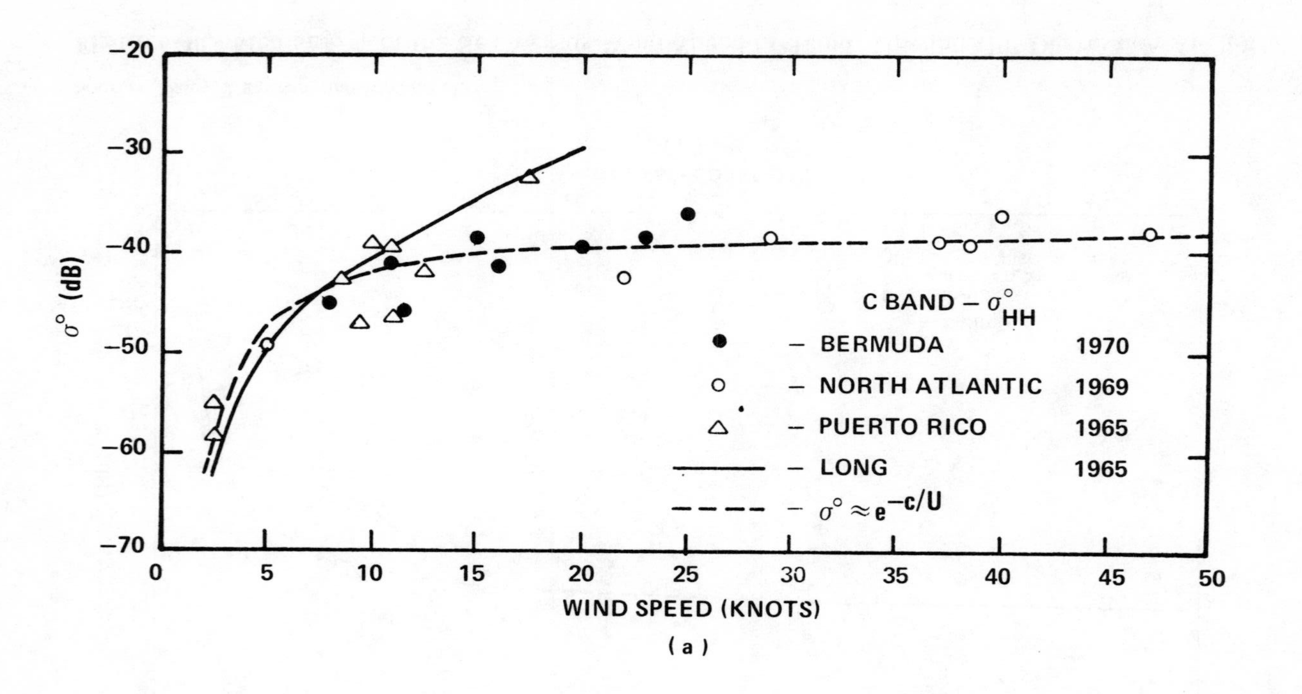

–20
–30
–40
–50
–60
–70
σ° (dB)
C BAND – σ°_{HH}
– BERMUDA 1970
– NORTH ATLANTIC 1969
– PUERTO RICO 1965
– LONG 1965
– $\sigma^{\circ} \approx e^{-c/U}$
0
5
10
15
20
25
30
35
40
45
50
WIND SPEED (KNOTS)

(a)

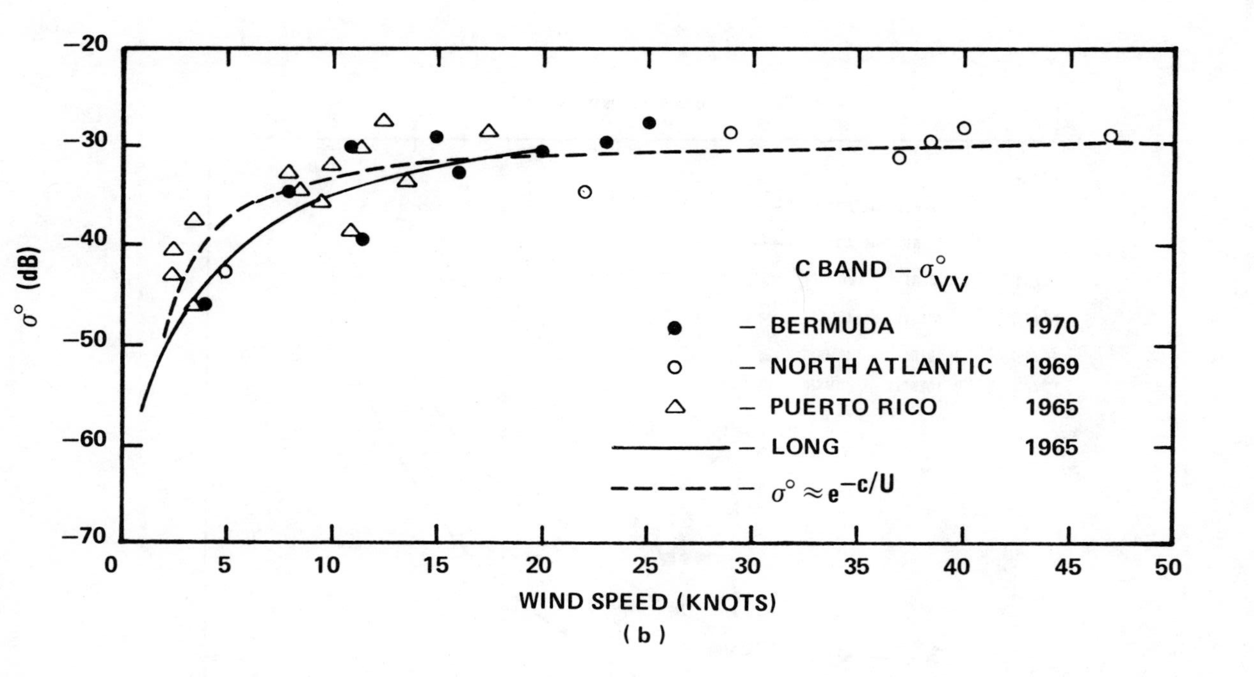

Figure 6-41. σ°_{HH} and σ°_{VV} Versus Wind Speed (*C* Band, Upwind). The dashed curves and the data points are from Daley, Ransone, and Burkett (1971). The solid curve is from Long (1965). The data from Long is for depression angles between 1.5° and 4°; the data from Daley et al. is for a 10° depression angle.

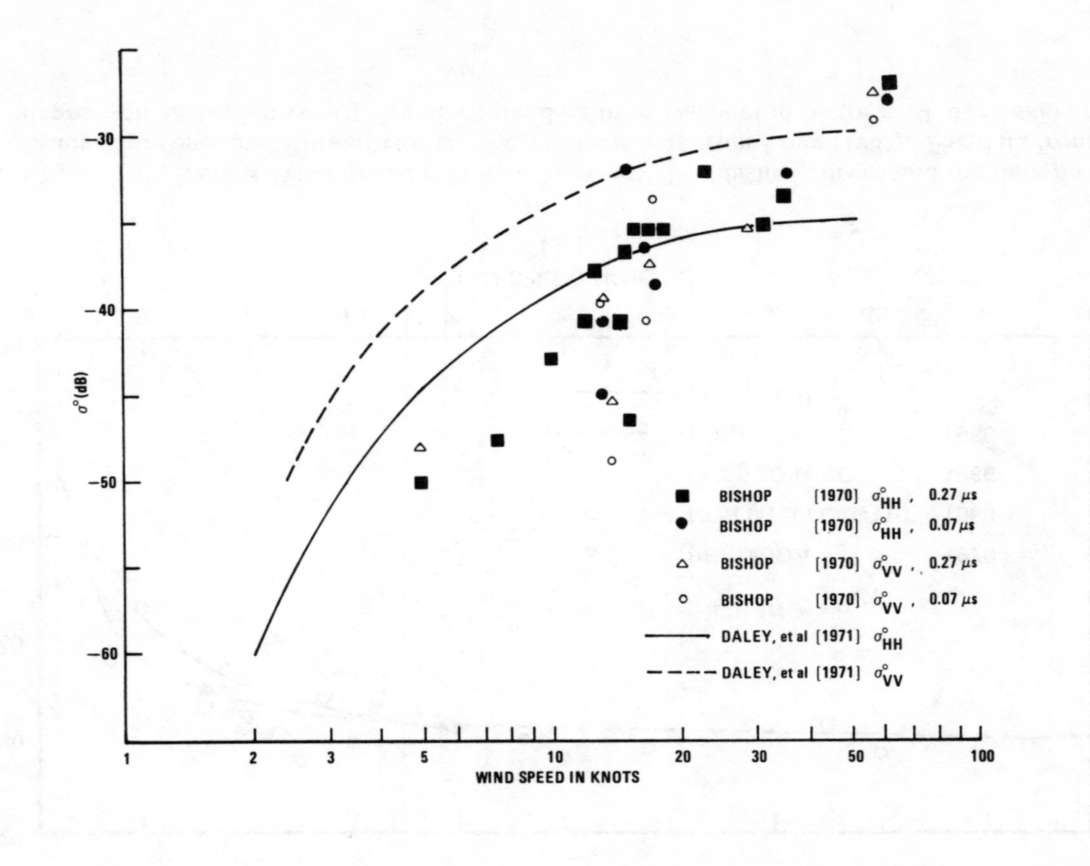
σ°(dB)
−30
−40
−50
−60
1
2
3
5
10
20
30
50
100
WIND SPEED IN KNOTS
BISHOP [1970] σ°HH , 0.27 μs
BISHOP [1970] σ°HH , 0.07 μs
BISHOP [1970] σ°VV , 0.27 μs
BISHOP [1970] σ°VV , 0.07 μs
DALEY, et al [1971] σ°HH
DALEY, et al [1971] σ°VV

Table 6-11
Wind Speed Dependence of $\sigma°$ for Vertical Incidence Reported by Daley, Ransone, and Davis (1973)

Band	*Wind Speed 0-20 knots*	*Wind Speed 20-50 knots*	*Comments*
X	$\sigma°$: 20 to 10 dB	$\sigma°$: between 5 and 10 dB	$\sigma°$ decreases with increases in wind speed
P	$\sigma°$: between −5 and +5 dB	$\sigma°$: between −5 and +5 dB	$\sigma°$ independent of wind speed

Daley et al. (1969) reported median $\sigma°$ at *X*, *C*, *L*, and *P* bands for *HH*, *VV*, *HV*, and *VH* polarizations. The ratios of the upwind-to-downwind and upwind-to-crosswind returns were investigated as functions of incidence angle, polarization, radar wavelength, wind velocity, and wind height. The upwind/downwind ratio was reported to decrease with increasing incidence angle and surface roughness. Horizontal polarization was found to be more sensitive to wind direction than was vertical polarization. The short wavelengths were more sensitive to wind direction than the long wavelengths. The longest wavelength (70 cm) was found to be essentially independent of wind direction at incidence angles greater than 10°.

6.19 Sea Echo Prediction

1. Predictions Based on Experiment. Empirical formulas fitted to experimental data (Long 1965) for 6.3 GHz indicate that $\sigma°$ varies approximately as the cube of the wind speed for horizontal polarization and as the square of the wind speed for vertical polarization. The data, which were for grazing angles between 1.5° and 4.0°, indicated that $\sigma°$ was not strongly dependent upon incidence angle. A small but statistically significant dependence on wave height ($\sigma°$ increases slightly with increase in wave height) was found for the *HH* echo. Essentially no dependence on wave height was observed for *VV* and *VH* echoes. No other polarizations were investigated.

Prediction equations for grazing angles between 1.5° and 4.0° were obtained by fitting the results of 198 measurements of σ°_{VV} and σ°_{HH} to empirical equations. At 4° (the largest angle for which measurements were made) these equations yield the following results:

$$10 \log \sigma^\circ_{VV} = 20.4 \log W + 7.45 \cos \beta - 64.0 \qquad (6.10)$$

$$10 \log \sigma^\circ_{HH} = 31.5 \log W + 9.6 \cos \beta - 81.4 \qquad (6.11)$$

where W = wind speed in knots

β = wind aspect angle in degrees

The wind aspect angle was defined as the angle between the line-of-sight and the wind direction, so that an antenna is pointing into the wind at a wind aspect angle of zero. Due to the shore location of the radar equipment, aspect angles larger than 90° were rare. The data were collected during an extended field operation at Boca Raton, Florida. The site was located where the ocean depth varies between 25 and 40 feet over a range interval of 350 to 1,500 yards. The data were for mean surface to peak wave heights up to 5 feet and for wind speeds up to 20 knots; therefore, the equations may not be valid for wind speeds much greater than 20 knots. In general, the surface of the sea was characterized by short-crested waves covered with chop; swell was observed on occasion. Local weather conditions and the shielding effect of offshore land masses normally precluded generation of deep sea waves within the radar range used. The limited radar range used minimized the effect of anomalous propagation conditions.

The solid curves in figure 6-41 (a) and (b) were calculated from equations (6.10) and (6.11) for the case of β equal to 0°. In accordance with the discussion on incidence angle dependence (sec. 6.15), curves for other plateau region angles presumably can be obtained by increasing σ° by roughly 1/4 dB for each degree of increase in θ above 4°. Figure 6-43 was prepared as a prediction of σ°_{VV} in the plateau region with 0° wind aspect angle and with wind speeds of 10, 16, and 30 knots. The given prediction equation was used for θ equal to 4° as the starting point, and slopes of 1/4 dB per degree were used in approximate agreement with the median of the data shown in figures 6-30 and 6-31. Notice the general agreement between the predictions of figure 6-43 and the measured data contained in figures 6-30 and 6-31. Since no correlation of σ°_{VV} with wavelength is apparent from figures 6-30 and 6-31, the prediction curves were used as a guide for estimating σ°_{VV} for all microwave frequencies (Long et al. 1965). Figure 6-43 was prepared some years before the 4-FR data were obtained by the Naval Research Laboratory. The author is now pleased to see the agreement between figures 6-43 and 6-47.

Curves similar to figure 6-43 but for σ°_{HH} can also be drawn; the curves would indicate general agreement with the measured σ°_{HH} data available in figure 6-32 for the plateau region. Since there is a peaking in the $\sigma^\circ_{VV}/\sigma^\circ_{HH}$ ratio for incidence angles near 20°, it is obvious that actual σ°_{VV} and σ°_{HH} versus θ curves are significantly different. Therefore, the straight line approximation for σ° versus θ used for figure 6-43 cannot be used to obtain valid comparisons between σ°_{VV} and σ°_{HH}.

It must be stressed that the prediction equations and curves are simply guesses of σ° based on limited data. The prediction curves were originally

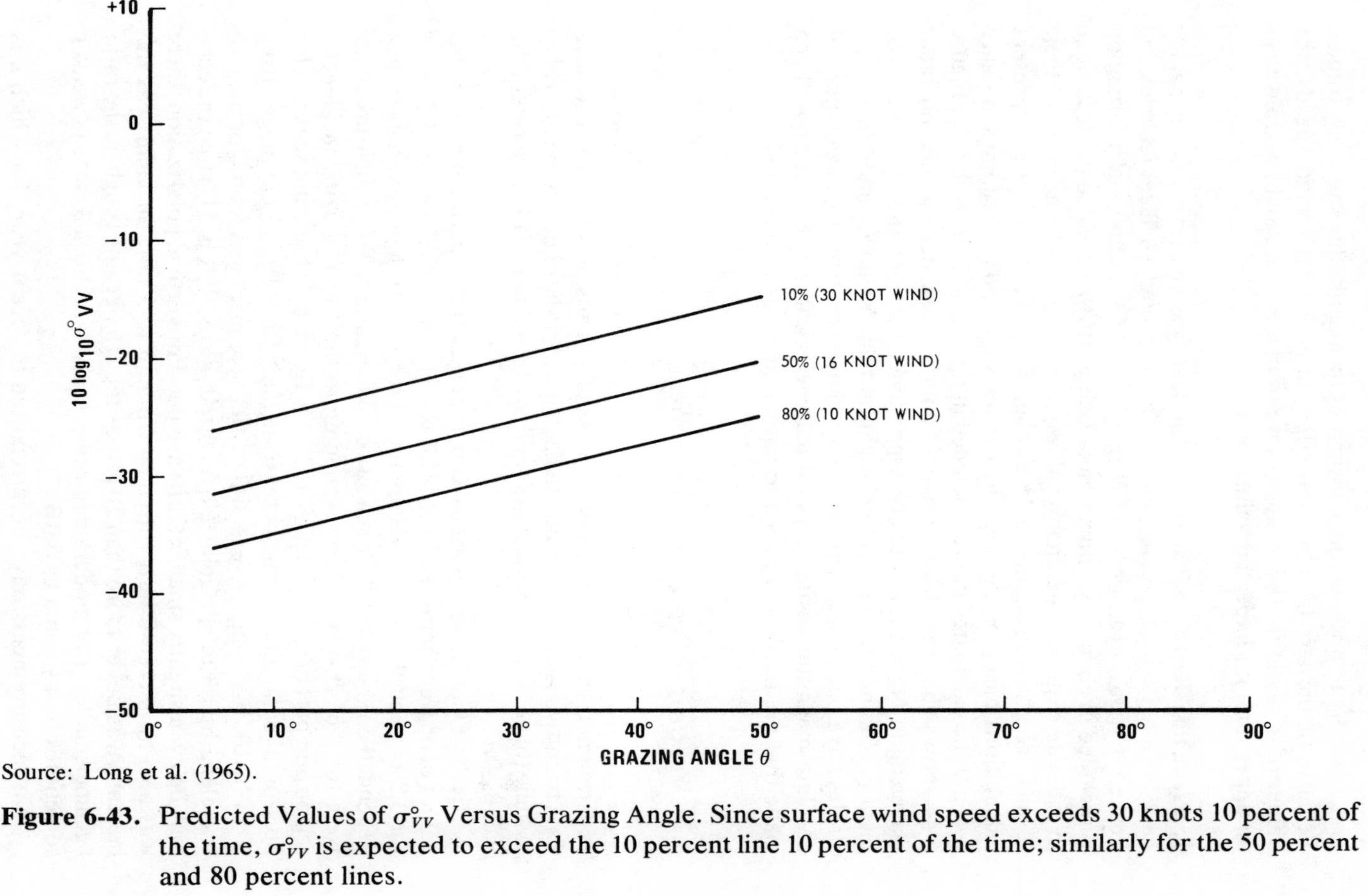

Source: Long et al. (1965).

Figure 6-43. Predicted Values of σ°_{VV} Versus Grazing Angle. Since surface wind speed exceeds 30 knots 10 percent of the time, σ°_{VV} is expected to exceed the 10 percent line 10 percent of the time; similarly for the 50 percent and 80 percent lines.

prepared *only* to provide rough estimates on magnitudes and likely trends. Because of the strong saturation effect in σ° for high wind speeds, the values of σ° given for the 30-knot curve of figure 6-43 should be viewed as an upper limit on expected values of σ°.

2. Prediction Based on Theory. In the last few years significant papers have been written on the composite theory for rough surfaces (Barrick and Peake 1968; Bass et al. 1968 and Wright 1966, 1968). Additional information is included in chapter 3. Guinard and Daley (1970) extended the technique to estimate expected maximum values of σ° for the sea, and some of their results are given in figures 6-33 through 6-36. The solid lines represent calculations based on theory. The narrow solid and dashed lines in figure 6-44 are the calculated curves included in figures 6-33 through 6-36. Figure 6-44 shows that the theory predicts only a weak dependence on radar wavelength. Notice a closer correspondence between theory and experiment is obtained for vertical polarization than for horizontal. Significant strides have been made with the theory, however, because calculations of absolute magnitude using physical parameters of the sea have now been made that are accurate at least to within an order of magnitude.

6.20 Wavelength Dependence for the Sea

To study the effect of wavelength on σ°, one generally compares the experimental values of σ° for two values of λ. Thus, it is necessary to have data from two radars calibrated to each other. It has been conventional to assume that σ° is proportional to λ^n and to express the results in terms of the value of n.

Herbert Goldstein reported (Kerr 1951) several experiments where the values of σ° were compared either for wavelengths of 1.2 cm and 3.2 cm or for 3.2 cm and 9 cm. The values found for n were highly variable; they depended on the state of the sea and on polarization. These measurements were for θ of 2° or less, corresponding to the "near grazing incidence" region and part of the "plateau" region in figure 6-1. For calm seas and for horizontal polarization, the increase in going from 10 to 3 cm ranged from 10 to 20 dB. For rougher seas, little difference in σ° was found for the two wavelengths, although Goldstein reported (Kerr 1951, p. 511) that the echo may be even slightly smaller (2dB) at 3 cm. For vertical polarization, there was also an increase in σ° in going from 10 to 3 cm; for calm seas the increase was never more than 10 dB; as the sea became rougher, the ratio became smaller. For moderate or rough seas, the echo at 3 cm was weaker sometimes by as much as 5 dB.

No measurements have been reported in recent years for which n is

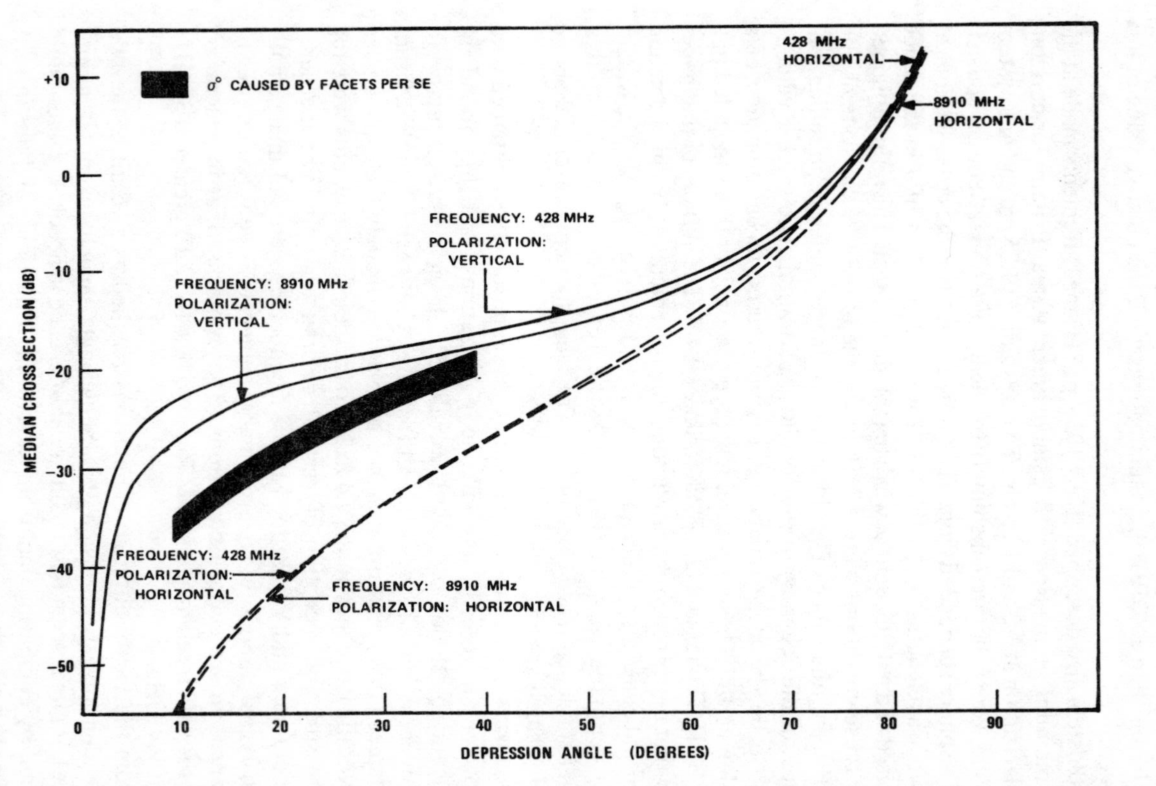

Figure 6-44. Calculations for σ° Versus θ. The narrow lines are from Guinard and Daley (1970) and show calculated σ° caused by ripples; the broad line is a guess for σ° caused by facets only.

positive,[c] as was reported by Goldstein for rough sea conditions. If the above measurements are considered to have a possible experimental error of ± 5 dB (Goldstein [1950] indicated that no wartime results can be trusted to be better than ± 5 dB), the dilemma regarding the positive value of n is removed.

Goldstein also described (Kerr 1951, p. 511) measurements made at the Telecommunications Research Establishment (Great Britain) for horizontal polarization at 3.2 and 1.25 cm. Two systems at 1.25 cm and one system at 3.2 cm were calibrated absolutely by means of a standard target consisting of a sphere suspended from a balloon. The ratio of the cross sections at the two wavelengths $\sigma°(1.25)/\sigma°(3.2)$ was independent of θ in the range measured and independent of wave height above 2 feet. The ratios obtained under these conditions were +3 dB using one of the 1.25 cm systems and +7 dB using the other. The difference of these values is an indication, perhaps, of the accuracy of the absolute calibration. The smaller value is stated to be the more reliable and is considered to be indicative of a variation as $1/\lambda$. These measurements are presumably those included in a paper by H. Davies and G.G. Macfarlane (1946). The authors of that paper stated that under rough sea conditions, measurements at 10 cm, 3 cm, and 1.2 cm indicated a λ^{-1} dependence on $\sigma°$. More recently Martin Katzin (1957) reported that measurements he reviewed also indicated a wavelength dependence close to λ^{-1}. The measurements described above were for small grazing angles.

J.C. Wiltse, S.P. Schlesinger, and C.M. Johnson (1957) obtained data for horizontal, vertical, and circular polarizations, and they found no significant correlation of $\sigma°$ with wavelength. Their measurements were made over the range 10 to 50 GHz and for angles between 15° and 90°. Grant and Yaplee (1957) measured the sea over approximately the same angles and over the frequency range 9.4 to 35 GHz; their measurements were for transmission and reception of vertical polarization only. The measurements of Grant and Yaplee for various frequencies were not made simultaneously and, therefore, are presumably subject to wide variations due to differences in sea surface conditions. They reported that in general $\sigma°$ increases with frequency and was found to be 8 to 12 dB greater at 35 GHz than at 9.4 GHz.

From the vastly different conclusions reached as a result of the two independent investigations (Wiltse, Schlesinger, and Johnson; Grant and Yaplee), it might seem that an obvious difference in absolute values could be found by comparing data from the two investigations, and this in turn would help pinpoint a possible calibration error. At one time this author reviewed the published data but he did not find any concrete suggestions

[c]This statement is true only for microwaves; it now seems that σ°_{VV} increases between P band and frequencies of a few megahertz (sec. 6.16).

regarding basic calibration problems. However, it is difficult to make meaningful comparisons of the data because detailed descriptions of the sea conditions are not given. There is a tendency to favor the data given by Wiltse, Schlesinger, and Johnson, because values of $\sigma°$ were measured simultaneously for the different frequencies. But, the data given by Wiltse, Schlesinger, and Johnson were collected on only three days, and it is risky to accept their conclusions because of the small amount of data and limited sea conditions investigated.

M.W. Long (1965) reviewed data on $\sigma°$ at 6.3 GHz and at 35.0 GHz and for grazing angles between 1.5° and 4.0°. The data were obtained by simultaneously receiving both horizontally and vertically polarized echo components for transmitted polarizations which were sequentially changed between horizontal and vertical. The results indicated that sea echo is caused primarily by two scattering mechanisms: (1) a wind-dependent fine structure of the sea (presumably ripples) that partly depolarizes and has a scattering cross section dependent on wavelength in accordance with λ^{-1}, and (2) gross structure of the sea (presumably smooth facets between the ripples) that does not depolarize and has a scattering cross section independent of wavelength. The cross sections for transmitting and receiving vertical polarization (σ°_{VV}) and for transmitting and receiving horizontal polarization (σ°_{HH}) are caused by the sum of the contributions from the two mechanisms. Therefore, although dependent on sea state and on polarization, it seems that σ°_{VV} and σ°_{HH} will tend to be independent of wavelength at the lower frequency end of the microwave spectrum and will tend to be dependent on wavelength in accordance with λ^{-1} at the higher end of the spectrum. The comparison of data at 6.3 GHz and at 35.0 GHz and for grazing angles between 1.5° and 4.0° indicated that wavelength dependencies of σ°_{VV} and σ°_{HH} are functions of sea state, but are greater than λ^0 and considerably less than λ^{-1}.

The lack of apparent correlation of $\sigma°$ with wavelength in figures 6-30, 6-31, and 6-32 suggests that σ°_{VV} and σ°_{HH} for the heavier seas change, at most, by several decibels throughout the microwave region. In other words, it appears that the spread in values of $\sigma°$ measured on different runs masks the differences caused by radar wavelength per se. The magnitudes of the curves in figure 6-43 were obtained from prediction equations from a study of experimental data not included in figures 6-30 and 6-31. The dependence of $\sigma°$ on grazing angle was drawn in figure 6-43 to correspond roughly to the data contained in figures 6-30 and 6-31. The general agreement between figure 6-43 and figures 6-30 and 6-31 further suggests that the dependence of σ°_{VV} on wavelength for the heavier seas is weak and is no larger than experimental errors and differences in $\sigma°$ caused by small changes in sea state.

There is always doubt, with good reason, regarding the validity of comparisons of absolute radar cross-section data for the sea. This is particularly true for σ°_{HH} (fig. 6-32) because the available data for the plateau region are very scanty. However, it is possible to gain some insight into the differences in wavelength dependence of σ°_{VV} and of σ°_{HH}. From data (fig. 6-39) on the relative magnitudes, it is clear that σ°_{VV} exceeds σ°_{HH} even for depression angles larger than the critical angle. The fact that X-band ratios are smaller than the corresponding L-band ratios suggests that σ°_{HH} increases faster with a decrease in wavelength than does σ°_{VV}.

Figures 6-45 through 6-48 show ranges in results reported by Guinard and Daley (1970) for *HH* and *VV* polarizations at 428 MHz and 9,310 MHz. The results are given in terms of median values of σ° and were prepared from six experimental runs with the following range of wind speeds:

24 m/s
5.0-7.5 m/s
4.0-6.0 m/s
4.0-5.5 m/s
3.5-6 m/s
0-0.5 m/s

Although the data are for median values of σ°, the relative values are expected to approximate those for averages of σ°. G.R. Valenzuela and M.B. Laing (1972) reported that observed sea echo distributions fall between Rayleigh and lognormal; the largest variance reported was 38 dB^2 and the smallest was 16 dB^2. Therefore, according to equation (6.1), the ratio of average σ° to median σ° is expected to be between that for Rayleigh (1.6 dB) and 4.4 dB, or nominally about 3 dB.

Figure 6-45 shows ranges in the values of median σ° given in figures 6-33 and 6-25 that are from Guinard and Daley (1970). Figures 6-46 through 6-48 include similar type data. The shaded areas were determined by extending straight lines between the smaller and larger values of σ°. Data points were not always obtained at a desired incidence angle, thereby causing some distortion in shape. However, the shaded areas obviously are useful for depicting general trends that exist in figures 6-33 through 6-36 for P and X bands.

The data for L and C bands (Guinard and Daley 1970) generally lie intermediate to those of P and X bands. The curves clearly depict results ascertained from earlier extrapolations, which follow:

1. For depression angles of 50° and greater, median σ°_{HH} and σ°_{VV} are essentially independent of frequency and polarization.
2. For depression angles between 5° and 90°, median σ°_{VV} is essentially independent of frequency.
3. For depression angles between 5° and 50°, median σ°_{VV} at any frequency is greater than median σ°_{HH} at any frequency

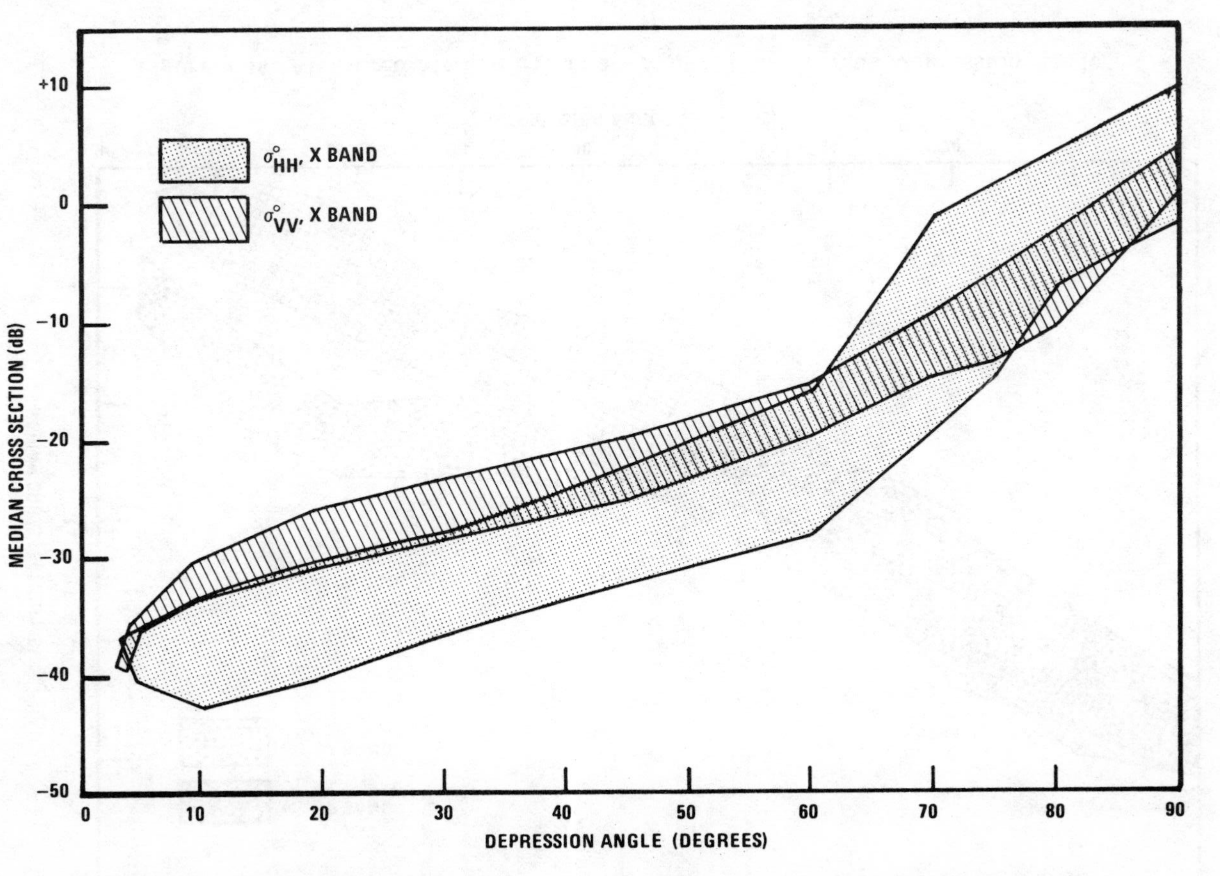

Figure 6-45. Measured Median σ°_{HH} and σ°_{VV} at X Band Versus Depression Angle.

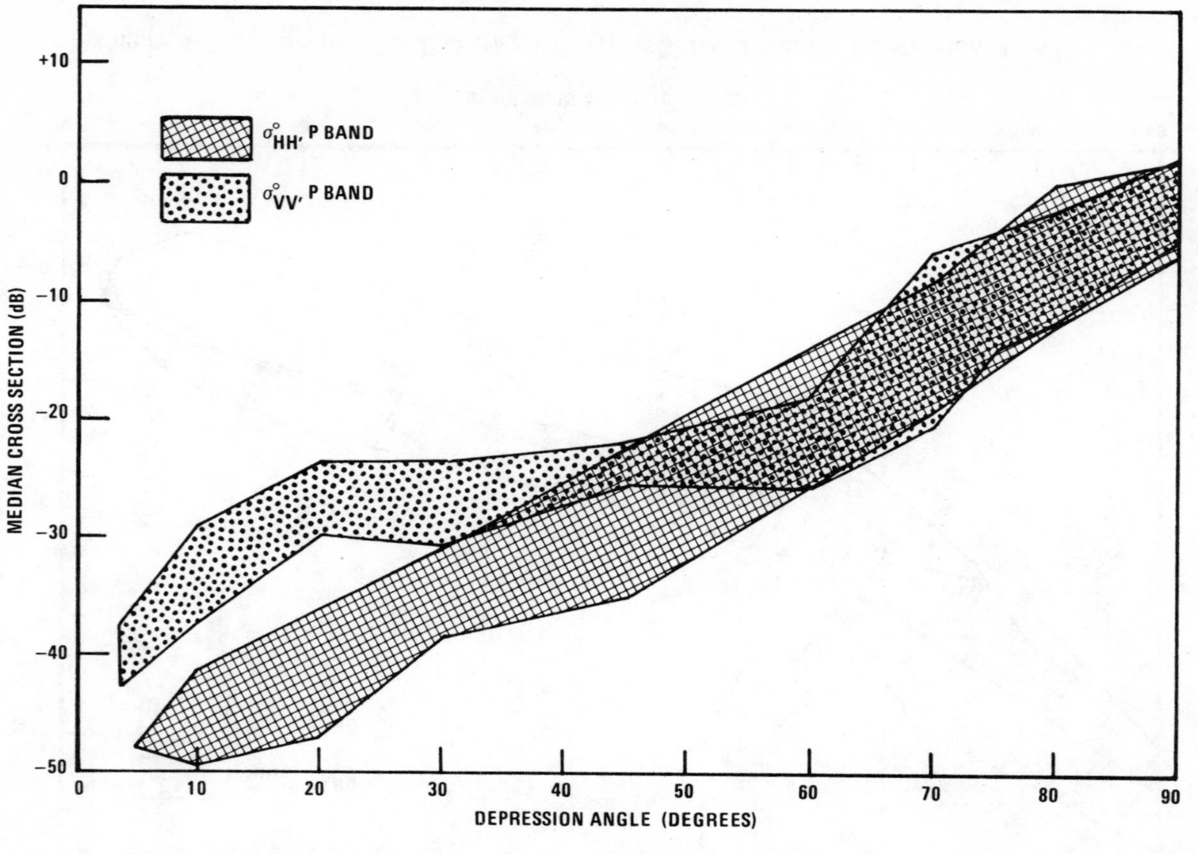

Figure 6-46. Measured Median σ°_{HH} and σ°_{VV} at P Band Versus Depression Angle.

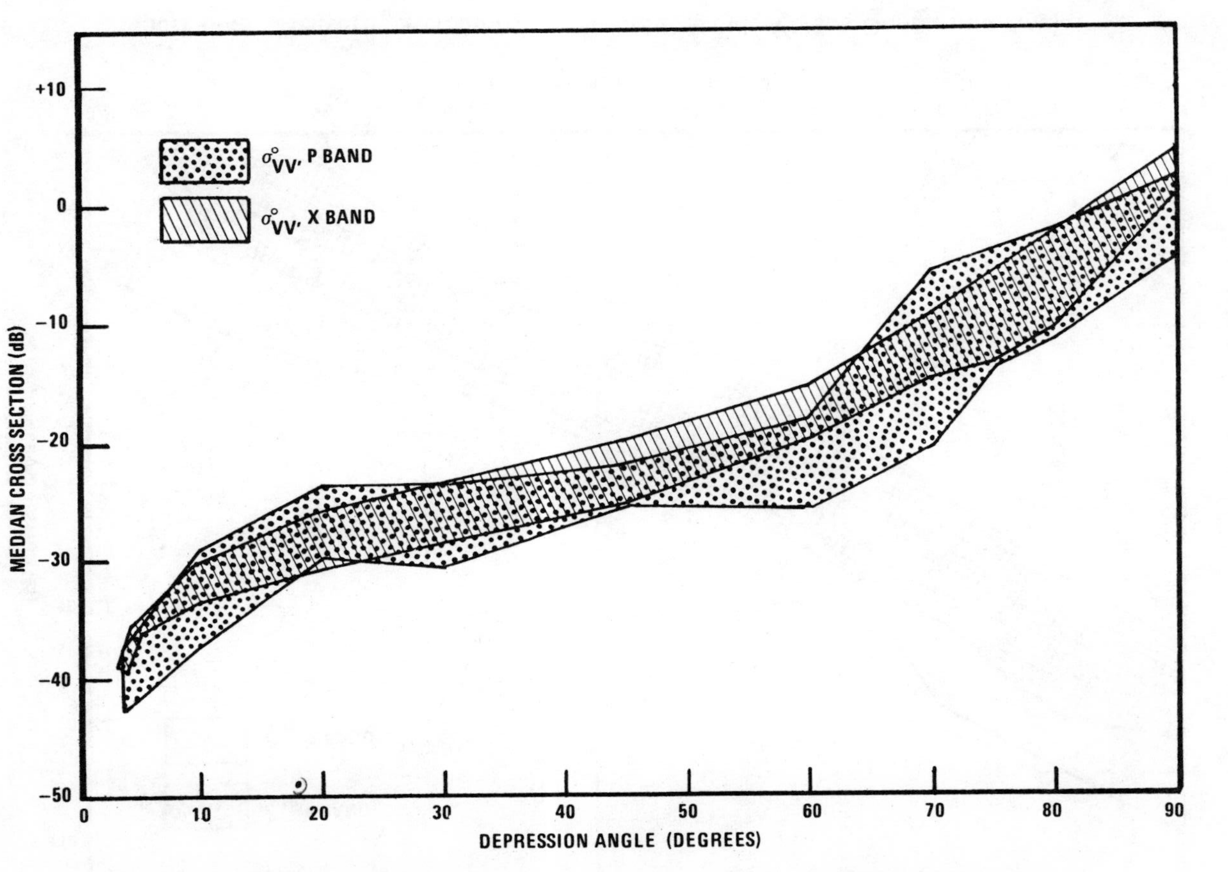

Figure 6-47. Measured Median σ°_{VV} at X and P Bands Versus Depression Angle.

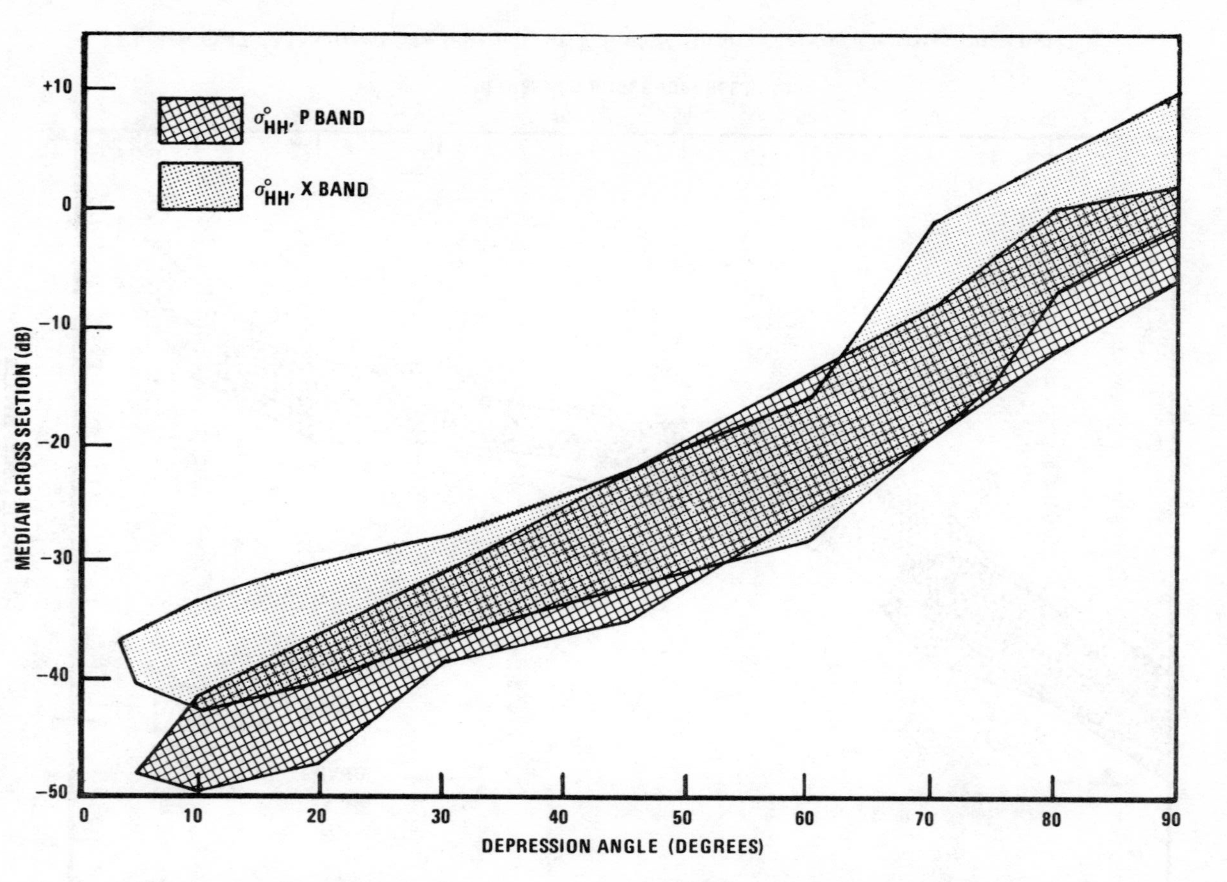

Figure 6-48. Measured Median σ°_{HH} at X and P Bands Versus Depression Angle.

4. For depression angles between 5° and 50°, median σ°_{HH} is larger at 9,310 MHz than at 428 MHz.

It should be recognized that the above statements are generalizations for which exceptions may exist when considering specific wind and wave conditions. However, the statements and curves are useful for describing matters that are, in fact, very complex.

The totality of available experimental data do not agree as to wavelength dependence of σ°. This is partly because the measurements are complicated by the many environmental factors that cannot be controlled. Further, in the low angle region where the interference effect is prominent, a λ^{-4} wavelength variation may be introduced and this complicates the analysis problem. However, in the plateau region (roughly 5° to 50° for microwaves) where the interference effect is minor, no clear correlation of σ° with wavelength is apparent. Recent experimental results and opinions indicate general agreement that the dependence on wavelength is between λ^{-1} and λ^0 within the plateau region. The lack of a good correlation of σ° measurements with wavelength can be accounted for by a number of factors: (1) the radar measurements required are extremely difficult; (2) the descriptions of the sea surface and other environmental conditions are inadequate and lack uniformity; (3) σ° is highly sensitive to fluctuations in environmental conditions (e.g., wind speed) and can change quite rapidly; and (4) the dependence of σ° on wavelength is weak.

Measurements on relative values for various polarizations show that σ°_{VV} usually exceeds σ°_{HH} even for grazing angles above the critical angle. Further, observations usually indicate that the ratio $\sigma^\circ_{VV}/\sigma^\circ_{HH}$ increases with increases in wavelength (fig. 6.39). Therefore, since σ°_{VV} is in essence independent[d] of wavelength for microwaves (see fig. 6-47), σ°_{HH} must decrease with increases in wavelength for both the near grazing and plateau regions. The data in section 6.18 support this conclusion for the near vertical region.

If this argument is extended, ultimately σ°_{HH} would exceed σ°_{VV} and, in fact, W.K. Rivers (1970) reported that at incidence angles of a few degrees σ°_{HH} exceeds σ°_{VV} at a 3-mm wavelength. This fact does not, however, require that either σ°_{HH} or σ°_{VV} is larger at 3 mm than at microwave frequencies.

For the data investigated by Rivers the median of $\sigma^\circ_{HH}/\sigma^\circ_{VV}$ was 4.5 dB, the average of the ratio was 4.8 dB, and the standard deviation of that ratio was 2.3 dB. The minimum ratio observed for those averages was 0 dB and the maximum was 8.5 dB. Rivers reasons that the reflection coefficient for an air-water interface at 3 mm for *VV* polarization relative to that for *HH*

[d]Average σ°_{VV} does, however, increase between *P* band and frequencies of a few megahertz (sec. 6.16).

polarization accounts for his observations that the various values of average σ°_{HH} exceed those for σ°_{VV}.

The wavelength dependence of σ° is only one of several dependencies which must be considered by a radar designer in choosing the best wavelength for a given application. Other characteristics of sea echo are affected by changes in wavelength; for example, short-term amplitude and frequency distributions. The designer also must consider the wavelength dependence of target cross sections, atmospheric attenuation, rain and cloud clutter, and so on. These questions, as well as those of available beamwidth, transmitter power, and receiver noise figure, are crucial to the choice of a wavelength.

6.21 Oil Slicks

Oil smooths rough water by modifying the effects of breaking waves, and by damping the small gravity and capillary waves; this smoothing effect permits oil slicks to be detected and monitored by radar. Slicks have been observed by Naval Research Laboratory personnel (Guinard 1971; Pilon and Purves 1973) with airborne radars at their four operating frequencies, i.e., 428 MHz, 1,228 MHz, 4,455 MHz, and 8,910 MHz. Oil films reduce sea echo more for *VV* than for *HH* polarization. For *VV* polarization and low sea states, the radars can reliably detect and map oil spills from the initial thickness of a polluting spill to thicknesses of 1 micron or less. K. Krishen (1972) reported that for depression angles between 40° and 75°, the vertically polarized 13.3-GHz sea echo strength was reduced 5 to 10 dB by the presence of oil.

6.22 Discussion on σ° for the Sea

Suppose the theoretical curves (fig. 6-44) accurately represent the scattering from ripples for both *HH* and *VV* polarizations. Then, there must be another mechanism to account for σ°_{HH} being as large as it is (see figs. 6-35 and 6-36). It has been suggested that scattering from the waves, per se (sec. 6.17), is a major contributor for *HH* polarization. Assume there is a mechanism that does not depolarize and which reflects equal power[e] for *VV* and *HH* polarizations. Let the broad, solid line in figure 6-44 be σ° for that mechanism. Compared to the theoretical curve, the solid line would contribute a relatively small amount to *VV* echo and would permit the $\sigma^\circ{}_{VV}/\sigma^\circ{}_{HH}$ ratio to peak in the vicinity of 20°. This additional mechanism, presumably

[e]The assumption of equal reflected power is made for simplicity. From figure 4-4 it can be seen that for a flat surface *HH* reflections will exceed those for *VV*.

due to the facets, could produce the low-frequency fluctuation that is a larger fraction of the total fluctuation spectrum for *HH* polarization than for *VV* polarization (see sec. 5.12). Irrespective of the mechanism that causes *HH* to be larger than the theoretical curve, the average σ° must increase inversely with wavelength because the $\sigma^\circ_{VV}/\sigma^\circ_{HH}$ ratio is smaller for the shorter radar wavelengths. Presumably, the contributing mechanism is the same that causes the echo for incidence angles near vertical incidence, and recall (see sec. 6.18) σ° does in fact increase with frequency for near vertical incidence.

Effects of multipath interference on reducing σ° of the facets, per se, should also be considered. Assume a significant wave height of about 4 feet or an rms surface roughness of 1 foot. Then, from equation (6.6) θ_c is about 10° at 428 MHz and less than 1/2° at *X* band. Brewster's angle at 428 MHz will be approximately 5°. Thus, the cross section reductions due to the classical interference effect under these conditions will be small for the angles considered in figure 6-44. It therefore seems that the differences between σ°_{HH} at 428 MHz and at *X* band for small depression angles (fig. 6-48) are caused by the local interference effect (sec. 4.18).

References

Ament, W.S., J.A. Burkett, F.C. Macdonald, and D.L. Ringwalt, "Characteristics of Radar Sea Clutter: Observations at 220 mc," Naval Research Laboratory Report 5218, November 19, 1958.

Ament, W.S., F.C. Macdonald, and R.D. Shewbridge, "Radar Terrain Reflections for Several Polarizations and Frequencies," *Transactions of the 1959 Symposium of Radar Return,* University of New Mexico, May 11-12, 1959.

Barrick, D.E., "First-Order Theory and Analysis of MF/HF/VHF Scatter from the Sea," *IEEE Transactions on Antennas and Propagation*, vol. AP-20, pp. 2-10, January 1972.

Barrick, D.E., and W.H. Peake, "A Review of Scattering from Surfaces with Different Roughness Scales," *Radio Science*, vol. 3 (new series), pp. 865-68, August 1968.

Barrick, D.E., J.M. Headrick, R.W. Bogle, and D.D. Crombie, "Sea Backscatter at HF: Interpretation and Utilization of the Echo," *Proceedings of the IEEE*, vol. 62, pp. 673-80, June 1974.

Bass, F.G., I.M. Fuks, A.I. Kalmykov, I.E. Ostrovsky, and A.D. Rosenburg, "Very High Frequency Radiowave Scattering by a Disturbed Sea Surface," *IEEE Transactions on Antennas and Propagation*, vol. AP-16, pp. 554-68, September 1968.

Beckmann, Petr and André Spizzichino, *The Scattering of Electromagnetic Waves from Rough Surfaces*, The MacMillan Company, New York, New York, 1963.

Bishop, Geoffrey, "Amplitude Distribution Characteristics of X-Band Radar Sea Clutter and Small Surface Targets," Royal Radar Establishment Memo No. 2348, Great Britain, 1970.

Campbell, J.P., "Backscattering Characteristics of Land and Sea at *X*-Band," *Transactions of the 1959 Symposium on Radar Return*, University of New Mexico, May 11-12, 1959.

Cosgriff, R.L., W.H. Peake, and R.C. Taylor, "Terrain Return Measurements and Applications," *Transactions of the 1959 Symposium on Radar Return*, University of New Mexico, May 11-12, 1959.

Cosgriff, R.L., W.H. Peake, and R.C. Taylor, "Terrain Scattering Properties for Sensor System Design (Terrain Handbook II)," Engineering Experiment Station Bulletin, vol. 29, no. 3, The Ohio State University, May 1960.

Cowan, E.W., "X-Band Sea-Return Measurements," Massachusetts Institute of Technology Radiation Laboratory Report No. 870, January 10, 1946.

Daley, J.C., W.T. Davis, J.R. Duncan, and M.B. Laing, "NRL Terrain Clutter Study, Phase II," Naval Research Laboratory Report 6749, October 21, 1968.

Daley, J.C., J.T. Ransone, Jr., J.A. Burkett, and J.R. Duncan, "Sea-Clutter Measurements on Four Frequencies," Naval Research Laboratory Report 6806, November 29, 1968.

Daley, J.C., J.T. Ransone, Jr., J.A. Burkett, and J.R. Duncan, "Upwind-Downwind-Crosswind Sea-Clutter Measurements," Naval Research Laboratory Report 6881, April, 14, 1969.

Daley, J.C., W.T. Davis, and N.R. Mills, "Radar Sea Return in High Sea States," Naval Research Laboratory Report 7142, September 25, 1970.

Daley, J.C., J.T. Ransone, Jr., and J.A. Burkett, "Radar Sea Return —JOSS I," Naval Research Laboratory Report 7268, May 11, 1971.

Daley, J.C., J.T. Ransone, Jr., and W.T. Davis, "Radar Sea Return —JOSS II," Naval Research Laboratory Report 7534, February 21, 1973.

Davies, H. and G.G. Macfarlane, "Radar Echoes from the Sea Surface at Centimeter Wave-Lengths," *Proceedings Physical Society,* vol. 58, pp. 717-29, 1946.

Dyer, F.B., and N.C. Currie, "Some Comments on the Characterization of Radar Sea Echo," *Digest of the International IEEE Symposium on Antennas and Propagation*, pp. 323-26, June 10-12, 1974.

Edison, A.R., R.K. Moore, and B.D. Warner, "Radar Return at Near-Vertical Incidence," *IRE Transactions on Antennas and Propagation*, vol. AP-8, pp. 246-54, May 1960.

Ericson, L.O., "Terrain Return Measurements with an Airborne X-Band Radar Station," Sixth Conference of the Swedish National Committee on Scientific Radio, March 13, 1963. (Translated paper dated October 1966.)

Flynt, E.R., F.B. Dyer, R.C. Johnson, M.W. Long, and R.P. Zimmer, "Clutter Reduction Radar," Engineering Experiment Station, Georgia Institute of Technology, Final Report, Contract NObsr-91024, December 1967.

Goldstein, Herbert, "Frequency Dependence of the Properties of Sea Echo," *Physical Review*, vol. 70, p. 938, 1946.

Goldstein, H., "A Primer of Sea Echo," U.S. Navy Electronics Laboratory Report NE 0506B, p. 157, August 7, 1950.

Goodyear Aircraft Corporation, "Radar Terrain Return Study, Final Report: Measurements of Terrain Back-Scattering Coefficients with an Airborne X-Band Radar," Contract NOas-59-6186-C, Report GERA-463, September 30, 1959.

Grant, C.R. and B.S. Yaplee, "Back Scattering from Water and Land at Centimeter and Millimeter Wavelengths," *Proceedings of the IRE*, vol. 45, pp. 976-82, July 1957.

Guinard, N.W., "The Remote Sensing of Oil Slicks," *Proceedings of the Seventh International Symposium on Remote Sensing of the Environment*, University of Michigan, May 1971.

Guinard, N.W. and J.C. Daley, "An Experimental Study of a Sea Clutter Model," *Proceedings of the IEEE*, vol. 58, pp. 543-50, April 1970.

Hayes, R.D., C.H. Currie, and M.W. Long, "An X-Band Polarization Measurements Program," *Record of the Third Annual Radar Symposium*, University of Michigan, February 1957.

Hayes, R.D. and F.B. Dyer, "Land Clutter Characteristics for Computer Modeling of Fire Control Radar Systems," Engineering Experiment Station, Georgia Institute of Technology, Technical Report No. 1, Contract DAAA 25-73-C-0256, May 1973.

Hayes, R.D., J.R. Walsh, D.F. Eagle, H.A. Ecker, M.W. Long, J.G.B. Rivers, and C.W. Stuckey, "Study of Polarization Characteristics of Radar Targets," Engineering Experiment Station, Georgia Institute of Technology, Final Report, Contract DA-36-039-sc-64713, October 1958.

Hoekstra, Pieter and Dennis Spanogle, "Backscatter from Snow and Ice Surfaces at Near Incident Angles," *IEEE Transactions on Antennas and Propagation*, vol. AP-20, pp. 788-90, November 1972.

Ingalls, R.P. and M.L. Stone, "Characteristics of Sea Clutter at HF," unpublished paper, Fall Meeting of the International Scientific Radio Union, (URSI), October 11, 1956.

Ivey, H.D., M.W. Long, and V.R. Widerquist, "Some Polarization Properties of K_a- and X_b-Band Echoes from Vehicles and Trees," *Record of the First Annual Radar Symposium*, University of Michigan, 1955.

Janza, F.J., R.K. Moore, and B.D. Warner, "Radar Cross Sections of Terrain Near Vertical Incidence at 415 Mc, 3800 Mc, and Extension of Analysis to X-Band," Technical Report EE-21, Engineering Experiment Station, University of Mexico, May 1959.

Katz, I., "Radar Reflectivity of the Earth's Surface," *APL Technical Digest*, pp. 11-17, January-February 1963.

Katz, Isadore and L.M. Spetner, "Polarization and Depression-angle Dependence of Radar Terrain Return," *Journal of Research of the National Bureau of Standards*, vol. 64D, pp. 483-86, September-October, 1960.

Katzin, Martin, "On the Mechanisms of Radar Sea Clutter," *Proceedings of the IRE*, vol. 45, pp. 44-54, January 1957.

Kerr, D.E., Ed., *Propagation of Short Radio Waves*, Massachusetts Institute of Technology Radiation Laboratory Series, vol. 13, McGraw-Hill Book Company, Inc., New York, New York, 1951.

Krishen, K., "Detection of Oil Spills Using a 13.3 GHz Radar Scatterometer," *Proceedings of the Eighth International Symposium on Remote Sensing of the Environment*, Environmental Research Institute of Michigan, October 1972.

Linell, T., "An Experimental Investigation of the Amplitude Distribution of Radar Terrain Return," Sixth Conference of the Swedish National Committee on Scientific Radio, March 13, 1963. (Translated paper dated October 1966.)

Long, M.W., "On the Polarization and the Wavelength Dependence of Sea Echo," *IEEE Transactions on Antennas and Propagation*, vol. AP-13, pp. 749-54, September 1965.

Long, M.W., "Polarization and Sea Echo," *Electronics Letters*, vol. 3, p. 51, February 1967.

Long, M.W., "On a Two-Scatterer Theory of Sea Echo," *IEEE Transactions on Antennas and Propagation*, vol. AP-22, pp. 667-72, September 1974.

Long, M.W., R.D. Wetherington, J.L. Edwards, and A.B. Abeling, "Wavelength Dependence of Sea Echo," Engineering Experiment Station, Georgia Institute of Technology, Final Report, Contract N62269-3019, July 1965.

Macdonald, F.C., "The Correlation of Radar Sea Clutter on Vertical and Horizontal Polarizations with Wave Height and Slope," *1956 IRE Convention Record*, Pt. 1, pp. 29-32, 1956.

Macdonald, F.C., "Radar Sea Return and Ocean Wave Spectra," *Proceedings of the Conference on Ocean Wave Spectra*, Prentice-Hall, Englewood Cliffs, New Jersey, pp. 323-29, 1963.

Maclusky, G.S.R. and H. Davies, *The Dependence of Sea Clutter on Angle of Elevation, Together with a Brief Note on the Rates of Fluctuation,* TRE Report No. T-1956, Great Britain, November 20, 1945.

Moore, R.K., "Radar Return from the Ground," Bulletin of Engineering No. 59, University of Kansas, 1969.

Müller, V., "Ergebnisse von Messungen der Ruckstreueigenschaften des Seeganges fur elektromagnetische Wellen bei Radarfrequenzen," *Ortung und Navigation*, Heft I, pp. 95-128, 1966.

Newton, R.W. and J.W. Rouse, Jr., "Experimental Measurements of 2.25-cm Backscatter from Sea Surfaces," *IEEE Transactions on Geoscience Electronics*, vol. GE-10, pp. 2-7, January 1972.

Peake, W.H. and T.L. Oliver, "The Response of Terrestrial Surfaces at Microwave Frequencies," Technical Report AFAL-TR-70-301, The Ohio State University, May 1971.

Pilon, R.O. and C.G. Purves, "Radar Imagery of Oil Slicks," *IEEE Transactions on Aerospace and Electronic Systems*, vol. AE2-9, pp. 630-36, September 1973.

Povejsil, D.J., R.S. Raven, and Peter Waterman, *Airborne Radar,* Principles of Guided Missile Design Series, vol. 8, D. Van Nostrand Company, Inc., Princeton, New Jersey, 1961.

Renau, Jacques and J.A. Collinson, "Measurements of Electromagnetic Backscattering from Known, Rough Surfaces," *The Bell System Technical Journal*, vol. 44, December 1965.

Rivers, W.K., "Low-Angle Radar Sea Return at 3-mm Wavelength," Engineering Experiment Station, Georgia Institute of Technology, Final Technical Report, Contract N62269-70-C-0489, November 1970.

Schooley, A.H., "Some Limiting Cases of Radar Sea Clutter Noise," *Proceedings of the IRE*, vol. 44, pp. 1043-47, August 1956.

Schwarz, D.E. and F. Caspall, "The Use of Radar in the Discrimination and Identification of Agricultural Land Use," *Proceedings of the Fifth Symposium on Remote Sensing of the Environment*, University of Michigan, April 1968.

Skolnik, M.I., *Radar Handbook*, McGraw-Hill Book Company, Inc., New York, New York, 1970.

Ulaby, F.T., "Radar Measurement of Soil Moisture Content," *IEEE Transactions on Antennas and Propagation*, vol. AP-22, pp. 257-65, March 1974.

Ulaby, F.T., "Radar Response to Vegetation," *IEEE Transactions on Antennas and Propagation,* vol. AP-23, pp. 36-45, January 1975.

Ulaby, F.T., R.K. Moore, R. Moe, and J. Holtzman, "On Microwave Remote Sensing of Vegetation," *Proceedings of the Eighth Symposium on Remote Sensing of the Environment,* Environmental Research Institute of Michigan, October 1972.

Ulaby, F.T., Josef Cihlar, and R.K. Moore, "Active Microwave Measurement of Soil Water Content," *Remote Sensing of the Environment*, vol. 3, pp. 185-203, 1974.

Ulaby, F.T., A. Sobti, J. Barr, and R.K. Moore, "Can Microwave Sensors Measure Soil Moisture from Space?" *Annual Meeting of International Union of Radio Science*, Boulder, Colorado, October 14-17, 1974.

Valenzuela, G.R. and M.B. Laing, "On the Statistics of Sea Clutter," Naval Research Laboratory Report 7349, December 30, 1971.

Valenzuela, G.R. and M.B. Laing, "Point-Scatterer Formulation of Terrain Clutter Statistics," Naval Research Laboratory Report 7459, September 27, 1972.

Waite, W.P. and H.C. MacDonald, "Vegetation Penetration with K-Band Imaging Radars," *IEEE Transactions on Geoscience Electronics*, vol. GE-9, pp. 147-55, July 1971.

Williams, C.S., Jr., C.H. Bidwell, and D.M. Bragg, "Radar Return from the Vertical for Ground and Water Surfaces," Sandia Corporation Report SCR-107, April 1960.

Wiltse, J.C., S.P. Schlesinger, and C.M. Johnson, "Back-Scattering Characteristics of the Sea in the Region from 10 to 50 KMC," *Proceedings of the IRE*, vol. 45, pp. 220-28, February 1957.

Wright, J.W., "Backscattering from Capillary Waves with Application to Sea Clutter," *IEEE Transactions on Antennas and Propagation*, vol. AP-14, pp. 749-54, November 1966.

Wright, J.W., "A New Model for Sea Clutter," *IEEE Transactions on Antennas and Propagation*, vol. AP-16, pp. 217-23, March 1968.

7 Interdependence of Polarization Characteristics

Introduction

7.1 General Observations

It is well known that the radar echo from targets usually depends on the polarizations used. The choice of polarization can sometimes help fill gaps in radar coverage. Circular polarization can improve the detection of aircraft in a heavy radar background of rain; horizontal polarization can be used to reduce average sea echo, but under certain conditions improvements can be made by changing to vertical polarization.

The discussion above applies to using the same transmitted and received polarizations. Of interest here also is depolarization—that part of the reradiated field with polarization different from that of the incident field. Usually, the rougher the surface the more depolarization generated. However, one must be cautious about this statement and not overgeneralize because the echo from a rough surface is not necessarily highly depolarized.

The study of the depolarization of echo is important to the understanding of the electromagnetic scattering properties of targets, and it has also proved useful for distinguishing classes of targets. For example, figure 7-1 shows previously unknown geological resources that had not been detected by investigations which included ground surveys, aerial photography, and radar without benefit of cross-polarized data. Notice that horizontal polarization is transmitted and both horizontal *HH* and vertical *HV* polarizations are received. Also notice that, in general, there is a marked similarity between the two images (which respond to average echo strength) as a function of range and azimuth. Even though there is a generally high correlation of echo strengths as a function of position, the echoes do not necessarily fluctuate together on a short-term basis. However, depolarized echo usually fluctuates at *rates* that are indistinguishable from those of echo polarized like that transmitted.

If echo from the same range and bearing are examined simultaneously for two polarizations, there may be a vast difference in the character of the fluctuations of the two polarizations. For example, if a target such as a vehicle, a sea wave, or a tree were to move through the radar beam, the

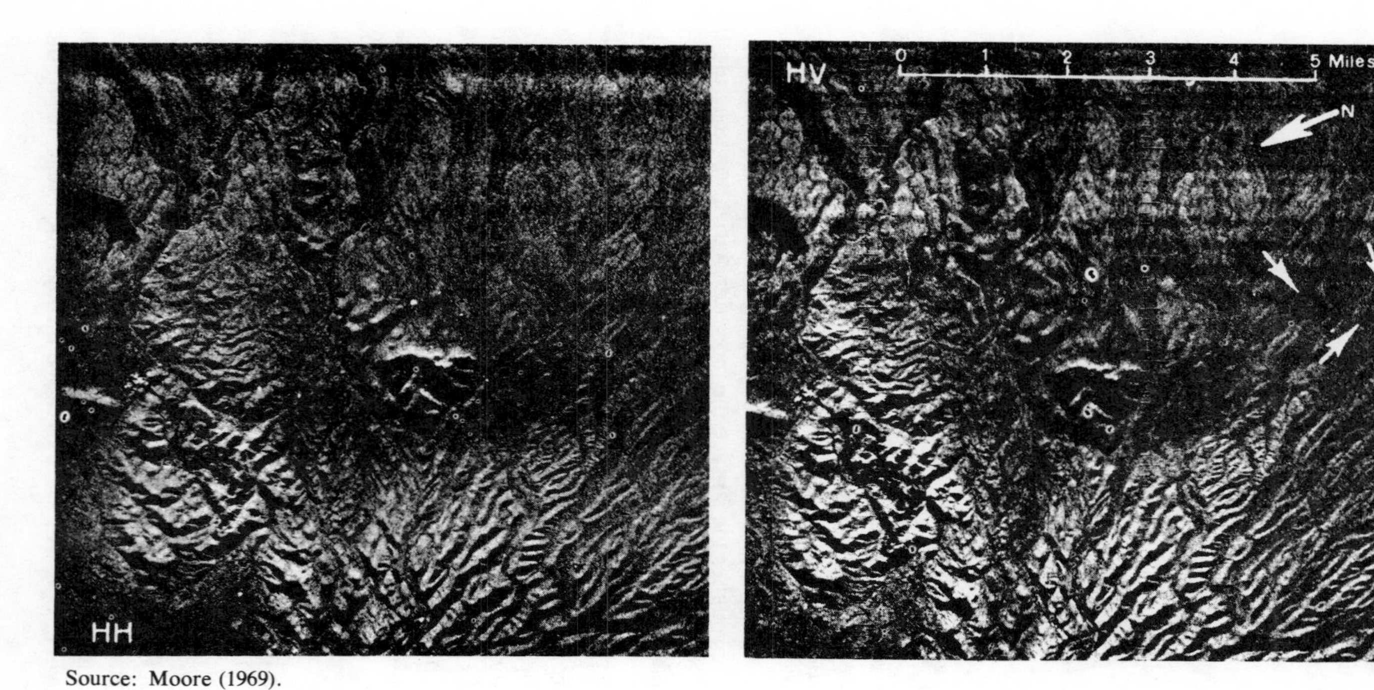

Source: Moore (1969).

Figure 7-1. Like-(*HH*) and Cross-polarized (*HV*) Radar Imagery, Twin Buttes Area, Arizona. Cross-polarized component clearly shows previously unknown outcrops of pyroxene rhyodacite, which had not been detected in field mapping, and examination of either aerial photographs or the like-polarized imagery.

echo strength would, *on the average*, increase for both polarizations. However, detailed examinations usually reveal little correlation between the fast fluctuations of the *polarized* and depolarized echo.

Examples, from chapter 5, of the lack of correlation between the fluctuations for different polarizations are reviewed here. Figure 5-1 (sec. 5.1) shows recordings of *HH* and *HV* echo for trees. These recordings were taken simultaneously and it is clear that the *HH* and *HV* echo strengths are not highly correlated. Figure 5-16 (sec. 5.5) shows recordings of *HH* and *VV* echo. The general character of the *VV* echo is different from that of the *HH* echo for this figure. The recordings of figure 5-16 were not made simultaneously. However, both measurements were made within a few minutes of one another and the general sea condition was unchanged. The echoes for circular polarizations have different characteristics than those for linear, but there is an interdependence between echoes for linear and circular polarizations that is considered later in this chapter.

7.2 Coherency, Statistical Independence, and Correlation

In general, the relative amplitudes and phases of the variously polarized echo components must be described in statistical terms. Two variables X and Y are statistically independent if the value of X is unrelated to the value of Y. Two variables are dependent if one of them is a function of the other. Coherence (incoherence) and correlation (noncorrelation) between waves are related to statistical dependence (independence).

Waves are said to be coherent if the phases of the various waves are fixed relative to each other; if the phases are randomly distributed over 2π, the waves are said to be incoherent. Consider the sum of a number of sinusoidal waves, all of the same frequency but of different amplitude and phase:

$$\sum_k A_k \cos(\omega t - \alpha_k) = \left(\sum_k A_k \cos\alpha_k\right)\cos\omega t + \left(\sum_k A_k \sin\alpha_k\right)\sin\omega t \qquad (7.1)$$

If all the phases are the same, for example, $\alpha_k = 0$, then the amplitudes of the cosine and sine terms will be $\Sigma_k A_k$ and 0, respectively. Then the amplitudes add, and the intensity is proportional to $(\Sigma_k A_k)^2$. Suppose the α's are completely independent of one another so that each α is equally likely to have any value between 0 and 2π, independent of the others. Then $\Sigma_k A_k \cos\alpha_k$ will be much less than $\Sigma_k A_k$ because there will be about as many terms with positive values of $\cos\alpha_k$ as with negative.

The square of the first summation on the right side of equation (7.1) is

$$\left(\sum_k A_k \cos\alpha_k\right)^2 = \sum_k A_k^2 \cos^2\alpha_k + \sum_{k\neq\ell} A_k A_\ell \cos\alpha_k \cos\alpha_\ell \tag{7.2}$$

It is easily seen that the average of equation (7.2) is $(1/2)\Sigma_k A_k^2$. The other summation in equation (7.1) gives an equal term so that the mean-square amplitude, or mean intensity, averaged over phases is the sum of the individual intensities. This is the state of complete incoherence in which for n waves of equal amplitude the intensity is n times the intensity of a single wave, rather than n^2 as for the coherent case.

If waves are statistically independent, they are incoherent but the converse is not always true. For example, waves of constant amplitude that differ by precise frequencies are dependent. However, they are incoherent because the second term on the right of equation (7.2) is zero. Roughly, two variables are correlated if the presence of one suggests the presence of the other. Thus, correlation and noncorrelation are also somewhat related to dependence and independence. If X and Y are independent, they are uncorrelated. However, as will be seen below, the converse is not always true.

The correlation coefficient[a] R_{XY} of variables X and Y is defined in terms of various averages. Let a bracket denote average; for example $\langle X\rangle$ means the average value of X. Then

$$R_{XY} = \frac{\langle XY\rangle - \langle X\rangle\langle Y\rangle}{\sqrt{(\langle X^2\rangle - \langle X\rangle^2)(\langle Y^2\rangle - \langle Y\rangle^2)}} \tag{7.3}$$

and it must lie between the limits

$$-1 \leq R_{XY} \leq 1$$

The reader may recognize the term $(\langle X^2\rangle - \langle X\rangle^2)$ as being the variance of X, and the term $(\langle XY\rangle - \langle X\rangle\langle Y\rangle)$ is the covariance of X and Y. If X and Y are independent, the covariance of X and Y is zero; therefore, R_{XY} is zero if X and Y are independent. If R_{XY} is positive, X and Y vary in harmony; if R_{XY} is negative, X and Y vary in opposition. For example, $R_{XY} = 1$ if $X = Y$; $R_{XY} = -1$ if $X = -Y$. If $X = \sin\theta$ and $Y = \cos\theta$, X and Y are dependent and yet $R_{XY} = 0$. It can be seen from equation (7.3) that $R_{XY} = 0$ because $\langle\sin\theta\rangle = \langle\cos\theta\rangle = \langle\cos\theta \sin\theta\rangle = 0$. Therefore, it is not necessarily true that X and Y are statistically independent if X and Y are uncorrelated. However if X and Y are independent, the variables X and Y are uncorrelated.

This section will use the relationships between the various linearly and circularly polarized echo components that are given in section 3.1. The symbols a_{ij} and $c_{k\ell}$ denote ratios of received-to-transmitted electric field

---

[a]See, for example, Petr Beckmann (1967, sec. 2.6).

strengths for transmitted-and-received-linear and transmitted-and-received-circular polarizations, respectively. From section 3.1, c_{ij} can be expressed in terms of the a_{ij} as follows:

$$|c_{11}| = \left| \frac{a_{xx} - a_{yy}}{2} + ja_{xy} \right|$$

$$|c_{12}| = \left| \frac{a_{xx} + a_{yy}}{2} \right|$$

and

$$|c_{22}| = \left| \frac{a_{xx} - a_{yy}}{2} - ja_{xy} \right| \tag{7.4}$$

The a's and c's are complex quantities that are proportional to electric field. To equate the various radar cross sections (proportional to electric field squared), care must be taken to account for phase appropriately in equations (7.4). It should be apparent to the reader that $\sigma_{ij} = K|a_{ij}|^2$ and $\sigma_{ij}/\sigma_{k\ell} = |a_{ij}|^2/|a_{k\ell}|^2$. Therefore,

$$\sigma_{11} = K \left| \frac{a_{xx} - a_{yy}}{2} + ja_{xy} \right|^2$$

$$\sigma_{12} = K \left| \frac{a_{xx} + a_{yy}}{2} \right|^2 \tag{7.5}$$

and

$$\sigma_{22} = K \left| \frac{a_{xx} - a_{yy}}{2} - ja_{xy} \right|^2$$

If a target area is composed of n scatterers with cross section for polarization ij of each scatterer denoted by σ_{ij}^k, the total cross section is

$$\sigma_{ij} = \left| \sum_{k=1}^{n} \sqrt{\sigma_{ij}^k} \exp (j\psi_{ij}^k) \right|^2 = |\sqrt{\sigma_{ij}}\, e^{j\psi_{ij}}|^2$$

or (7.6)

$$\sigma_{ij} = \frac{1}{K} \left| \sum_{k=1}^{n} a_{ij}^k \right|^2 = \frac{1}{K} \left| |a_{ij}|\, e^{j\psi_{ij}} \right|^2$$

The echo (σ_{ij}) will fluctuate if either the individual magnitudes σ_{ij}^k or the phases ψ_{ij}^k fluctuate. It is also clear that ψ_{ij}, the phase of a_{ij}, will change if there is a change in any of the σ_{ij}^k or ψ_{ij}^k. Therefore, for a complex target consisting of many scatterers, there is a change in the resultant echo phase associated with a change in echo intensity.

Now consider the relative phases for two polarizations, ψ_{gh} and ψ_{ij}, if the amplitudes ($|a_{gh}|$ and $|a_{ij}|$) or the intensities, σ_{gh} and σ_{ij}, are independent. As noted above, changes in the echo magnitudes will be caused by either changes in the phases or in the magnitudes of the individual scatterers, and these changes will also cause changes in the resultant phases. In other words, if $|a_{gh}|$ and $|a_{ij}|$ (or σ_{gh} and σ_{ij}) are independent, then the ψ_{gh} and ψ_{ij} are expected to be independent. Therefore, if the magnitudes of the echoes for various polarizations are independent, the echoes are incoherent.

For computing average values, certain simplifications can be made for targets for which the a's are statistically independent (waves are incoherent). Let a bar denote a time average. If a_{xx}, a_{yy}, and a_{xy} are statistically independent of one another, then (from eqs. (7.5)) average cross section can be expressed as

$$\overline{\sigma}_{11} = \overline{\sigma}_{22} = \frac{\overline{\sigma}_{xx}}{4} + \frac{\overline{\sigma}_{yy}}{4} + \overline{\sigma}_{xy}$$

and (7.7)

$$\overline{\sigma}_{12} = \overline{\sigma}_{21} = \frac{\overline{\sigma}_{xx}}{4} + \frac{\overline{\sigma}_{yy}}{4}$$

For scatterers of this type, it is obvious that $\overline{\sigma}_{11}$ would exceed $\overline{\sigma}_{12}$.

7.3 A Simplified Polarization Model for Rough Terrain

In principle, it is possible to calculate radar cross section for any polarization if all information contained in the polarization matrix is known. The various elements of the matrix, in general, fluctuate with time. Therefore, a voluminous amount of relative amplitude and phase data would be necessary to calculate expected average values. There is a need to calculate averages for various polarizations when only a few data on averages exist. To this end, equations are developed by using a model consisting of (1) scatterers that reradiate coherently and do not depolarize and (2) scatterers that may or may not depolarize and reradiate incoherently. The model is applicable to angles near vertical incidence; that is, angles that are often of interest for altimeters, satellite-borne radar, and radar astronomy.

In general, the field backscattered by a rough surface has a constant (or slowly varying) component and a random component. A field is called coherent if the phase is constant or varies in a deterministic manner. A field is called incoherent if the phase is random and distributed over an interval of phase 2π. The mean-power density of the sum of incoherent fields is the arithmetic sum of the separate power densities; the total power density of coherent fields is obtained by summing the individual fields vectorially, and determining the total power from the resultant total field.

In conformity with above, the a's will be expressed as

$$a_{ij} = a^c_{ij} + a^i_{ij}$$

where the superscripts c and i denote coherent and incoherent, respectively.

Since by definition the various a^c and a^i are incoherent of one another, the average cross sections $\overline{\sigma}_{11}$ and $\overline{\sigma}_{12}$ can be obtained through use of the equations in (7.5) as follows:

$$\begin{aligned}\overline{\sigma}_{11} = K\overline{|c_{11}|^2} &= K\overline{\left|\frac{a^c_{xx} - a^c_{yy}}{2} + ja^c_{xy}\right|^2} \\ &+ K\overline{\left|\frac{a^i_{xx}}{2}\right|^2} + K\overline{\left|\frac{a^i_{yy}}{2}\right|^2} + K\overline{|a^i_{xy}|^2} \qquad (7.8) \\ \overline{\sigma}_{12} = K\overline{|c_{12}|^2} &= K\overline{\left|\frac{a^c_{xx} + a^c_{yy}}{2}\right|^2} + K\overline{\left|\frac{a^i_{xx}}{2}\right|^2} + K\overline{\left|\frac{a^i_{yy}}{2}\right|^2}\end{aligned}$$

Therefore, it is seen that by comparison of equations (7.8) with equations (7.5) and (7.7)

$$\begin{aligned}\overline{\sigma}_{11} &= \overline{\sigma}^c_{11} + \overline{\sigma}^i_{11} \\ \overline{\sigma}_{12} &= \overline{\sigma}_{21} = \overline{\sigma}^c_{12} + \overline{\sigma}^i_{12}\end{aligned}$$

Assume that the *depolarized* echo is caused by a large collection of scatterers for which $\overline{\sigma}_{xx} = \overline{\sigma}_{yy}$. Also assume that there are large surfaces with small scale roughness that is smooth with respect to wavelength, that these surfaces do not depolarize, and that the scattering is equal for xx and for yy polarizations. With these assumptions the coherent and incoherent contributions are from the large surfaces and from the depolarizing scatterers, respectively. Therefore, the model yields the following constraints:

$$\begin{aligned}\overline{a^{i\,2}_{xx}} &= \overline{a^{i\,2}_{yy}} \\ a^c_{xy} &= a^c_{yx} = 0 \qquad (7.9) \\ a^c_{xx} &= a^c_{yy}\end{aligned}$$

Since a^c_{xy} is equal to zero for this model, $\overline{\sigma}_{11}$ is equal to $\overline{\sigma}_{22}$ (see σ_{11} and σ_{22} in eqs. (7.5) and $\overline{\sigma}_{11}$ in eqs. (7.8)). Recall that $\sigma_{ij} = K|a_{ij}|^2$. Then the equations in (7.8) with the constraints of the model give

$$\begin{aligned}\sigma^c_{11} &= 0 \\ \sigma^c_{12} &= \sigma^c_{21} = \sigma^c_{xx} = \sigma^c_{yy} \\ \overline{\sigma}_{11} &= \overline{\sigma}_{22} = \overline{\sigma}^i_{xy} + (1/2)\overline{\sigma}^i_{xx} \\ \overline{\sigma}_{12} &= \overline{\sigma}_{21} = \overline{\sigma}^c_{xx} + (1/2)\overline{\sigma}^i_{xx} \qquad (7.10)\end{aligned}$$

where

$$\overline{\sigma}_{xy} = \overline{\sigma}^i_{yx}, \quad \overline{\sigma}^i_{xx} = \overline{\sigma}^i_{yy} \quad \text{and} \quad \overline{\sigma}^c_{xx} = \overline{\sigma}^c_{yy}$$

Contained within the assumptions used thus far is the restriction that whatever depolarization exists is caused by the incoherent scatterers. On the other hand, the energy scattered with the same polarization as transmitted is caused by both the coherent and the incoherent scatterers. The following equations express the words above:

$$\overline{\sigma}_{xy} = \overline{\sigma}^i_{xy} = \overline{\sigma}^i_{xx}/r$$

$$\overline{\sigma}_{xx} = \overline{\sigma}^c_{xx} + \overline{\sigma}^i_{xx}$$

Depolarized scattering from natural surfaces is the result of current flowing in all possible directions. We expect any one direction of current, on the average, to be as likely as any other direction. Thus, we expect the ratio $\overline{\sigma}^i_{xx}/\overline{\sigma}^i_{xy}$ to be 3, the same as $\overline{\sigma}_{xx}/\overline{\sigma}_{xy}$ for a collection of randomly oriented dipoles (sec. 3.5). Therefore, for an extended area consisting of randomly oriented depolarizing scatterers and nondepolarizing scatterers that scatter coherently:

$$\begin{aligned} \overline{\sigma}_{xy} &= \overline{\sigma}_{yx} = (1/3)\overline{\sigma}^i_{xx} \\ \overline{\sigma}_{xx} &= \overline{\sigma}_{yy} = \overline{\sigma}^c_{xx} + \overline{\sigma}^i_{xx} \\ \overline{\sigma}_{12} &= \overline{\sigma}_{21} = \overline{\sigma}^c_{xx} + (1/2)\overline{\sigma}^i_{xx} \\ \overline{\sigma}_{11} &= \overline{\sigma}_{22} = (5/6)\overline{\sigma}^i_{xx} \end{aligned} \tag{7.11}$$

where $\overline{\sigma}^i_{xx}$ and $\overline{\sigma}^c_{xx}$ are the average cross sections with xx polarization for the depolarizing and the nondepolarizing scatterers, respectively. By using the various constraints, the equations in (7.11) can be expressed as

$$\begin{aligned} \overline{\sigma}_{12} &= \overline{\sigma}_{21} = \overline{\sigma}_{xx} - (3/2)(\overline{\sigma}_{xy}) \\ \overline{\sigma}_{11} &= \overline{\sigma}_{22} = (5/2)(\overline{\sigma}_{xy}) \\ \overline{\sigma}_{xy} &= \overline{\sigma}_{yx} = (2/5)(\overline{\sigma}_{11}) \\ \overline{\sigma}_{xx} &= \overline{\sigma}_{yy} = \overline{\sigma}_{12} + (3/5)(\overline{\sigma}_{11}) \end{aligned} \tag{7.12}$$

Other useful equations follow:

$$\frac{\overline{\sigma}_{12}}{\overline{\sigma}_{22}} = \frac{\overline{\sigma}_{21}}{\overline{\sigma}_{22}} = \frac{\overline{\sigma}_{21}}{\overline{\sigma}_{11}} = \frac{\overline{\sigma}_{12}}{\overline{\sigma}_{11}} = \frac{\overline{\sigma}_{xx}/\overline{\sigma}_{xy} - 3/2}{5/2} = \frac{2}{5}\frac{\overline{\sigma}_{xx}}{\overline{\sigma}_{xy}} - \frac{3}{5} \tag{7.13}$$

$$\frac{\overline{\sigma}_{yy}}{\overline{\sigma}_{yx}} = \frac{\overline{\sigma}_{yy}}{\overline{\sigma}_{xy}} = \frac{\overline{\sigma}_{xx}}{\overline{\sigma}_{yx}} = \frac{\overline{\sigma}_{xx}}{\overline{\sigma}_{xy}} = \frac{5}{2}\frac{\overline{\sigma}_{12}}{\overline{\sigma}_{11}} + \frac{3}{2} \tag{7.14}$$

$$\frac{\overline{\sigma}^c_{xx}}{\overline{\sigma}^i_{xx}} = \frac{1}{3}\frac{\overline{\sigma}_{xx}}{\overline{\sigma}_{xy}} - 1 \tag{7.15}$$

and

$$\frac{\overline{\sigma^c_{xx}}}{\overline{\sigma^i_{xx}}} = \frac{1}{3}\left[\frac{5\overline{\sigma}_{12}}{2\overline{\sigma}_{11}} + \frac{3}{2}\right] - 1 = \frac{5\overline{\sigma}_{12}}{6\overline{\sigma}_{11}} - \frac{1}{2} \tag{7.16}$$

Effects of the ratio $\overline{\sigma^c_{xx}}/\overline{\sigma^i_{xx}}$ are illustrated in table 7-1 and figure 7-2. In interpreting the table and the figure, recall that without a superscript σ denotes total cross section:

$$\overline{\sigma_{xx}} = \overline{\sigma^c_{xx}} + \overline{\sigma^i_{xx}}$$

In practice, the depolarizing scatterers are small (ripples, twigs, leaves) and are moved briskly by the wind, but the nondepolarizing scatterers tend to be large (waves, tree trunks, and limbs) and they tend to have less movement. Therefore, the ratio $\overline{\sigma^c_{xx}}/\overline{\sigma^i_{xx}}$ that gives cross section of the nondepolarizing scatterers relative to that for the depolarizing scatterers tends to be a ratio of cross sections for the constant or slower fluctuating echoes relative to the fast fluctuations.

Notice that with this model the ratio $\overline{\sigma}_{xy}/\overline{\sigma}_{xx}$ becomes increasingly small for increasingly large fractions of coherent (quasi-specular) scattering, $\overline{\sigma}_{11}$ always exceeds $\overline{\sigma}_{xy}$ by 4 dB, and $\overline{\sigma}_{11}$ may be either less than or greater than $\overline{\sigma}_{12}$.

Equations (7.13) through (7.16) are used in section 7.4 to study reflections from the moon versus incidence angle, in section 7.8 to study trees, and in section 7.12 to study the sea.

7.4 Use of the Polarization Model for the Moon

The lunar surface is an excellent target for testing theories of scattering by rough surfaces. It is especially useful for studying the simple polarization model because its circular symmetry allows the various constraints of that model to be satisfied. Further, because the rapid relative motion between the moon's surface and that of the earth, averages are obtained rapidly.

A radar pulse transmitted from earth illuminates a finite area (range ring) of the moon. As the range ring recedes with the propagation of the pulse, its circumference increases and its width diminishes. In fact, the two effects exactly compensate for each other so that the illuminated area remains constant. To illustrate, let θ be the angle between the direction of propagation and the tangent plane at the moon's surface (see fig. 7-3), let a be the moon's radius (1738 km), let τ be the radar pulse length, and let c be the velocity of propagation (3×10^8 m/sec). Then, from simple geometry, the radius of the range ring is $a \cos \theta$, and its width (see sec. 2.3 for

Table 7-1
Calculated Polarization Ratios as a Function of Ratio of Coherent to Incoherent Cross Sections

$\frac{\overline{\sigma^c_{xx}}}{\overline{\sigma^i_{xx}}}$	$\frac{\overline{\sigma_{xy}}}{\overline{\sigma_{xx}}}$ (dB)	$\frac{\overline{\sigma_{11}}}{\overline{\sigma_{xx}}}$ (dB)	$\frac{\overline{\sigma_{12}}}{\overline{\sigma_{xx}}}$ (dB)	$\frac{\overline{\sigma_{12}}}{\overline{\sigma_{11}}}$ (dB)
0	− 4.8	− 0.8	−3.0	−2.4
1/3	− 6.0	− 2.0	−2.0	0
1	− 7.8	− 3.8	−1.2	+2.6
10	−15.2	−11.2	−0.2	11.0
20	−18.0	−14.0	−0.1	13.9
30	−19.7	−15.7	−0.1	15.6
100	−24.8	−20.8	~0	20.8

discussion on range resolution) is $(c\tau/2)\sec\theta$. Therefore, the illuminated area is $a\pi c\tau$.

For backscatter from the moon, the incidence angle varies through all values in less than 12 msec. Therefore, a measurement of echo power as a function of delay time t_d (from the time that the pulse first reaches the moon) is a measurement of echo power as a function of incidence angle. Again from simple geometry, the sine of θ in terms of the time delay t_d is $1 - ct_d/2a$. Values of the grazing angle θ versus t_d for the moon are included in figure 7-3.

Average circularly polarized echo power $\overline{P}_{12}$ drops rapidly as the angle θ departs from 90° (fig. 7-3). For much smaller angles (longer delay times), the variation is slower and tends to approach *a* $\sin\theta$ dependence (Evans and Hagfors 1968). From this it seems that for near vertical incidence the scatterers tend to be primarily quasi-specular, and for small values of θ they tend to be diffuse (the $\sin\theta$ dependence implies that $\sigma^\circ \propto \sin\theta$ because illuminated moon area is independent of incidence angle). J.V. Evans and Tor Hagfors (1968) reported that $\overline{P}_{11}$ varies as $\cos\eta$ (fig. 7-4), where η is the complement of θ. Since η is the complement of θ, figure 7-4 illustrates that $\bar{\sigma}^\circ_{11}$ varies directly with $\sin\theta$ for θ at least as small as 8° ($\sin 8° = 0.14$). This θ dependence is characteristic of rough (diffuse) terrestrial surfaces (secs. 3.6 and 6.8).

Few moon experiments have been made with linear polarization because of problems caused by polarization rotation due to the Faraday effect. However, Evans and Hagfors (1966) reported both linearly and circularly polarized measurements that were made at 23 cm. Results for linear polarization are given in figure 7-5 and some of their circular polarization data have already been discussed (figs. 7-3 and 7-4).

If the polarization model is valid and the experimental data of figures 7-3

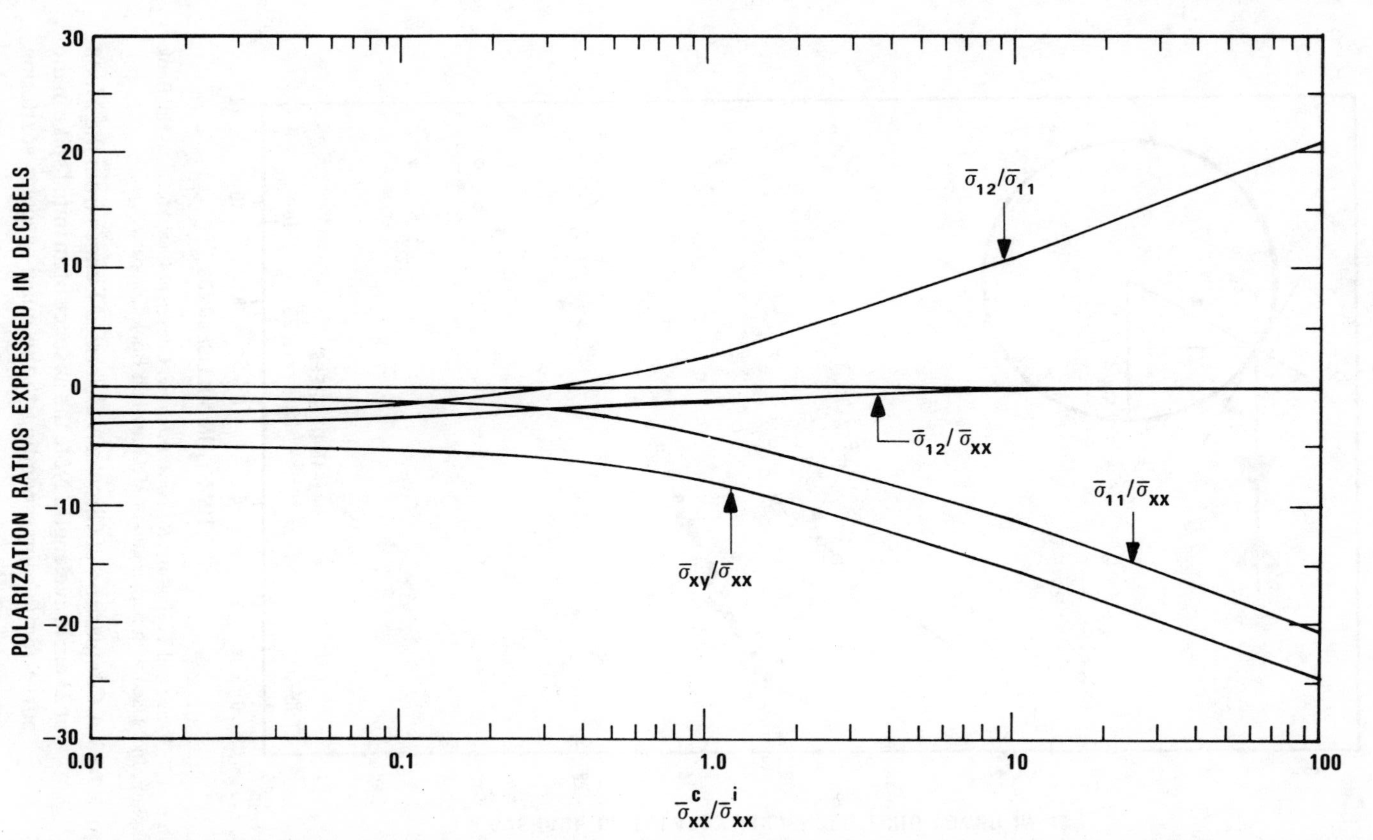

Figure 7-2. Polarization Ratios Versus Cross Section of Quasi-specular Reflectors Relative to That for Diffuse Scatterers Versus Polarization Ratio.

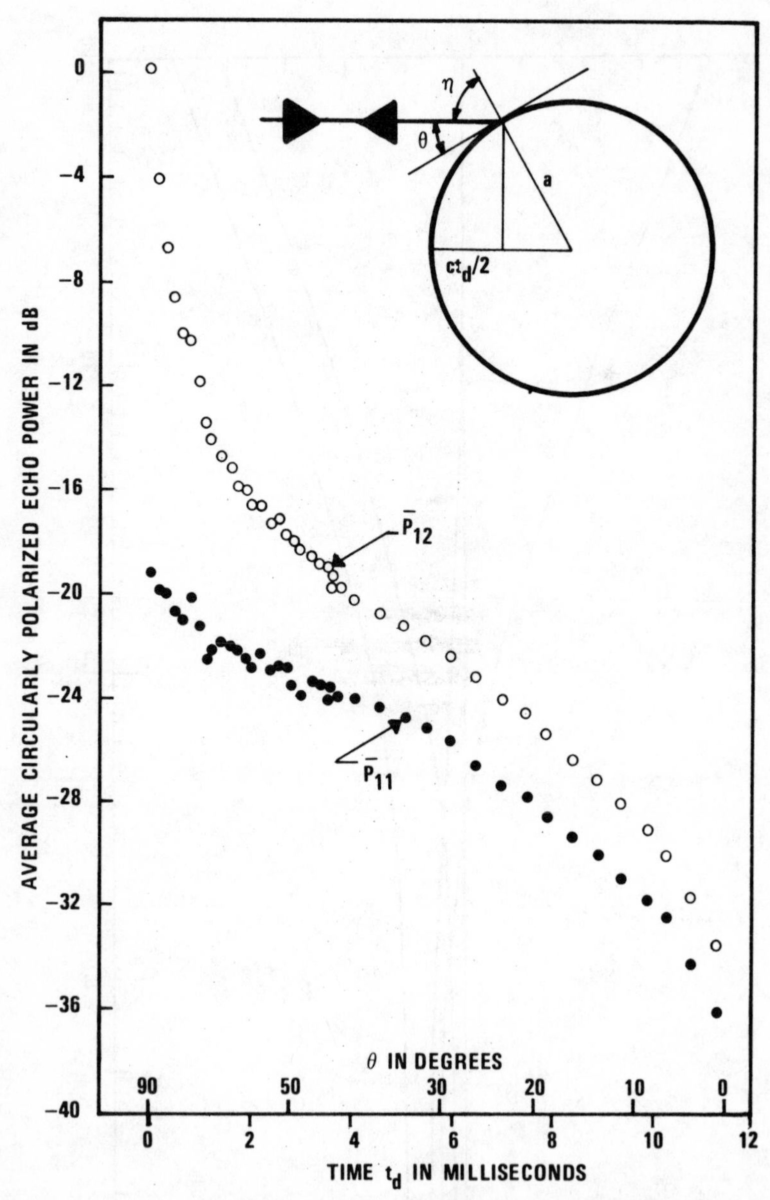

Source: J.V. Evans and T. Hagfors, *Radar Astronomy*. Copyright 1968, McGraw-Hill Book Company, Inc. Used with permission of McGraw-Hill Book Company.

Figure 7-3. A Comparison of Circularly Polarized Echoes from the Moon at 23 cm Wavelength. Note the absence of an initial spike in the curve for $\overline{P}_{11}$ corresponding to the quasi-specular scattering observed in $\overline{P}_{12}$.

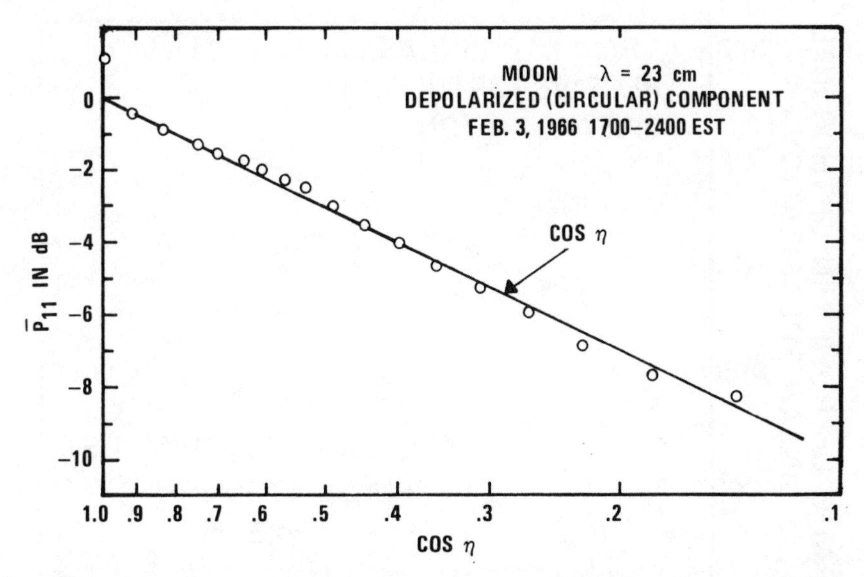

Source: J.V. Evans and T. Hagfors, *Radar Astronomy*. Copyright 1968, McGraw-Hill Book Company, Inc. Used with permission of McGraw-Hill Book Company.

Figure 7-4. The Angular Scattering Law Observed for $\bar{P}_{11}$ Plotted in Figure 7-3.

and 7-5 were without error, equations (7.15) and (7.16) should each yield the same ratio for power from specular scatterers to that from diffuse scatterers. Because of the scatter of data, especially large for linear polarization, the range of delay times of 5 to 10 msec will be considered. From figure 7-3, the applicable range of σ_{12}/σ_{11} is 2.6 to 3.5 dB; equation (7.16) indicates that $\sigma^c_{xx}/\sigma^i_{xx}$ is 1.0 to 1.4. The median is 3.1 dB and this implies that $\sigma^c_{xx}/\sigma^i_{xx}$ is 1.2. For the 5 to 10 msec region of figure 7-5, the range for σ_{xx}/σ_{xy} is roughly 5 to 10, indicating that $\sigma^c_{xx}/\sigma^i_{xx}$ is between 0.6 and 2.3. The median of σ_{xx}/σ_{xy} is 6.9 (8.4 dB), which indicates that $\sigma^c_{xx}/\sigma^i_{xx}$ is 1.3. Therefore, within the accuracy of the data, the measurements using linear and circular polarizations support the polarization model, and the results indicate there is a substantial contribution to σ_{12} and σ_{xx} by quasi-specular reflections from the moon, even for large delay times (small grazing angles).

Another test of the model can be made by calculating σ_{xx}/σ_{xy} expected for a given value of σ_{12}/σ_{11}. For example, by using the 3.1 dB median for σ_{12}/σ_{11}, equation (7.14) indicates that σ_{xx}/σ_{xy} will be 6.6 (8.2 dB). Recall that the median of σ_{xx}/σ_{xy} from figure 7-5 was 6.9 (8.4 dB) for delay times of 5 to 10 msec. Therefore, from figures 7-3 and 7-5 it seems that the model provides a valid method for calculating σ_{xx}/σ_{xy} in terms of σ_{12}/σ_{11}.

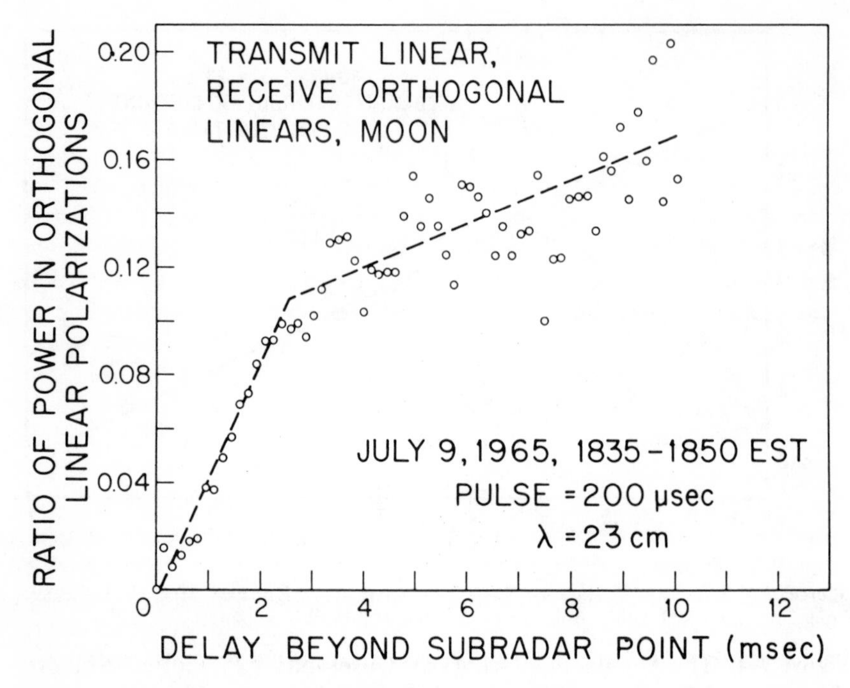

Source: J.V. Evans and T. Hagfors, *Radar Astronomy*. Copyright 1968, McGraw-Hill Book Company, Inc. Used with permission of McGraw-Hill Book Company.

Figure 7-5. The Ratio of the Power Observed between the Expected Linear Component and One Orthogonal to it. These measurements were made at a wavelength of 23 cm and complement the results obtained with circularly polarized waves shown in Figure 7-3.

As stated previously, it seems plausible that the constraints of equation (7.9) are valid for the moon. However, the assumption $\overline{\sigma^i_{xy}} = (1/3)\,\overline{\sigma^i_{xx}}$ is a relationship for dipoles in free space which may not be valid for the present application. Nonetheless, the calculations in this section seem to support the validity of equations (7.11) for the moon.

Echo from Land, Principally Trees

Sections 7.5 through 7.8 contain data on the relationships between the horizontally, vertically and circularly polarized echo components from terrain. The interdependencies of amplitude fluctuations, relative phases, and average amplitudes are considered. By using the results of section 7.3, the relative power backscattered by depolarizing and nondepolarizing scatterers is deduced.

7.5 Amplitude Fluctuations

In general, echo from ground contains a constant component caused by the "fixed" targets plus fluctuating echoes. Theoretically, for a large number of independently moving scatterers, the probability density function of received power is exponential and is the well known "Rayleigh" distribution. Results from World War II studies (sec. 5.2) showed that heavily wooded terrain with strong winds is approximately Rayleigh distributed, and the distribution for the same target area under less windy conditions was peaked. Radar echo is, therefore, sometimes peaked at a value near the amplitude of the constant component caused by the fixed scatterers.

A number of measurements reported by H.D. Ivey, M.W. Long, and V.R. Widerquist (1955) showed that for both pine and deciduous trees the statistical distributions approximated the Rayleigh shape at X_b and K_a bands under windy conditions. Later, other Georgia Tech personnel (Hayes and Walsh 1959) studied vegetation at X band and found from repeated measurements that distributions with heavy winds were approximately Rayleigh distributed for all polarizations available: *HH, VV, HV, VH, RL, RR, LR,* and *LL*. The distributions were sometimes peaked if there was no wind. Figure 5-4 is typical of the distributions observed by Hayes and Walsh, and figure 5-5 shows a peaking that was sometimes observed.

N.W. Guinard et al. (1967) studied echo distributions at *P, L, C,* and *X* bands measured simultaneously with airborne equipment for horizontal and vertical polarizations. Normalized radar cross section was reported in terms of the 10, 50 (median), and 90 percentile values of its distribution. Two types of statistical processes were determined for the cross-section data—the Rayleigh distribution between 10 and 90 percentile values, and one that departed radically from this distribution. The one that departed radically from the Rayleigh distribution was identified as a Ricean distribution produced by a strongly reflecting point target plus random "Rayleigh" scatterers. In other words, statistical characteristics obtained with airborne equipment at four radar bands and for horizontal and vertical polarizations agree with that reported much earlier (sec. 5.2).

From above it seems that independent of polarization and radar frequency, amplitude distributions can be described as the sum of a constant echo plus fluctuating echoes. Although a sensitivity of amplitude distribution to polarization has not been reported, or at any rate clearly underscored, such a sensitivity does exist in principle. For example, scatterers that cause depolarization when linear polarization is transmitted also contribute to the nondepolarized echo. By definition, the nondepolarizing echo only contributes to the polarization transmitted. Therefore, with sufficiently precise measurements it should be possible to detect differences, although they might be quite small for trees, between the distributions for

VV and *VH* or between *HH* and *HV* echoes. Obviously differences can also exist between *VV* and *HH* echoes; for example, multipath transmission caused by ground reflections will cause a difference between the various polarizations. Circular-polarized echo is describable in terms of *VV, HH*, and *HV* echo (see eqs. (7.5)). Therefore, in principle, statistical distributions for circular polarizations are different than those for linear.

Figures 5-1 and 5-2 are samples of pen recorder playouts of parallel and cross returns from deciduous and pine trees observed by R.D. Hayes et al. (1958). The data represent echo from a fixed range. In addition to illustrating the amplitude variation of return obtained for range-gated samples, the figures show an increase in fluctuation rate which generally accompanies an increase in wind speed. The radar (Hayes, Currie, and Long 1957) was operating at 9,375 MHz with a 0.25 microsecond pulse width and equal horizontal and vertical beamwidths of 1.86°. Notice that, in general, neither P_{HH} and P_{HV} nor P_{VV} and P_{VH} fluctuate together. A similar lack of correlation in fluctuations is observed for trees when transmitting and receiving circular polarizations.

Figures 5-1 and 5-2 show echo strength as a function of time, but they do not depict a dependency of frequency spectrum on polarization. As in the case of amplitude distributions, the author does not know of any clearly describable differences of frequency spectra for trees as a function of polarization.

7.6 Interdependence of Amplitude and Phase of Orthogonally Polarized Echoes

Figures 7-6 and 7-7 are examples of scatter diagrams showing the relationship between parallel- and cross-polarized amplitudes from trees as read from pen recordings with short-term smoothing such as those in figures 5-1 and 5-2. Those diagrams illustrate that there is little correlation in the short-term fluctuations of either P_{HH} and P_{HV} or P_{RR} and P_{RL}.[b] A similar lack of correlation in the short-term fluctuations was observed for trees when comparing P_{VV} with P_{VH} and when comparing P_{LL} with P_{LR}.

H.D. Ivey, M.W. Long, and V.R. Widerquist (1955) reported correlation coefficients that they calculated between *HH* and *HV* echoes and between *VV* and *VH* echoes for trees. Their results, which indicate there is also little correlation in the short-term fluctuations of echo amplitudes at 6.3 GHz and 35 GHz, are given in table 7-2. Scatter diagrams reported for trees at 6.3 GHz and 35 GHz are similar (Ivey, Long, and Widerquist 1955) to those of figures 7-6 and 7-7 for *X* band.

[b]The letters *L* and *R* represent left circular and right circular, respectively.

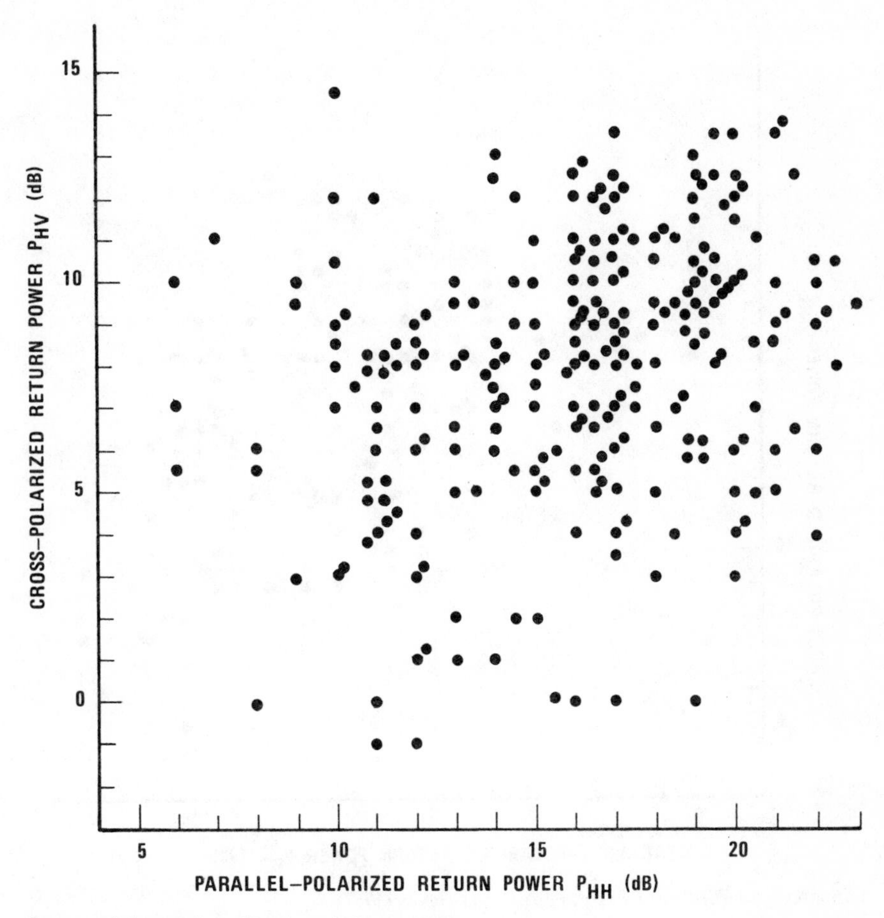

Source: Adapted from Harris and Hayes (1957).

Figure 7-6. Polarization Scatter Diagram of Dry Pine Tree Return with Horizontal Transmission.

The relative phase between two linearly polarized echo components has also been measured. A number of measurements were made on the phase between *HH* and *HV* echoes and between *VV* and *VH* echoes (Hayes and Long 1957). Examples are given in figures 7-8 and 7-9. The measurements were made at *X* band and were taken in a heavily wooded area. Usually the distributions were uniform; that is, any phase difference was as likely as any other. Sometimes there tended to be a peaking (fig. 7-9) that was roughly Gaussian in shape, but even in those cases the distribution was clearly random. One would expect that the peaking was caused by "fixed" targets, perhaps tree trunks. No discernible differences in prevailing weather conditions were associated with the distributions being either

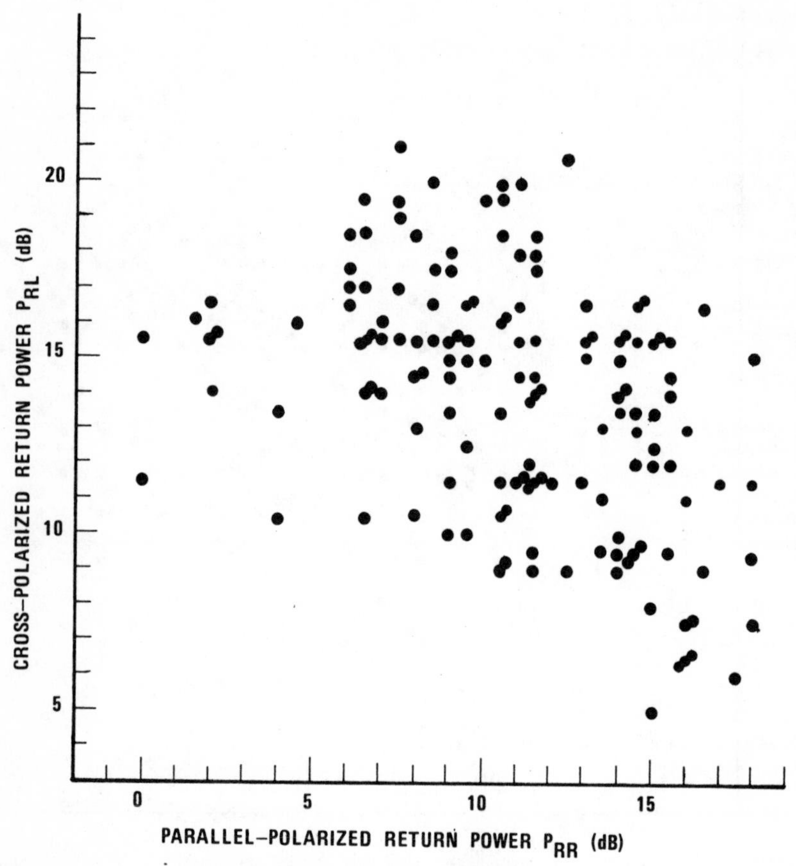

Source: Adapted from Eagle, Harris, and Hayes (1957).

Figure 7-7. Polarization Scatter Diagram of Dry Deciduous Tree Return with Right-Circular Transmission.

roughly uniform or peaked. For that matter, there were no results which indicated that any particular site would cause a more peaked distribution in phase than any other site. The result of random phase is interesting because it provides another random variable, in addition to amplitude, that must be used to describe the values of depolarized echo relative to that for receiving the same polarization that is transmitted. The randomness of the relative phase between *polarized* and depolarized echo components, in addition to the uncorrelated amplitudes of the relative components, provides strength to assumptions used later that the average of $|a_{xx} + a_{xy}|^2$ and the average of $|a_{xx} - a_{xy}|^2$ equal the average of $[|a_{xx}|^2 + |a_{xy}|^2]$. For examples, refer to section 3.2.

Table 7-2
Table of Linear Correlation Coefficients[a]

Linear Correlation Coefficients for Tree Clutter (35 GHz)

Tree Type	*Wind Speed (mph)*	*Transmission Polarization*	*Correlation Coefficient*
Deciduous	5-8	Horizontal	0.11
Deciduous	5-8	Horizontal	0.08
Deciduous	10	Horizontal	0.05
Deciduous	10	Horizontal	0.06
Coniferous	20	Horizontal	0.03
Deciduous	5-6	Vertical	0.13
Deciduous	5-8	Vertical	0.04
Deciduous	5-8	Vertical	0.08
Deciduous	8	Vertical	0.04
Deciduous	8	Vertical	0.00
Deciduous	8	Vertical	0.02
Deciduous	10	Vertical	0.08
Deciduous	10	Vertical	0.12
Coniferous	15	Vertical	0.10

Linear Correlation Coefficient for Tree Clutter (6.3 GHz)

Tree Type	*Wind Speed (mph)*	*Transmission Polarization*	*Correlation Coefficient*
Damp pine trees	0-5	Horizontal	−0.05
Damp pine trees	0-5	Horizontal	−0.22
Pine trees	0-5	Horizontal	0.08
Pine trees	0-5	Horizontal	−0.03
Pine trees	0-5	Horizontal	0.02
Pine trees	0-5	Horizontal	0.03
Pine trees	0-5	Vertical	0.02
Pine trees	0-5	Vertical	−0.01
Pine trees	0-5	Vertical	−0.02
Pine trees	0-5	Vertical	0.01
Deciduous trees[b]	0-5	Vertical	−0.09
Deciduous trees[b]	0-5	Vertical	0.17

[a]From Ivey, Long, and Widerquist (1955).
[b]Brown leaves, rain.

7.7 Average and Median Value Data, and Depression Angle Dependence

Various averaging methods are used to describe relative echo strength as a function of polarization. Most authors report the ratio of average (mean)

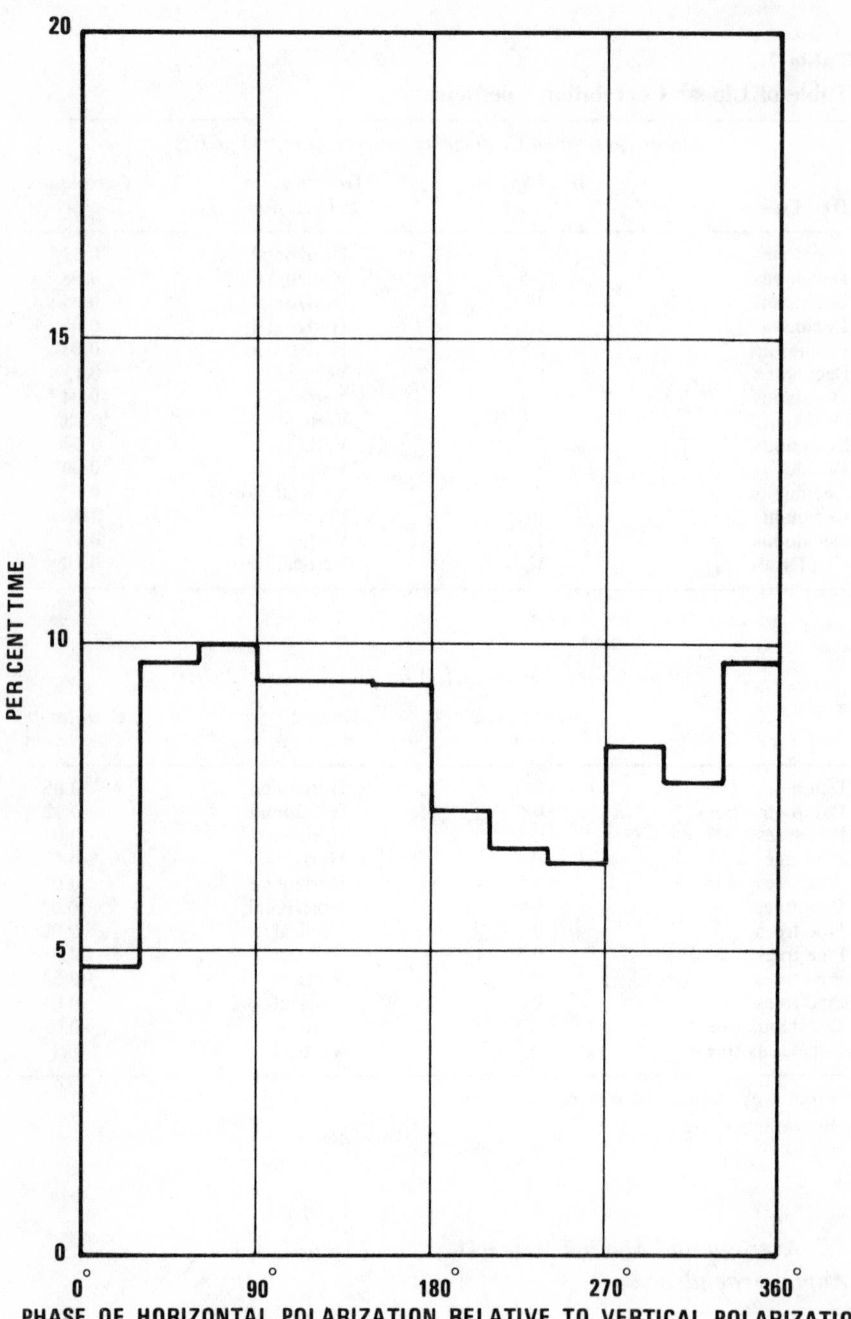

Source: Hayes and Long (1957).

Figure 7-8. Phase Distribution of *X*-Band Deciduous Tree Return. Vertically polarized transmission.

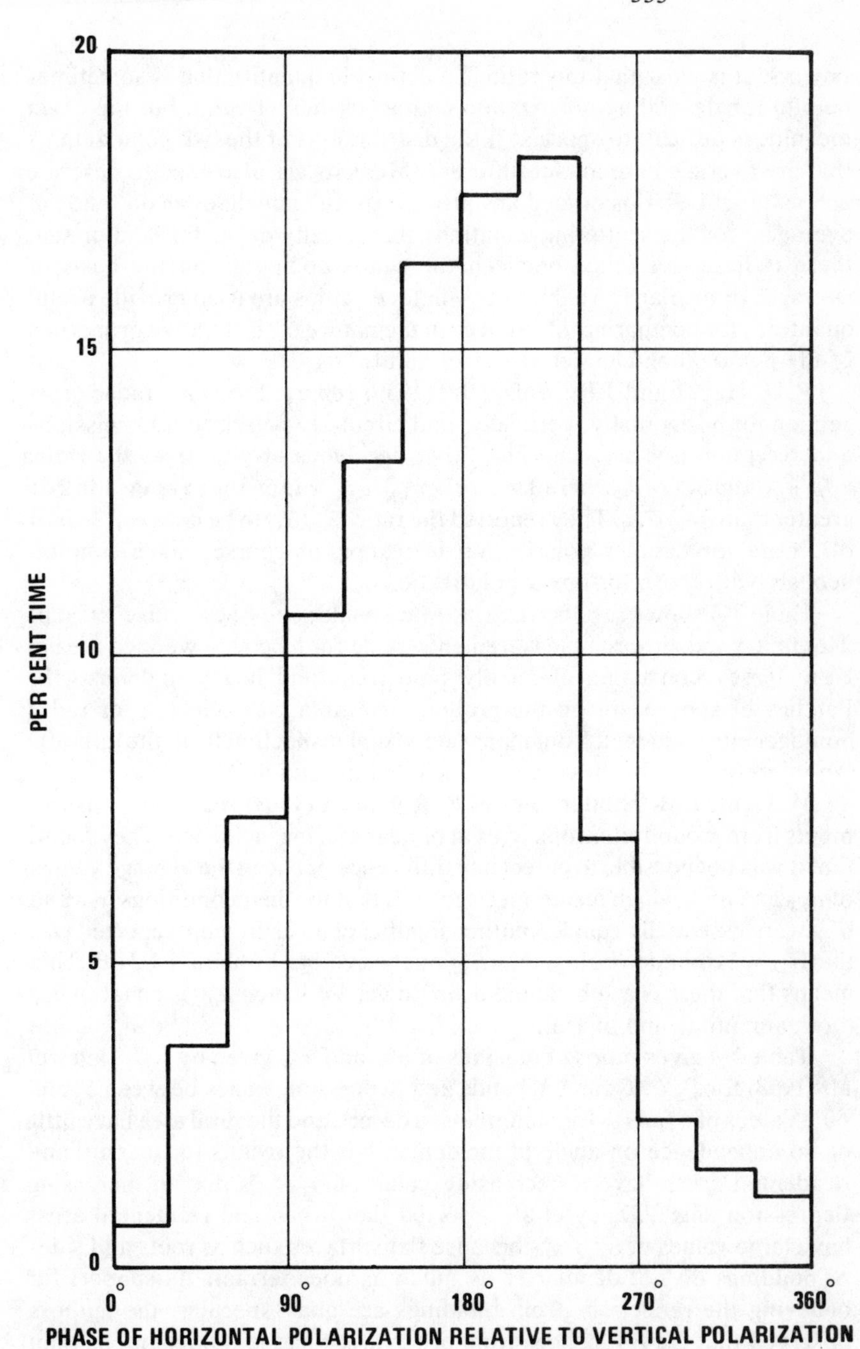

Source: Hayes and Long (1957).

Figure 7-9. Phase Distribution of *X*-Band Deciduous Tree Return. Vertically polarized transmission.

powers. It is clear that this ratio is a definable quantity that is sometimes helpful for describing polarization characteristics of echo, but the exact meaning is difficult to appraise if the distributions of the two polarizations that are averaged are in fact different. Medians are also used to describe echo strength. Ratios of medians provide useful guidelines as do ratios of averages; for the scattering situations that usually occur for land or sea, there is little difference between the ratios computed on the basis of averages or medians. At any rate, whatever ratios are used provide useful quantities for comparing differences in the nature of the relative properties of the polarization characteristics of natural targets.

R.D. Hayes and J.R. Walsh, Jr. (1959) reported average radar cross section for horizontally, vertically, and circularly polarized transmissions and reception (see sec. 6.8). For either deciduous or pine trees the ratios σ_{HH}/σ_{HV} and σ_{VV}/σ_{VH} where 3 to 10 dB; σ_{HH}/σ_{HV} was on the average 1 to 2 dB greater than $\overline{\sigma}_{VV}/\overline{\sigma}_{VH}$. They reported the ratio $\overline{\sigma}_{12}/\overline{\sigma}_{11}$ to be between 2 and 4 dB. Data for circular polarization were not, of course, taken simultaneously with those for linear polarizations.

Table 7-3 shows results (Ament, Macdonald, and Shewbridge 1959, p. 246) of X-band airborne measurements made for a heavily wooded area in New Jersey consisting of mostly pine trees and heavy undergrowth. Patches of snow were on the ground. The area was selected for radar homogeneity as judged from maps and visual inspection from the aircraft. An interpretation of these data is included in section 7.8.

M. Gent, I.M. Hunter, and N.P. Robinson (1963) reported measurements from ground sites for angles at or near grazing incidence. They found that it was not possible to detect any difference between the average values of σ_{RV} and σ_{RH}, which means (see sec. 3.2) that for their conditions $\overline{\sigma}_{VV}$ and $\overline{\sigma}_{HH}$ were essentially equal. Another significant measurement reported was that $\overline{\sigma}_{RR}$ was found to be less than $\overline{\sigma}_{RL}$ on the average by about 1 1/2 dB. This means that there is a substantial echo power for which a_{HH} is equal to a_{VV} (both amplitude and phase).

Table 7-4 gives ratios of medians of σ°_{VV} and σ°_{HV} given by J.C. Daley et al. (1968) for P, L, C, and X bands and depression angles between 5° and 60°. Values of $\sigma^\circ_{VV}/\sigma^\circ_{HV}$ for mountains, a desert, and the rural area have little or no dependence on angle of incidence, but the results for marshy and residential areas have a decreasing value of $\sigma^\circ_{VV}/\sigma^\circ_{HV}$ for an increasing depression angle. Daley et al. reported that urban and residential areas have large values of $\sigma^\circ_{VV}/\sigma^\circ_{HV}$ because flat surfaces such as roofs and sides of buildings do not depolarize as much as does terrain. As support for believing the reflections from buildings are quasi-specular, the authors observed that the *HV* echoes from urban areas had less fluctuations than the corresponding *VV* echoes. *SS* in the tables for Phoenix refers to the number of data points (see sec. 6.9) for a measurement.

Table 7-3
Ratios of Median Power Expressed in Decibels[a] (New Jersey Woods, *X* Band, 2.5 μs Pulse)

Depression Angle	*HH/VV*	*HH/HV*	*VV/VH*	*HV/VH*
10°	−1.0	+10.2	+12.3	+1.1
20°	−0.9	+10.4	+11.6	+0.3
30°	−1.5	+ 8.0	+ 7.0	−2.3
40°	+0.1	+ 7.5	+ 6.8	−0.6
50°	+0.4	+ 8.2	+ 7.1	−0.7
60°	+1.7	+ 9.9	+ 6.4	−1.8
70°	+2.4	+11.5	+ 9.8	+0.8
80°	+0.4	+12.0	+11.0	−0.6
90°	−2.9	+12.5	+10.9	−4.5
Avg.	−0.1	+10.0	+ 9.2	−0.9

[a]From Ament, Macdonald, and Shewbridge (1959).

7.8 Relative Magnitude of Coherent and Incoherent Scattering from Trees

It is seen that for trees neither σ_{HH} and σ_{HV} nor σ_{VV} and σ_{VH}, in general, fluctuate together. This is not surprising because scatterers that depolarize contribute to σ_{VH} (equals σ_{HV}) and to σ_{VV} and σ_{HH}. However, nondepolarizing scatterers contribute to σ_{HH} and σ_{VV} only.

The experimental data also indicate that σ_{11} and σ_{12} do not fluctuate together. This is not surprising when it is recalled that σ_{11} contains a contribution from a_{HV}, but σ_{12} does not. Since a_{HH} and a_{VV} appear to be statistically independent of a_{HV} (and a_{VH}), equations (7.5) reduce to

$$\overline{\sigma_{11}} = \sigma_{22} = (K/4)\overline{|a_{HH} - a_{VV}|^2} + K\overline{|a_{HV}|^2}$$

and

$$\overline{\sigma_{12}} = \overline{\sigma_{21}} = (K/4)\overline{|a_{HH} + a_{VV}|^2}$$

Let δ be defined as the instantaneous phase angle between a_{HH} and a_{VV}. Then

$$\overline{\sigma_{11}} = \overline{\sigma_{22}} = (K/4)[\overline{|a_{HH}|^2 - 2|a_{HH}||a_{VV}|\cos\delta + |a_{VV}|^2}] + K\overline{|a_{HV}|^2}$$

and

$$\overline{\sigma_{12}} = \overline{\sigma_{21}} = (K/4)[\overline{|a_{HH}|^2 + 2|a_{HH}||a_{VV}|\cos\delta + |a_{VV}|^2}]$$

Since the various σ's are proportional to $K|a|^2$, we have

$$\overline{\sigma_{11}} = (1/4)[\sigma_{HH} - \overline{2\sqrt{\sigma_{HH}\sigma_{VV}}\cos\delta} + \sigma_{VV}] + \overline{\sigma_{HV}}$$

Table 7-4
Cross Polarization Ratios $\sigma^{\circ}_{VV}/\sigma^{\circ}_{HH}$[a]

θ	X VV/HV	C VV/HV	L VV/HV	P VV/HV
Desert				
5°	+ 7½	+ 8	—	—
8°	—	—	+ 4½	+ 4½
12°	+ 6	+ 7	+ 5½	+ 6
30°	+ 6	+ 7½	—	—
50°	+ 7½	+ 9	+ 7½	—
60°	—	—	—	—
75°	—	—	—	—
Marsh				
5°	—	—	—	—
8°	+ 9½	+ 7	+11½	None
12°	+ 7½	+ 7	+ 9	None
15°	+ 6½	+ 5	—	None
30°	+ 6½	+ 5	+10	None
50°	+ 5	+ 4	+ 7½	None
60°	+4½	+ 4	+ 7½	None
75°	—	—	—	—
Phoenix; SS > 25,000				
5°	—	—	+ 9	+ 2½
8°	+10½	+ 9½	+11½	+10½
12°	+13	+15½	+10	+10½
15°	—	—	—	—
30°	+ 6	+ 9	+ 6½	+ 5
50°	+ 6½	+ 9½	—	—
60°	—	—	—	—
75°	+ 9	+13	—	—
New Jersey Residential				
8°	+12½	+ 7	+12½	None
12°	+11	+ 8½	+ 9	None
15°	+ 9	+ 9	+ 8	None
30°	+10½	+ 5	+ 7½	None
50°	—	—	+ 7½	None
60°	+ 8½	+ 7½	+ 6½	None
Mountains				
5°	—	—	—	—
8°	+ 6	+ 5½	+ 7	+ 5
12°	+ 6½	+ 8	+ 5	+ 6
30°	+ 6	+ 6	+ 6	+ 9
50°	+ 6	+ 6	+ 5½	+ 9½
60°	—	—	—	—
75°	—	—	—	—
Phoenix; SS = 1000				
5°	+12½	+16	+ 9½	+ 6
8°	+11½	+ 8½	+ 7½	+15½
12°	+13½	+12	+10½	+13
15°	—	—	—	—
30°	+ 7	+10	+ 8½	+ 6½
50°	+ 6	+11½	—	—
60°	—	—	—	—
75°	+ 9	+11½	—	—
New Jersey Rural				
5°	—	—	—	—
8°	+ 7	+ 4	+ 5½	None
12°	+ 7	+ 5	+ 6½	None
15°	+ 8	+ 4½	+ 7	None
30°	+ 4	+ 3	+ 6	None
50°	+ 4½	+ 4	+ 6	None
60°	+ 5	+ 3½	+ 6½	None
75°	—	—	—	—

[a]From Daley et al. (1968).

and

$$\overline{\sigma}_{12} = (1/4)[\sigma_{HH} + 2\sqrt{\sigma_{HH}\sigma_{VV}}\cos\delta + \sigma_{VV}]$$

For almost all ground clutter data reported, $\overline{\sigma}_{11} < \overline{\sigma}_{12}$. Therefore, the effective value of $\cos\delta$ must be (on the average) positive, and the effective phase angle between a_{HH} and a_{VV} is consequently less than 90°.

From the discussion above, it is apparent that there is a discernible coherent component in the fluctuating echoes from trees for both *HH* and for *VV* polarizations. The coherent scattering, it would seem, is caused by quasi-specular reflections that are usually large for angles near normal incidence, for which the phase and amplitude of the reflection coefficients (for *HH* and *VV*) of a surface (tree trunks, limbs, etc.) approach equal values.

From above, it is evident that for trees a substantial portion of the echoes for *HH, VV*, and *HV* (or *VH*) polarizations are statistically independent. However, there are also contributions to the *HH* and *VV* echoes that are equal both in amplitude and phase.

At this point, insight can be gained by using the concept of coherent and incoherent fields discussed in section 7.3 for cases that $\overline{\sigma}_{HH}$ and $\overline{\sigma}_{VV}$ are nearly equal. M. Gent, I.M. Hunter, and N.P. Robinson (1963) reported essentially no difference between $\overline{\sigma}_{RV}$ and $\overline{\sigma}_{RH}$, which means (see sec. 3.2) there was little difference between $\overline{\sigma}_{HH}$ and $\overline{\sigma}_{VV}$. They also reported a $\overline{\sigma}_{12}/\overline{\sigma}_{11}$ value of 1 1/2 dB to an accuracy of 1/2 dB. Therefore, it appears from equation (7.16) that the ratio of coherent to incoherent scatterers $\overline{\sigma^c_{xx}}/\overline{\sigma^i_{xx}}$ was between 0.6 and 0.8.

The results of Hayes and Walsh (1959) and of Daley et al. (1968) on ratios of averages and medians for various polarizations ($\overline{\sigma}_{HH}/\overline{\sigma}_{VV}$, etc.) vary over wide ranges, indicating apparent effects of multipath interference. Effects of reflections from a ground plane, if averaged over several interference lobes, increase radar cross section. On the average, the effects of ground reflections will increase σ_{HH} most, σ_{HV} (and σ_{VH}) somewhat less, and will increase σ_{VV} least. Therefore, the expected effect (on the average) of the ground is to increase $\overline{\sigma}_{HH}/\overline{\sigma}_{HV}$ and to decrease $\overline{\sigma}_{VV}/\overline{\sigma}_{VH}$. Thus, the result of ground reflections, if many sites and observations are included, is to spread the composite range of σ_{HH}/σ_{HV} and σ_{VV}/σ_{VH} (given as 3-10 dB by Hayes and Walsh) by more than would exist for trees in complete absence of ground reflections. It should be recalled, however, that values of σ_{HH}/σ_{HV} and σ_{VV}/σ_{VH} can vary widely depending on the orientations of tree trunks, limbs, stumps, and other specular reflectors.

To obtain a feel for the polarization ratio that might typically exist for trees at small depression angles and in absence of ground reflections, the mid-ranges of 7 dB and 3 dB for $\overline{\sigma}_{HH}/\overline{\sigma}_{HV}$ and $\overline{\sigma}_{12}/\overline{\sigma}_{11}$, respectively, are chosen. According to equations (7.15) and (7.16), these values of $\overline{\sigma}_{HH}/\overline{\sigma}_{HV}$ and $\overline{\sigma}_{12}/\overline{\sigma}_{11}$ indicate ratios for $\overline{\sigma^c_{xx}}/\overline{\sigma^i_{xx}}$ of 0.7 and 1.2, respectively. Therefore, the reported ratios $\overline{\sigma}_{HH}/\overline{\sigma}_{HV}$, $\overline{\sigma}_{VV}/\overline{\sigma}_{VH}$, and $\overline{\sigma}_{12}/\overline{\sigma}_{22}$, when used with the scattering model of section 7.3, indicate that for trees the coherent (quasi-specular) scatterers contribute *roughly* the same echo power as do the incoherent (diffuse) scatterers.

Polarization ratios for large depression angles are available from table 7-3 where some of the results indicate that the averages for *HH* and *VV* echoes are equal. From experience with highly accurate and repeatable measurements on vegetation made for incidence angles between 10° and 80° (see figs. 6-7, 6-12, and 6-13), it is expected that the averages for *HH* and *VV* echoes are, in fact, equal. Hence, those measurements in table 7-3 for which *HH/VV* is 1.0 dB or greater will be rejected. From reciprocity, *HV* and *VH* echoes are equal. Then, if the medians of *HH* and *VV* echoes are equal, the ratios *VV/VH* and *HH/VH* are, in principle, equal. It therefore seems that *HH/HV* and *VV/VH* are within the ranges 6.8 to 12.0 dB for depression angles between 20° and 80°. From equation (7.15), these ratios indicate that the ratio of coherently to incoherently scattered power is between 0.6 and 4.3 for *HH* and *VV* echoes. The large variability in percentage of coherent scattering indicated by the simplified polarization model is compatible with the fact that the amplitude distributions reported for trees are sometimes Rayleigh and sometimes Ricean in character.

Sea Echo

7.9 Fluctuations of Orthogonally Polarized Components

The characteristics of sea echo fluctuations for *HH* and *VV* polarizations are discussed in chapter 5. Density functions of received power can often be approximated by the Rayleigh model. Some experimental curves, principally for small depression angles and small illuminated areas, have a break in the slope suggesting two sources of fluctuation. The deviation from Rayleigh is more pronounced for *HH* than for *VV* polarization, and the distributions for cross-polarized return (*HV* or *VH*) seem to be intermediate between the shapes of the distributions for *HH* and *VV* polarizations.

Differences in the statistical distributions for *HH* and *VV* echo have been attributed to a "spiky" character generally associated with the gross wave structure of the sea. Spikes are observed for a wide variety of sea conditions and polarizations, and they are especially prominent when there is a well-ordered swell structure on an otherwise calm sea.

Each spike usually extends approximately one pulse length in range and one beamwidth in azimuth, as would the return from a single dominant scatterer or a closely grouped collection of scatterers. The spikes are less depolarized than the more uniformly distributed return; ratios of parallel to cross-polarized return between 20 and 30 dB have been measured. These high polarization ratios suggest reflection from a smooth surface. Spikes have been observed to remain in excess of 10 seconds and fixed in position except for a slow drift in the direction of propagation of swell and sea waves.

Some understanding of sea echo can be obtained from the relative magnitudes of orthogonally polarized echoes received simultaneously. Consider the probability distribution of the instantaneous ratio obtained by simultaneously receiving horizontal and vertical polarizations when one of these polarizations is transmitted. The depolarized linearly polarized echo is caused by ripples and other asymmetric scatterers, but *HH* and *VV* echoes are caused by any type of scatterer. In other words two orthogonal, linearly polarized echoes usually will not fluctuate together. Therefore, as a first approximation it will be assumed that orthogonal, linearly polarized echoes are Rayleigh distributed and statistically independent.

Consider two statistically independent Rayleigh density functions for power

$$W(P)\,dP = \frac{1}{\overline{P}} e^{-P/\overline{P}}\,dP$$

and

$$W(Q)\,dP = \frac{1}{\overline{Q}} e^{-Q/\overline{Q}}\,dQ$$

The joint probability density function of two independent density functions is the product $W(P) \cdot W(Q)$. Therefore, the probability that the power Q is instantaneously less than some multiple of P (e.g., aP) is

$$W(aP > Q)\,dP = W(P)\,dP \int_0^{aP} W(Q)\,dQ$$

if the power P is between P and $P + dP$. Therefore, the probability that Q is less than aP for any value of P and of Q is

$$\text{Prob}\,\{aP > Q\} = \int_0^{\infty} W(aP > Q)\,dP \tag{7.17}$$

After evaluating the integrals, equation (7.17) is found to be

$$\text{Prob}\,\{aP > Q\} = 1 - \frac{1}{1 + (a\overline{P}/\overline{Q})} \tag{7.18}$$

Assume that the average values of the two distributions are equal, that is, $\overline{P} = \overline{Q}$. Then it is seen, as expected, from equation (7.18) that the probability of the power P exceeding the power Q is 1/2.

From equation (7.18) one can readily derive the probability density function $p(S)$ for the ratio P/Q expressed in decibels. Let $S = 10 \log (P/Q)$ and let $\overline{S}$ denote $10 \log (\overline{P}/\overline{Q})$. Then[c]

$$p(S) = \frac{\ln 10}{10} \frac{\log^{-1}\left(\dfrac{S - \overline{S}}{10}\right)}{\left[1 + \log^{-1}\left(\dfrac{S - \overline{S}}{10}\right)\right]^2}$$

or

[c]The symbol ln represents the natural logarithm; therefore, ln 10 is equal to 2.3026.

$$p(S) = \frac{\ln 10}{40} \operatorname{sech}^2 \left[\frac{\ln 10}{20} (S - \overline{S}) \right] \tag{7.19}$$

From equation (7.19) it may be seen that the mean and median of S are equal to $\overline{S}$ and the shape (width) of $p(S)$ is independent of $\overline{S}$.

Figure 7-10 gives a histogram of the instantaneous ratio of σ_{HH} and σ_{HV} that was taken from E.R. Flynt et al. (1967). The data were obtained with an X-band radar, a depression angle of a few degrees, a pulse length of 0.25 microsecond, and a resolution cell of approximately 10^3 square meters. For comparison, a histogram is also included that shows the theoretical predition using equation (7.19) with $\overline{S} = 11.4$ dB. Wave height and wind speed were given as 3 feet and 12 mph. The fit between the two histograms seems remarkably good, but data were not reported for other sea conditions.

The equation for the density function for the ratio of two orthogonally polarized circular waves is more difficult to obtain than that for the linear polarizations considered above. As previously discussed (see eqs. (7.4)), circularly polarized waves can be considered as being comprised of the sum (includes phase) of linearly polarized components. Since the amplitude distributions for linear polarizations are more or less Rayleigh and the orthogonally polarized linear components are essentially statistically independent, the circularly polarized echo powers are also expected to be Rayleigh distributed.

Suppose $\sigma_{VV}^\circ >> \sigma_{HH}^\circ$ and $\sigma_{HH}^\circ > \sigma_{VH}^\circ$. Then, from equations (7.5) and the relations that exist between the a's and the σ's, σ_{11}° and σ_{12}° would both be nearly equal to $\sigma_{VV}/4$. Therefore, the width of the probability density function depends on relative values of various cross sections and will be quite narrow if $\sigma_{VV}^\circ >> \sigma_{HH}^\circ > \sigma_{VH}^\circ$.

The derivation for the probability density function for the instantaneous ratio of two *dependent* random variables which are Rayleigh distributed is complicated, but it has been accomplished by H.A. Ecker and J.W. Cofer (1969). Equation (7.20) is the density function for σ_{12}/σ_{11} under the assumption that σ_{VV}, σ_{VH}, and σ_{HH} are statistically independent and each is Rayleigh power distributed:

$$\begin{aligned} p(T) = {} & \frac{\ln 10}{20} \times \log^{-1}\left(\frac{T}{10}\right) \left[\frac{\sigma_{HH}^\circ}{\sigma_{VV}^\circ} + \frac{\sigma_{VH}^\circ}{\sigma_{VV}^\circ}\left(\frac{\sigma_{HH}^\circ}{\sigma_{VV}^\circ} + 1\right)\right] \\ & \times \left[\left(\frac{\sigma_{HH}^\circ}{\sigma_{VV}^\circ} + 4\frac{\sigma_{VH}^\circ}{\sigma_{VV}^\circ} + 1 \right) \log^{-1}\left(\frac{T}{10}\right) + \frac{\sigma_{HH}^\circ}{\sigma_{VV}^\circ} + 1 \right] \\ & \div \left\{ \frac{1}{4}\left[\left(\frac{\sigma_{HH}^\circ}{\sigma_{VV}^\circ} + 4\frac{\sigma_{VH}^\circ}{\sigma_{VV}^\circ} + 1 \right) \log^{-1}\left(\frac{T}{10}\right) + \frac{\sigma_{HH}^\circ}{\sigma_{VV}^\circ} + 1 \right]^2 \right. \\ & \left. - \log^{-1}\left(\frac{T}{10}\right)\left(\frac{\sigma_{HH}^\circ}{\sigma_{VV}^\circ} - 1\right)^2 \right\}^{3/2} \end{aligned} \tag{7.20}$$

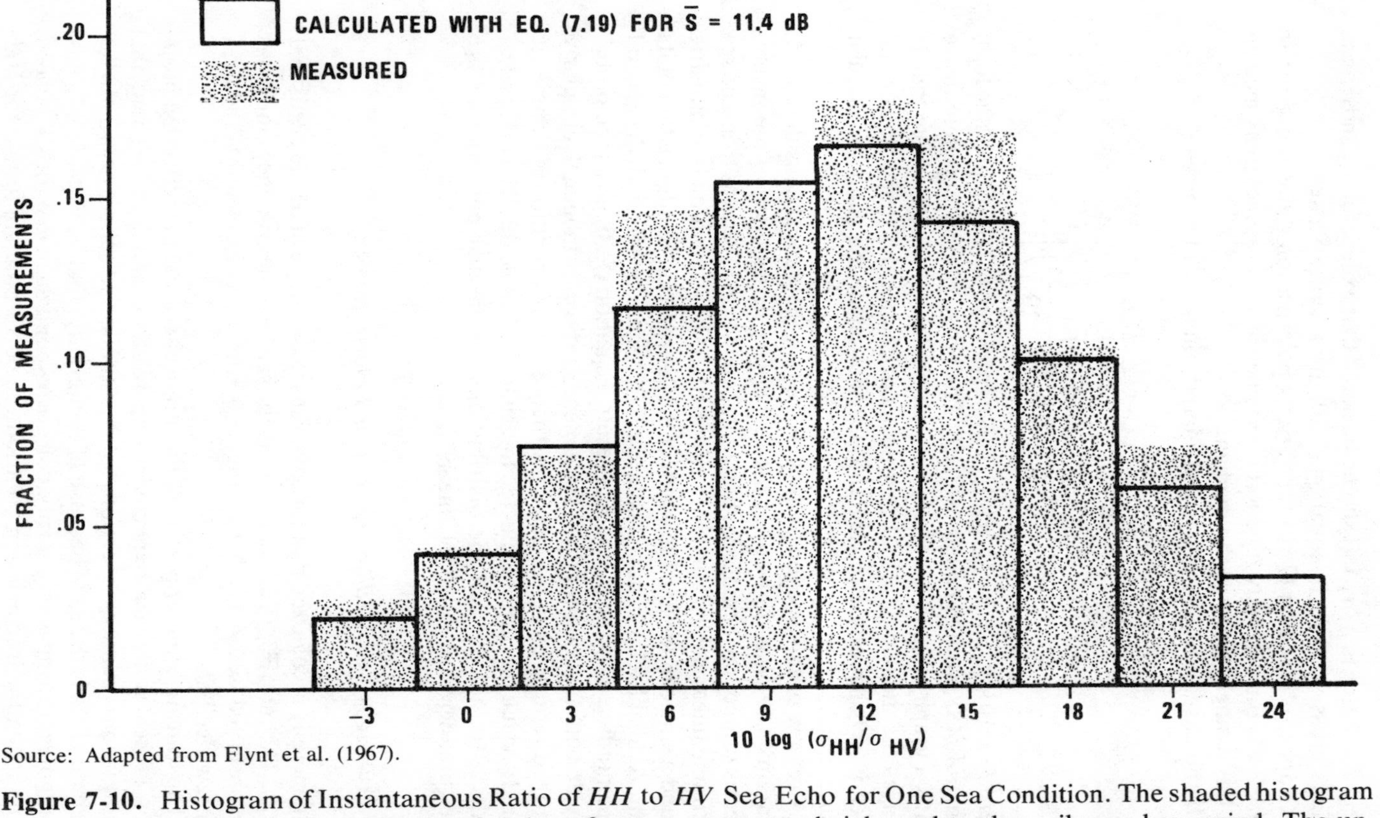

Source: Adapted from Flynt et al. (1967).

Figure 7-10. Histogram of Instantaneous Ratio of *HH* to *HV* Sea Echo for One Sea Condition. The shaded histogram represents measurements for three-foot average wave height and twelve-mile-per-hour wind. The unshaded histogram represents a theoretical distribution function for two independent echoes with ratio of the average powers equal to 11.4 dB.

where $T = 10 \log (\sigma_{12}/\sigma_{11})$. It should be noted that except for σ_{12} and σ_{11} the various cross sections in equation (7.20) are average values.

Graphs of equation (7.20) for various polarization ratios are given in figure 7-11. Each density function is symmetric with respect to its median and has a single mode at the median.

If the case of $\sigma^\circ_{HH} = \sigma^\circ_{VV}$ is considered, then $p(T)$ becomes

$$p(T) = \frac{\ln 10}{10} \frac{\log^{-1}\left(\dfrac{T + T'}{10}\right)}{\left[1 + \log^{-1}\left(\dfrac{T + T'}{10}\right)\right]^2} \tag{7.21}$$

where $T' = 10 \log [1 + 2(\sigma^\circ_{VH}/\sigma^\circ_{VV})]$. This expression $p(T)$ is of the same form as that of $p(S)$ given in equation (7.19). The broadest distribution in figure 7-11 is a graph of equation (7.21) for a particular value of T'. Note that T' changes the median of $p(T)$, but it does not change the shape. Ecker and Cofer (1969) have shown that the distribution of the ratio for circular polarization is never broader than that for linear polarization.

A review of this section follows. Based on theoretical considerations, equation (7.19) was derived for the probability distribution of the instantaneous ratio of the orthogonal horizontal and vertical returns from the sea when one linear polarization is transmitted. The distribution is symmetric around a single mode and the width of the distribution is independent of its mean. Available experimental data (fig. 7-10) agree closely with the general shape predicted by the theoretical result. Equation (7.20) gives the probability density function for the instantaneous ratio of orthogonal, circularly polarized returns from the sea. This theoretical distribution is symmetric around a value that is not expected to depart far from 0 dB. The theoretical results also indicate that the distributions for circular polarizations are never broader than those for linear.

7.10 Averages and Medians for Linear Polarization

It is generally recognized that for calm seas the average of σ°_{VV} exceeds that of σ°_{HH}. The ratio $\overline{\sigma^\circ_{VV}}/\overline{\sigma^\circ_{HH}}$ decreases as the sea becomes rougher, and it has been observed (sec. 6.17) that average σ°_{HH} sometimes exceeds average σ°_{VV} by a few decibels.

The transmission properties of the atmosphere and the reflection properties of sea water are expected to be bilateral and linear. Then, the instantaneous radar cross sections for the depolarized echoes (σ°_{VH} and σ°_{HV}) are equal. This is the case because of reciprocity, that is, because σ°_{VH} has an outgoing transmission contribution corresponding to that for *VV* echoes and an incoming transmission contribution corresponding to that for *HH* echoes and conversely for σ°_{HV}.

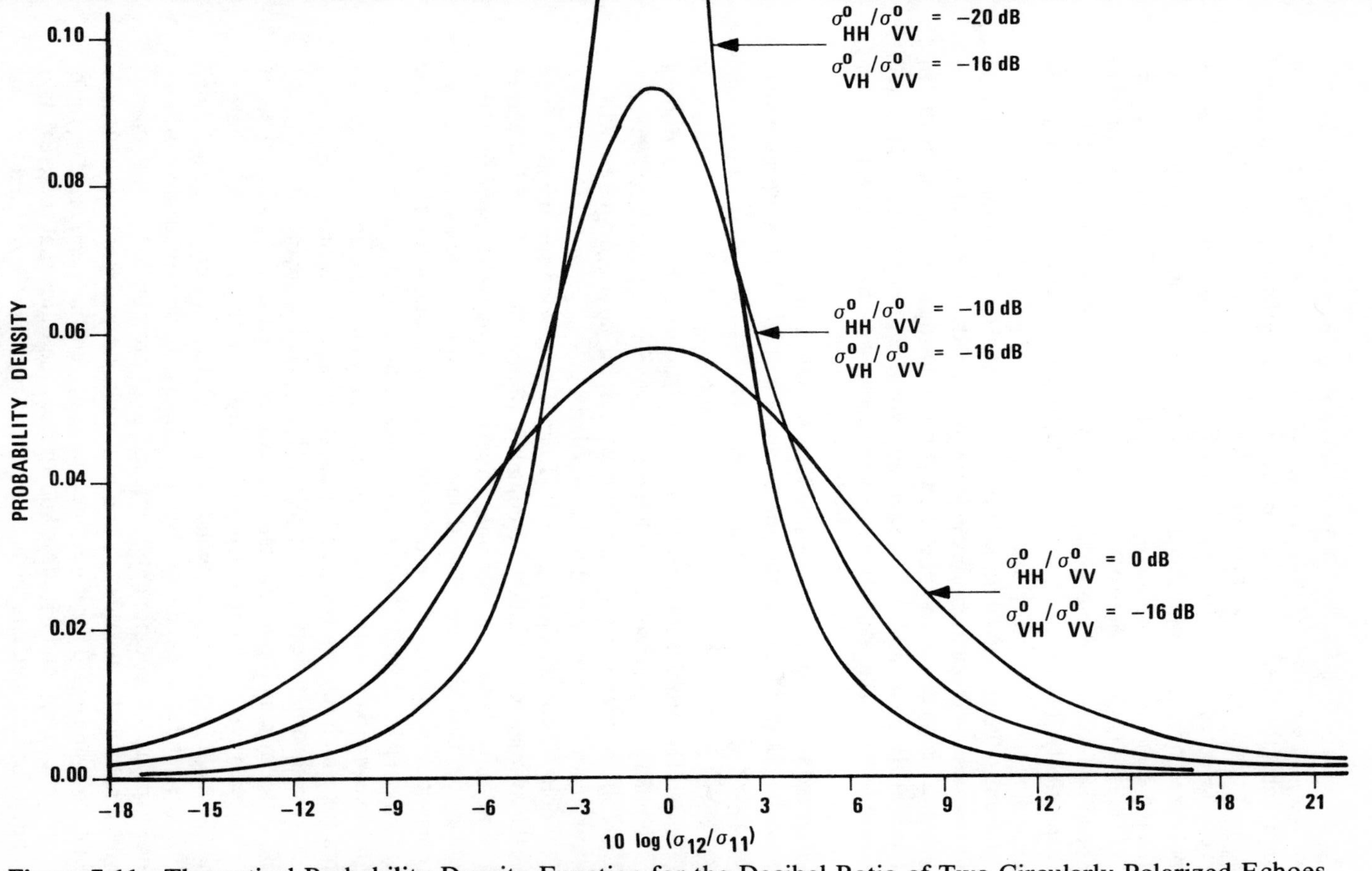

Figure 7-11. Theoretical Probability Density Function for the Decibel Ratio of Two Circularly Polarized Echoes.

Figure 7-12 illustrates the variation of $\sigma^\circ_{VV}/\sigma^\circ_{VH}$ with σ°_{VH} and $\sigma^\circ_{HH}/\sigma^\circ_{HV}$ with σ°_{HV} at 6.3 GHz for grazing angles between 1.5° and 4.0°. Each point represents the mean of a group of 10 runs; the lengths of the lines are equal to two times the standard deviation for each group. Data for figure 7-12 were selected from 10-minute runs for which there was a vertically polarized transmission followed by a horizontally polarized transmission for the same antenna beam direction and radar range. The data for figure 7-12 are for a wide variety of wind and wave conditions; the ranges of wind speed and average wave height were 3 to 20 knots and 0.3 to 4.1 feet, respectively. The reader is referred to M.W. Long (1965) for further details regarding the sea surface and other operating conditions.

Long (1965) reported for 6.3 GHz and for 35.0 GHz that the average σ° for all observed polarizations is strongly dependent on wind speed. The dependence of σ° on wave height was also investigated (see sec. 6.19). In these analyses a small but statistically significant dependence on wave height (σ° increases slightly with increases in wave height) was found for the *HH* echo at 6.3 GHz. Essentially no dependence on wave height was observed for the other polarizations at 6.3 GHz, and no dependence on wave height was found in the limited amount of 35.0-GHz data collected.

The strong observed dependence on wind speed suggests that wind-generated ripples are the major cause of the depolarized echo. Also, surfaces with radii small compared with a wavelength (such as the sharp edges of ripples) are expected to depolarize (Beckman and Spizzichino 1963, pp. 152-75); the echo from a surface with radii large compared with a wavelength is not expected to be depolarized.

Table 7-5 contains data given by Long (1965). Wind speed was defined as the average measured over a complete 10-minute run. The wind was in general moving toward the radar. Average height was estimated with a wave pole near the surf. The data for the two frequencies and the two transmitted polarizations were not obtained simultaneously. However, for each vertical transmission run there was a horizontal transmission run taken sequentially with the same antenna pointing angle and radar range, and for which the sea conditions had not greatly changed.

By comparing the two columns of table 7-5, it may be seen that σ°_{HH} and σ°_{VV} differ by a multiplicative factor between λ^0 and λ^{-1}; σ°_{VH} and σ°_{HV} seem to have a wavelength dependence close to λ^{-1}. With good reason, there is always doubt regarding the validity of absolute radar cross section data. However, the fact that the ratios of σ°_{VV} and σ°_{HH} to their respective orthogonally polarized components (σ°_{VH} and σ°_{HV}) are smaller at the shorter wavelengths indicates that σ°_{VV} and σ°_{HH} are less dependent on wavelength than is σ°_{VH} and σ°_{HV}. Therefore, since the scatterers that depolarize will also contribute to σ°_{HH} and σ°_{VV}, it is known from the ratios that there exists a nondepolarizing mechanism that contributes power to the *HH* and *VV* echo

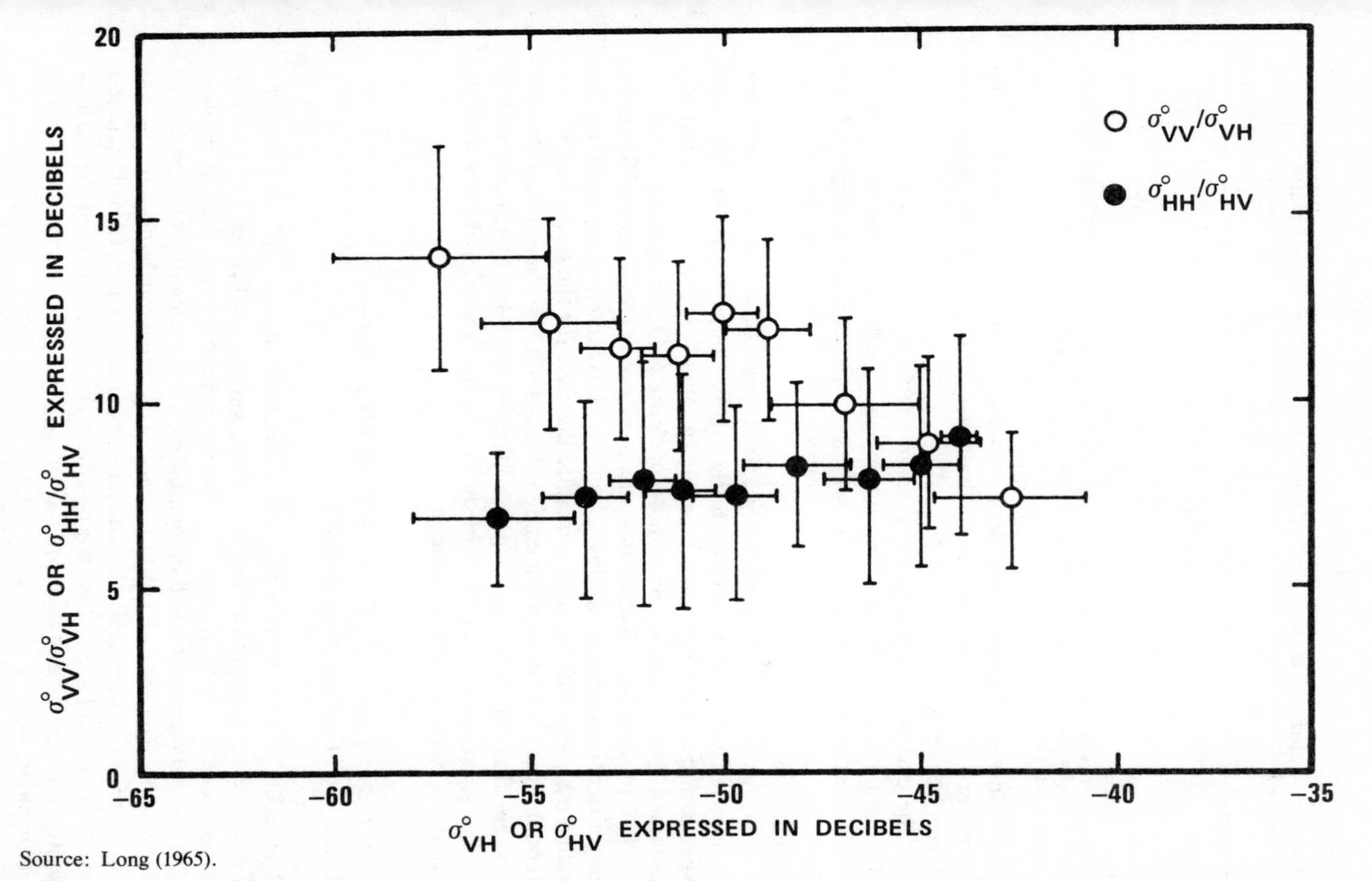

Source: Long (1965).

Figure 7-12. A Comparison of Average σ° at 6.3 GHz for Various Polarizations and a Wide Variety of Sea States.

Table 7-5
Average Values of σ° and Prevailing Conditions for Culled Data[a]

	6.3 GHz	35.0 GHz
σ°_{VV}	-36.8 ± 1.9 dB	-33.6 ± 1.6 dB
σ°_{HH}	-40.8 ± 2.4 dB	-36.8 ± 3.1 dB
$\sigma^\circ_{HV} = {}^\circ_{VH}$	-48.5 ± 3.0 dB	-40.6 ± 1.9 dB
Wind speed	11.1 ± 2.2 knots	12.6 ± 2.6 knots
Grazing angle	$2.5 \pm 0.6°$	$3.3 \pm 0.6°$
Average wave height	1.1 ± 0.5 feet	1.0 ± 0.4 feet

[a]From Long (1965).

that is less wavelength dependent than are the cross sections of the depolarizers. Thus far the discussion in this section has been for depression angles between 1.5° and 4.0°.

The curves for average and median values of σ°_{HH}, σ°_{VV}, σ°_{HV}, and σ°_{VH} depend on sea conditions, transmitter wavelength, direction of the radar beam relative to wind and wave directions, and depression angle. Therefore, the ratios of these averages and medians also depend on those various parameters.

Curves for $\sigma^\circ_{VV}/\sigma^\circ_{HH}$ versus depression angle for a number of operating conditions are given in chapter 6 (fig. 6-39). For convenience, the wind speeds and wave heights are repeated in table 7-6. Data on σ°_{HH} and σ°_{VH} obtained simultaneously with those of figure 6-39 are included in figures 7-13, 7-14, and 7-15. Notice that σ°_{VH} does not increase as rapidly at the larger depression angles as do σ°_{HH} and σ°_{VV}. It is therefore apparent that the ratios $\sigma^\circ_{HH}/\sigma^\circ_{HV}$ and $\sigma^\circ_{VV}/\sigma^\circ_{HV}$ generally increase for the larger angles. Although figure 7-15 does not include values of σ°_{VH} for angles larger than 60° (except that the February 11 graph includes an X-band measurement at 75°), it seems that the ratios of medians $\sigma^\circ_{HH}/\sigma^\circ_{HV}$ and $\sigma^\circ_{VV}/\sigma^\circ_{VH}$ are possibly 30 dB or more at vertical incidence.

7.11 Interdependence of Averages and Medians for Linear and Circular Polarizations

The facets of a sea surface that are comparable in size to a wavelength or larger are not expected to depolarize (a_{VH} and a_{HV} negligible), but would in general contribute to a_{HH} and a_{VV}. Asymmetric scatterers, such as ripples, that have dimensions small compared with a wavelength are expected to depolarize. Ripples will, of course, also generate contributions to a_{VV} and a_{HH}. Since relative magnitudes and phases of the a's for a given illuminated patch of sea will depend on the positions and orientations of the various scatterers (ripples and waves), a_{HV} (or a_{VH}) is expected to be statistically independent of a_{HH} and a_{VV}.

Table 7-6
Wind Speed and Wave Height for Measurements Reported by Daley, Davis, and Mills (1970)

Date (1969)	*Wind Speed (knots)*	*Average Wave Height (feet)*
February 6	40	15
February 10	30-33	11.5-13.1
February 11	46-48	21.3
February 13	35-39	23
February 14	37-40	23-26
February 17	5	3
February 18	22	9.8
February 20	29	16.4

If a_{HH}, a_{VV}, and a_{HV} were statistically independent of one another, then (from eqs. (7.7)) average cross section could be expressed as

$$\overline{\sigma}_{11} = \overline{\sigma}_{22} = \frac{\overline{\sigma}_{HH}}{4} + \frac{\overline{\sigma}_{VV}}{4} + \overline{\sigma}_{HV}$$

and (7.22)

$$\overline{\sigma}_{12} = \overline{\sigma}_{21} = \frac{\overline{\sigma}_{HH}}{4} + \frac{\overline{\sigma}_{VV}}{4}$$

For scatterers of this type, $\overline{\sigma}_{11}$ exceeds $\overline{\sigma}_{12}$.

It is well known that $\overline{\sigma}_{VV}$ is greater than $\overline{\sigma}_{HH}$ for a calm sea at small or intermediate depression angles. If $\overline{\sigma}_{VV}$ is much greater than $\overline{\sigma}_{HH}$ and $\overline{\sigma}_{HV}$, equations (7.22) indicate that

$$\overline{\sigma}_{11} = \overline{\sigma}_{22} = \overline{\sigma}_{12} = \overline{\sigma}_{VV}/4 \tag{7.23}$$

For this case, the various cross sections for circular polarization would be equal and would be 6 dB less than $\overline{\sigma}_{VV}$. J.M. Hunter and T.B.A. Senior (1966) reported that for low sea states σ_{11} and σ_{12} are equal on an instantaneous basis; this is consistent with equation (7.23).

Hunter and Senior (1966) reported relative values of $\overline{\sigma}_{12}$ and $\overline{\sigma}_{11}$ at X band for depression angles between 0.3° and 3°. The ratio of these averages, $\overline{\sigma}_{12}/\overline{\sigma}_{11}$, was between −1 and +2 dB for all available sea states and wind conditions (the experimental error was given as 1/2 dB). For high seas the average values of σ_{11} and σ_{12} were found to be about equal, but instantaneous values were not. Thus, the equations in (7.22) appear to be approximately valid for high sea states and small depression angles; therefore, much of the echo contributing to σ_{HH}, σ_{VV}, and σ_{HV} is statistically indepen-

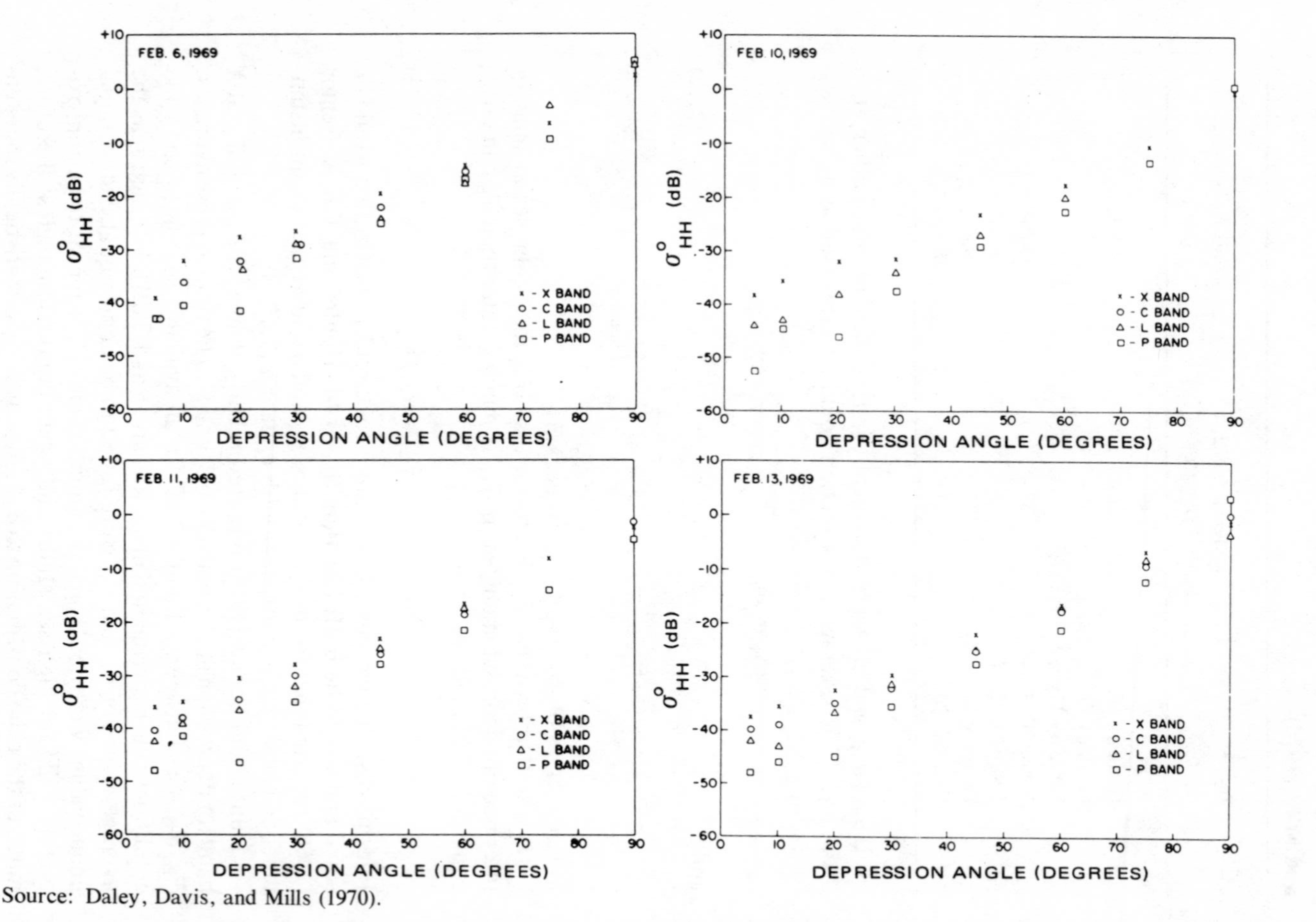

Source: Daley, Davis, and Mills (1970).

Figure 7-13. Median σ°_{HH} of the Ocean Versus Depression Angle. Data are for upwind conditions and obtained on

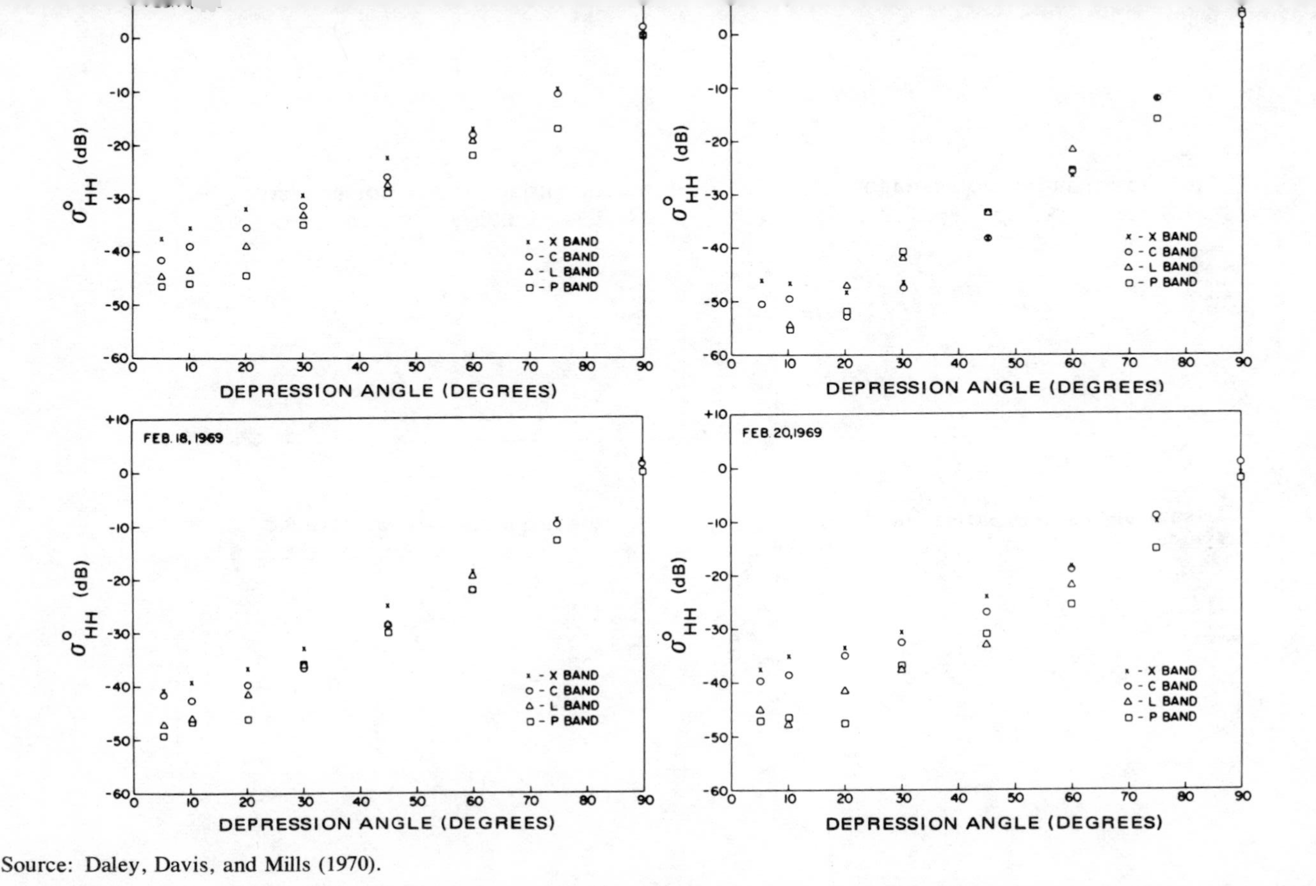

Source: Daley, Davis, and Mills (1970).

Figure 7-14. Median σ°_{HH} of the Ocean Versus Depression Angle. Data are for upwind conditions and obtained on February 14, 17, 18, and 20, 1969.

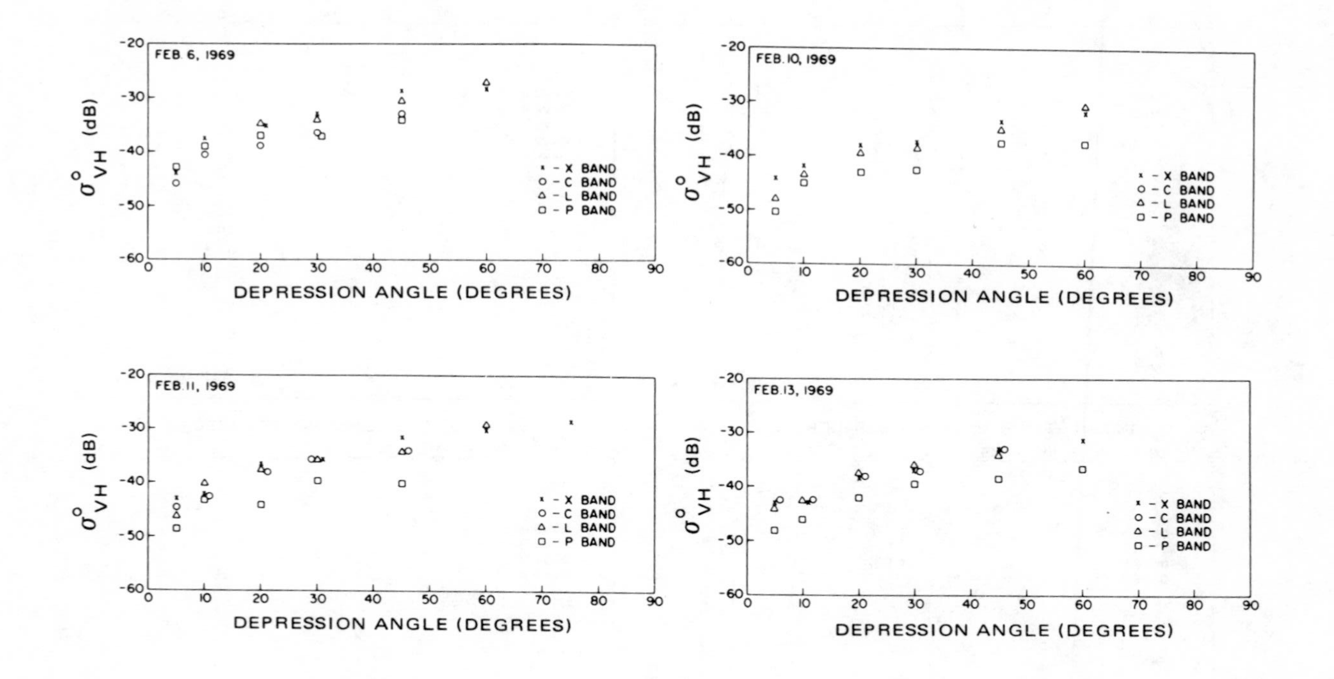
FEB. 6, 1969
FEB. 10, 1969
FEB. 11, 1969
FEB. 13, 1969
x - X BAND
o - C BAND
Δ - L BAND
□ - P BAND
σ^{o}_{VH} (dB)
DEPRESSION ANGLE (DEGREES)

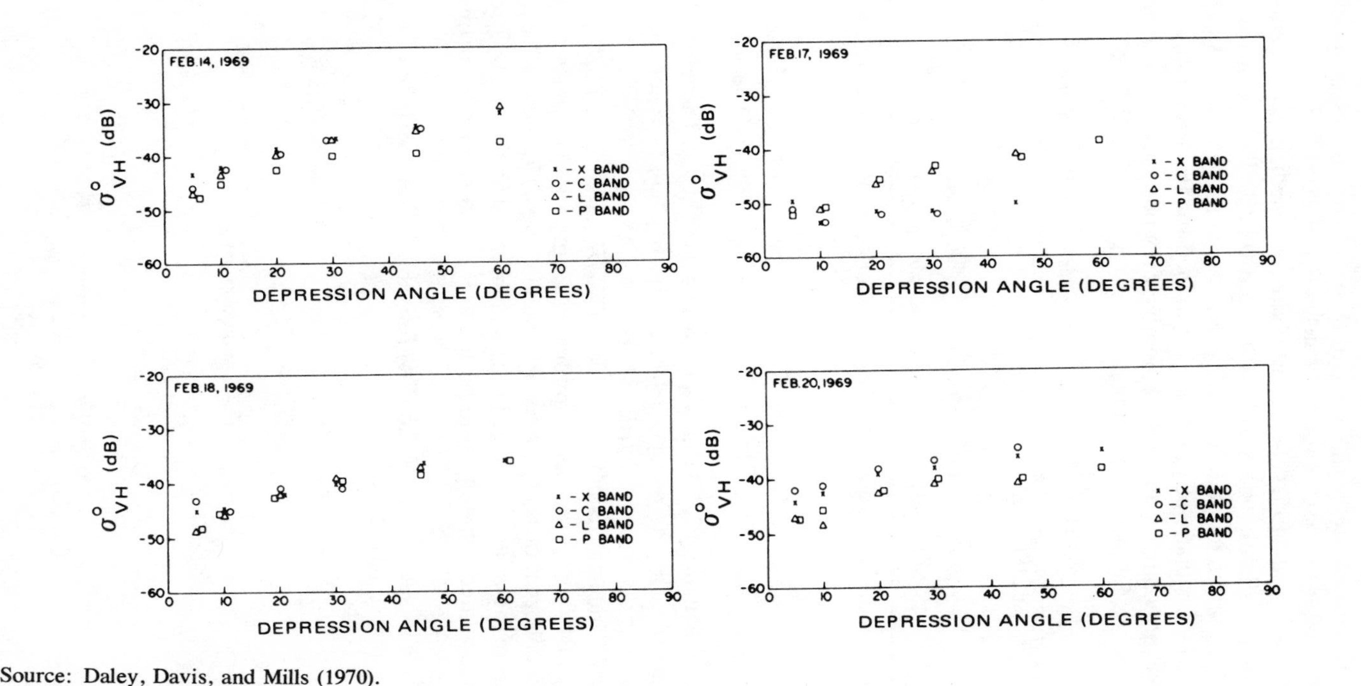

Source: Daley, Davis, and Mills (1970).

Figure 7-15. Median σ°_{VH} of the Ocean Versus Depression Angle. Data are for upwind conditions and obtained on February 6, 10, 11, 13, 14, 17, 18, and 20, 1969.

dent. As discussed later in this section, there is also a significant amount of coherent scattering that frequently causes $\overline{\sigma}_{12}$ to exceed $\overline{\sigma}_{11}$.

E.R. Flynt et al. (1967) also reported the ratio median σ_{12} to median σ_{11} to be centered slightly above zero—at about 1/2 dB. Their measurements were at X band and for depression angles in the range of 1° to 4°. As a result of comparing medians of σ_{12} and σ_{11} obtained for the wide range of sea conditions prevailing over several months, they observed ratios in accordance with the following percentages.

Median σ_{12} to Median σ_{11}	*Percentage of Measurements*
0 dB to 1 dB	52
−1 dB to 2 dB	73
−2 dB to 3 dB	94
−3 dB to 4 dB	99

From section 7.3, radar cross section can be expressed as the sum of coherent and incoherent cross sections as follows:

$$\begin{aligned} \overline{\sigma}_{11} &= \overline{\sigma}^{i}_{11} + \overline{\sigma}^{c}_{11} \\ \overline{\sigma}_{12} &= \overline{\sigma}^{i}_{12} + \overline{\sigma}^{c}_{12} \end{aligned} \tag{7.24}$$

Recall from equation (7.22) that $\overline{\sigma}^{i}_{11}$ is never less than $\overline{\sigma}^{i}_{12}$. Therefore, when $\overline{\sigma}_{11}$ is less than $\overline{\sigma}_{12}$ as observed frequently, $\overline{\sigma}^{c}_{12}$ must exceed $\overline{\sigma}^{c}_{11}$. This condition was discussed for trees in section 7.8.

Table 7-7 includes calculated values of $\overline{\sigma}_{12}/\overline{\sigma}_{11}$, $\overline{\sigma}_{HH}/\overline{\sigma}_{VV}$, $\overline{\sigma}_{HH}/\overline{\sigma}_{HV}$, and $\overline{\sigma}_{VV}/\overline{\sigma}_{VH}$ based on various assumptions regarding relative values of coherent and incoherent cross sections. For values of $\overline{\sigma}_{12}/\overline{\sigma}_{11}$ near 1/2 dB (the mid-range of the values obtained both by Hunter and Senior and by Flynt et al.), the calculated coherent contribution to $\overline{\sigma}_{HH}$ is about half the incoherent contribution.

The equations used (see sec. 7.8) for table 7-7 are included below in equations (7.25) and (7.26).

$$\begin{aligned} \overline{\sigma}_{HH} &= \overline{\sigma}^{i}_{HH} + \overline{\sigma}^{c}_{HH} \\ \overline{\sigma}_{VV} &= \overline{\sigma}^{i}_{VV} + \overline{\sigma}^{c}_{VV} \\ \overline{\sigma}_{VH} &= \overline{\sigma}_{HV} = \overline{\sigma}^{i}_{HV} = \overline{\sigma}^{i}_{VH} \\ \overline{\sigma}_{11} &= \overline{\sigma}^{i}_{11} + \overline{\sigma}^{c}_{11} \\ \overline{\sigma}_{12} &= \overline{\sigma}^{i}_{12} + \overline{\sigma}^{c}_{12} \\ \overline{\sigma}^{i}_{11} &= \overline{\sigma}^{i}_{HH}/4 + \overline{\sigma}^{i}_{VV}/4 + \overline{\sigma}^{i}_{VH} \\ \overline{\sigma}^{i}_{12} &= \overline{\sigma}^{i}_{HH}/4 + \overline{\sigma}^{i}_{VV}/4 \end{aligned} \tag{7.25}$$

Table 7-7
Calculated Values of Relative Cross Section for Coherent and Incoherent Scatterers[a]

Assumptions				*Calculations*			
$\frac{\sigma^i_{HV}}{\sigma^i_{HH}}$	$\frac{\sigma^i_{VV}}{\sigma^i_{HH}}$	$\frac{\sigma^c_{HH}}{\sigma^i_{HH}}$	$\frac{\sigma^c_{VV}}{\sigma^i_{VV}}$	$\frac{\sigma_{12}}{\sigma_{11}}$	$\frac{\sigma_{HH}}{\sigma_{VV}}$	$\frac{\sigma_{HH}}{\sigma_{HV}}$	$\frac{\sigma_{VV}}{\sigma_{VH}}$
−7 dB	1	0	0	−1.5 dB	0 dB	7.0 dB	7.0 dB
−7 dB	1	1/2	1/4	0.9 dB	0.8 dB	8.8 dB	8.0 dB
−7 dB	1	1	1/2	2.3 dB	1.3 dB	10.0 dB	8.7 dB
−5 dB	1	1/2	1/4	0.2 dB	0.8 dB	6.8 dB	6.0 dB
−7 dB	2	0	0	−1.0 dB	−3.0 dB	7.0 dB	10.0 dB
−7 dB	2	1/2	1/8	0.7 dB	−1.5 dB	8.8 dB	10.3 dB
−7 dB	2	1	1/4	1.8 dB	−0.5 dB	10.0 dB	10.5 dB
−5 dB	2	1/2	1/8	0.2 dB	−1.5 dB	6.8 dB	8.3 dB
−7 dB	4	0	0	−1.7 dB	−6.0 dB	7.0 dB	13.0 dB
−7 dB	4	1/2	1/16	0.5 dB	−4.4 dB	8.8 dB	13.2 dB
−7 dB	4	1	1/8	1.3 dB	−3.2 dB	10.0 dB	13.2 dB
−5 dB	4	1/2	1/16	0.1 dB	−4.4 dB	6.8 dB	11.2 dB

[a]All values of σ in this table are averages.

For small depression angles, it seems that the coherent echo is caused by the smooth fronts of waves and that these wave fronts do not depolarize. Therefore, it is assumed that (1) $\overline{\sigma^c_{HV}}$ and $\overline{\sigma^c_{VH}}$ are zero, and (2) the phase change on reflection is the same for *HH* and *VV* echoes (see figures 4-3 and 4-5). Equations that express these assumptions follow:

$$\overline{\sigma^c_{HV}} = \overline{\sigma^c_{VH}} = 0$$

$$\overline{\sigma^c_{11}} = (1/4)\overline{[\sigma^c_{HH} - 2\sqrt{\sigma^c_{HH}\sigma^c_{VV}} + \sigma^c_{VV}]} \tag{7.26}$$

$$\overline{\sigma^c_{12}} = (1/4)\overline{[\sigma^c_{HH} + 2\sqrt{\sigma^c_{HH}\sigma^c_{VV}} + \sigma^c_{VV}]}$$

We expect σ^c_{HH} and σ^c_{VV} to be directly proportional to ρ^2_H and ρ^2_V (see fig. 4-4). The ratio $\overline{\sigma^c_{HH}/\sigma^c_{VV}}$ is assumed to be 2 for the calculations, but the results are not highly sensitive to this value. If $\overline{\sigma^c_{HH}/\sigma^c_{VV}}$ equals 2, equations (7.26) give

$$\overline{\sigma^c_{11}} = 0.02\overline{\sigma^c_{HH}}$$

$$\overline{\sigma^c_{12}} = 0.73\overline{\sigma^c_{HH}}$$

From table 7-7, ratios of $\overline{\sigma^c_{HH}/\sigma^i_{HH}}$ and $\overline{\sigma^c_{VV}/\sigma^i_{VV}}$ are 1/2 and 1/16, respectively, for typically observed relative values for $\overline{\sigma}_{12}$, $\overline{\sigma}_{11}$, $\overline{\sigma}_{HH}$ $\overline{\sigma}_{VV}$, and $\overline{\sigma}_{HV}$. For example, the third row measured from bottom has relative values for $\overline{\sigma}_{HH}$, $\overline{\sigma}_{VV}$, and $\overline{\sigma}_{HV}$ similar to those in table 7-5 for 6.3 GHz. Although perhaps simply fortuitous, it is of interest that the power ratios of slow to fast fluctuations reported in section 5.12 correspond roughly to the values of $\overline{\sigma^c_{HH}}/\overline{\sigma^i_{HH}}$ and $\overline{\sigma^c_{VV}}/\overline{\sigma^i_{VV}}$ given in table 7-7. The model and the extrapolation to account for fluctuations are oversimplifications. For example, σ^c_{HV} is assumed to be zero—certainly *HV* echo has both slow and fast fluctuations.

Figure 7-16 shows results given by J.C. Wiltse, S.P. Schlesinger, and C.M. Johnson (1957) for radar cross section using linear and circular polarizations for a wide range of depression angles. The data were obtained at 24 GHz for a moderately heavy sea and a 26-knot wind speed. In order to illustrate experimental error, each curve in figure 7-16 was drawn with a width of 5 dB to illustrate the estimated probable error of 2.5 dB given by the authors.

For small depression angles, the polarization dependence of average σ° indicated by figure 7-16 is consistent with the echoes being caused principally by incoherent scatterers. The equations in (7.22) are valid for incoherent scatterers. At the small depression angles, average σ°_{VV} is larger than average σ°_{HH}, and average σ°_{HH} is larger than average σ°_{HV}. From equations (7.22) and (7.23), average σ°_{12} may be as much as 6 dB less than average σ°_{VV}. From equation (7.22), average σ°_{11} is greater than average σ°_{HV}, and the averages of σ°_{11} and σ°_{12} are nearly equal. Therefore, the equations in (7.22) seem to be in general agreement with the small depression angle data of figure 7-16.

By comparing equations (7.5) with figure 7-16 at large depression angles, a new insight is obtained. It is expected that average σ°_{VV} and average σ°_{HH} are equal for near vertical incidence, and average σ°_{HV} is less than average σ°_{VV}. From equation (7.5) average σ°_{11} is greater than average σ°_{HV}, which is consistent with the figure. However, also from equation (7.5), if average σ°_{11} is consistently less than average σ°_{HH} and average σ°_{VV}, the amplitudes and phases of the *HH* and *VV* echoes (should they exist simultaneously) must be nearly equal on the average. For this condition to exist, the equations require the averages of σ°_{12}, σ°_{HH}, and σ°_{VV} to be approximately equal, as reported in figure 7-16. Therefore, a comparison of the equations and the figure shows that the preponderance of the *HH* and *VV* echoes at large depression angles have, instantaneously, equal phase and amplitude. Therefore it seems the echo is quasi-specular; that is, the echo is caused primarily by the air-water interfaces that are oriented near-normal to the radar beam.

7.12 Coherent and Incoherent Scattering from the Sea

The relative magnitudes at small depression angles for fast and slow echo fluctuations are discussed in chapter 5. Based on measured autocorrelation

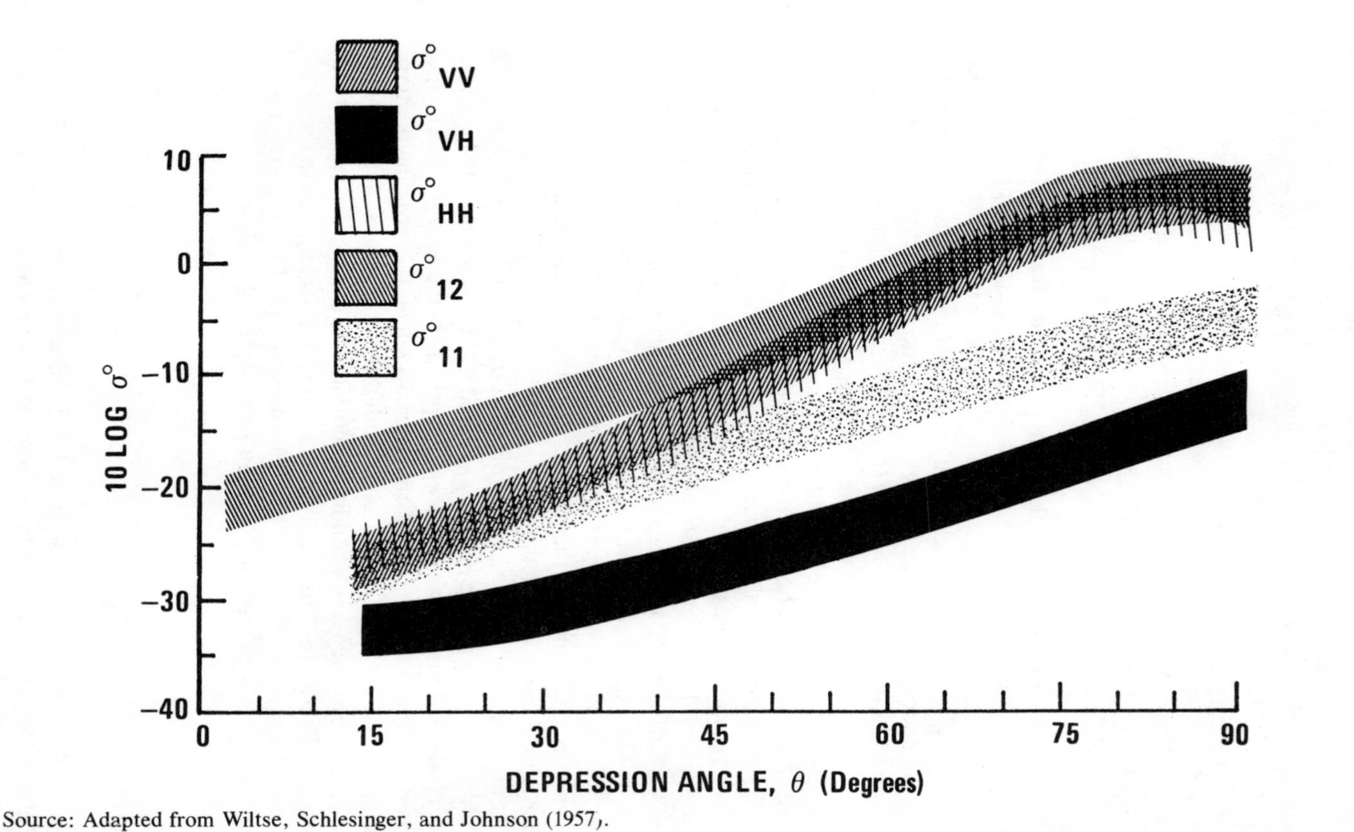

Source: Adapted from Wiltse, Schlesinger, and Johnson (1957).

Figure 7-16. Average σ° Versus Depression Angle for Vertical, Horizontal, Circular, and Cross Polarization at 24 GHz. The data were taken on November 22, 1955. Sea moderately heavy, wind speed 26 knots.

functions for *HH* polarization, it was estimated that the power of the slow fluctuations can be approximately that of the fast fluctuations. For *VV* polarization, the power of the fast fluctuations is sometimes as much as 10 times that of the slow fluctuations.

It was also observed in chapter 5 that the source of the fast fluctuations is area extensive and that the slow fluctuations are usually spiky and appear to be generated principally by wave fronts. That is, at small depression angles the fast fluctuations appear to be caused by incoherent scattering and much of the slow fluctuations seem to be caused by coherent scattering.

In section 7.13 estimates of the relative magnitudes of the coherent and incoherent scattering are obtained by comparing data for circular and linear polarizations, and by making certain assumptions regarding the coherent scattering. The relative values of coherent and incoherent scattering obtained in this way seem to be nearly equal to the power ratios reported for the slow and fast fluctuations of *HH* and *VV* echoes. Coherent scatterers are not, of course, the sole cause for slow fluctuations, and incoherent scatterers are not the sole cause for the fast fluctuations.

For angles near vertical incidence, the simplified polarization model (sec. 7.3) is useful for calculating echo strengths for various polarizations and for determining the relative echo strength from coherent and incoherent scatterers. From figure 7-16 the ratios $\overline{\sigma^\circ_{HH}}/\overline{\sigma^\circ_{HV}}$ and $\overline{\sigma^\circ_{12}}/\overline{\sigma^\circ_{11}}$ are approximately 18 dB and 12 dB, respectively. The errors (uncertainties) in the ratios would exceed those of the individual measurements and would be about 3.5 dB. In view of uncertainties in the ratios, table 7-1 indicates that the coherent component of the backscattered power is at least 10 times that of the incoherent component.

At near-vertical incidence and for a sea that is smooth with respect to the radar wavelength, the echo is expected to be caused mainly by quasi-specular reflections. However, it was surprising to the author when he first learned that the scattering at 1.2 cm radar wavelength is principally quasi-specular ($\overline{\sigma^c_{HH}} >> \overline{\sigma^i_{HH}}$), even for a moderately rough sea and a 26-knot wind speed. For such a sea condition, both the fine and gross sea surfaces are rough. Therefore, this example illustrates that statements to the effect that a surface is smooth simply because one or more polarization ratios are large should be accepted only if they pass a critical review.

References

Ament, W.S., F.C. Macdonald, and R.D. Shewbridge, "Radar Terrain Reflections for Several Polarizations and Frequencies," *Transactions of the 1959 Symposium of Radar Return*, The University of New Mexico, May 11-12, 1959.

Beckmann, Petr, *Elements of Applied Probability Theory*, Harcourt, Brace and World, Inc., New York, New York, 1967.

Beckmann, Petr and André Spizzichino, *The Scattering of Electromagnetic Waves from Rough Surfaces*, The MacMillan Company, New York, New York, 1963.

Daley, J.C., W.T. Davis, J.R. Duncan, and M.B. Laing, "NRL Terrain Clutter Study, Phase II," Naval Research Laboratory Report 6749, October 21, 1968.

Daley, J.C., W.T. Davis, and N.R. Mills, "Radar Sea Return in High Sea States," Naval Research Laboratory Report 7142, September 25, 1970.

Eagle, D.F., M.D. Harris, and R.D. Hayes, "Study of Polarization Characteristics of Radar Targets," Engineering Experiment Station, Georgia Institute of Technology, Quarterly Report No. 9, Contract DA 36-039 sc-64713, October 1957.

Ecker, H.A. and J.W. Cofer, "Statistical Characteristics of the Polarization Power Ratio for Radar Return with Circular Polarization," *IEEE Transactions on Aerospace and Electronic Systems*, vol. AES-5, pp. 762-69, September 1969.

Evans, J.V. and Tor Hagfors, "Study of Radio Echoes for the Moon at 23 Centimeter Wavelength," *Journal of Geophysical Research*, vol. 71, pp. 4871-99, 1966.

Evans, J.V. and Tor Hagfors, *Radar Astronomy*, McGraw-Hill Book Company, Inc., New York, New York, 1968.

Flynt, E.R., F.B. Dyer, R.C. Johnson, M.W. Long, and R.P. Zimmer, "Clutter Reduction Radar," Engineering Experiment Station, Georgia Institute of Technology, Final Report, Contract NObsr-91024, December 1967.

Gent, M., I.M. Hunter, and N.P. Robinson, "Polarization of Radar Echoes, Including Aircraft, Precipitation and Terrain," *Proceedings of the IEE*, vol. 110, pp. 2139-48, 1963.

Guinard, N.W., J.T. Ransone, Jr., M.B. Laing, and L.E. Hearton, "NRL Terrain Clutter Study, Phase I," Naval Research Laboratory Report 6487, May 10, 1967.

Harris, M.D., and R.D. Hayes, "Study of Polarization Characteristics of Radar Targets," Engineering Experiment Station, Georgia Institute of Technology, Quarterly Report No. 8, Contract DA-36-039 sc-64713, July 1957.

Hayes, R.D., C.H. Currie, and M.W. Long, "An X-band Polarization

Measurements Program," *Record of the Third Annual Radar Symposium*, University of Michigan, February 1957.

Hayes, R.D., J.R. Walsh, D.F. Eagle, H.A. Ecker, M.W. Long, J.G.B. Rivers, and C.W. Stuckey, "Study of Polarization Characteristics of Radar Targets," Engineering Experiment Station, Georgia Institute of Technology, Final Report, Contract DA-36-039 sc-64713, October 1958.

Hayes, R.D. and M.W. Long, "Study of Polarization Characteristics of Radar Targets," Engineering Experiment Station, Georgia Institute of Technology, Quarterly Report No. 6, Contract DA-36-039 sc-64713, January 1957.

Hayes, R.D. and J.R. Walsh, Jr., "Some Polarization Properties of Targets at X-Band," *Transactions of the 1959 Symposium on Radar Return*, University of New Mexico, May 11-12, 1959.

Hunter, I.M. and T.B.A. Senior, "Experimental Studies of Sea-Surface Effects on Low-Angle Radars," *Proceedings of the IEE*, vol. 113, p. 1931, 1966.

Ivey, H.D., M.W. Long, and V.R. Widerquist, "Some Polarization Properties of K_a- and X_b-Band Echoes from Vehicles and Trees," *Record of the First Annual Radar Symposium*, University of Michigan, 1955.

Long, M.W., "On the Polarization and the Wavelength Dependence of Sea Echo," *IEEE Transactions on Antennas and Propagation*, vol. AP-13, pp. 749-54, September 1965.

Moore, R.K., "Radar Return from the Ground," Bulletin of Engineering No. 59, University of Kansas, 1969.

Wiltse, J.C., S.P. Schlesinger, and C.M. Johnson, "Back-Scattering Characteristics of the Sea in the Region from 10 to 50 KMC," *Proceedings of the IRE*, vol. 45, pp. 220-28, February 1957.

APPENDIX A

The Weibull Distribution

The Weibull distribution has properties that lie between those of the Rayleigh (exponential) and the lognormal distributions. As may be seen from figure 5-22(b), measured data can also lie between those distributions. Robert R. Boothe was, it seems, the first person to use the Weibull function for describing clutter (Boothe, 1969). He noted that the curves in figures 5-6 and 5-7, over the limited range of measured amplitudes, can be described with Weibull statistics equally as well as was previously done by Linell with lognormal statistics.

The Weibull density function is

$$W(\sigma) = \frac{1}{\bar{\sigma}} b \sigma^{(b-1)} \exp\left(\frac{-\sigma^{b}}{\bar{\sigma}}\right) \tag{A.1}$$

It is of interest to notice that for b equals unity, equation (A-1) reduces to the Rayleigh (exponential) density function. Therefore the Rayleigh function is a member of the Weibull family.

For the Weibull function, the ratio of average RCS, $\bar{\sigma}$, to median RCS, σ_m, is given (Schleher 1976; 1977) by

$$\frac{\bar{\sigma}}{\sigma_m} = \frac{\Gamma\left(1 + \frac{1}{b}\right)}{(\ln 2)^{1/b}}$$

where

$\Gamma(1 + 1/b)$ is the Gamma function of the argument $(1 + 1/b)$ and

$\ln 2$ is the natural logarithm of 2.

Figure A-1 includes several Weibull cumulative distributions for various values of the parameters b of the equation (A-1). It should be noted that each of these distributions is graphed with respect to its average value $\bar{\sigma}$. These curves were taken from a research report by Rivers (1970). Weibull cumulative distributions that are graphed with respect to their medians are included in Schleher (1976).

Figure A-2 is included for comparing the lognormal and the Weibull functions. It includes cumulative distributions, drawn relative to average value, for lognormal density functions with standard deviation S of 3, 4, 5, 6, 7, and 8 dB. These curves were also reproduced from Rivers (1970). It should be noted that each cumulative distribution of figure A-2 is a straight line, consistent with the fact that normal probability graph paper is used and the amplitude unit (abscissa) is logarithmic instead of linear.

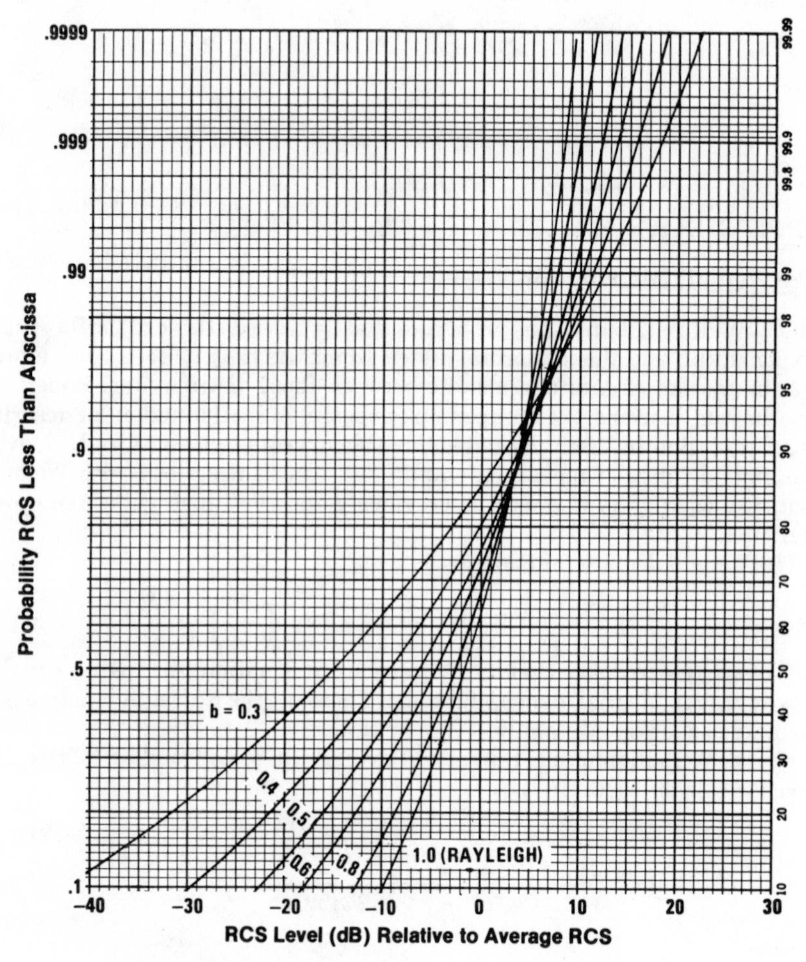

Figure A-1. Weibull Cumulative Distributions.
Source: Rivers (1970).

The lognormal distribution is useful for describing clutter of the type that attains very large RCS values, e.g., 20 dB above its median, for higher percentages of time than does clutter that obeys Weibull statistics. To illustrate, the reader should compare the $S = 8$ dB curve of figure A-2 with the $b = 0.6$ curve of figure A-1. First, notice that for probability less than 0.9 and probability greater than 0.1, i.e., for 80% of the time, the ranges of RCS values for the two curves are approximately equal. Even so, the median for the Weibull is approximately 3 dB larger than the median for the lognormal curve and yet each curve is normalized to 0 dB average value.

It is also apparent that, for this case of the two curves with equal averages and nearly equal ranges for 80% of the time, large RCS values are far more

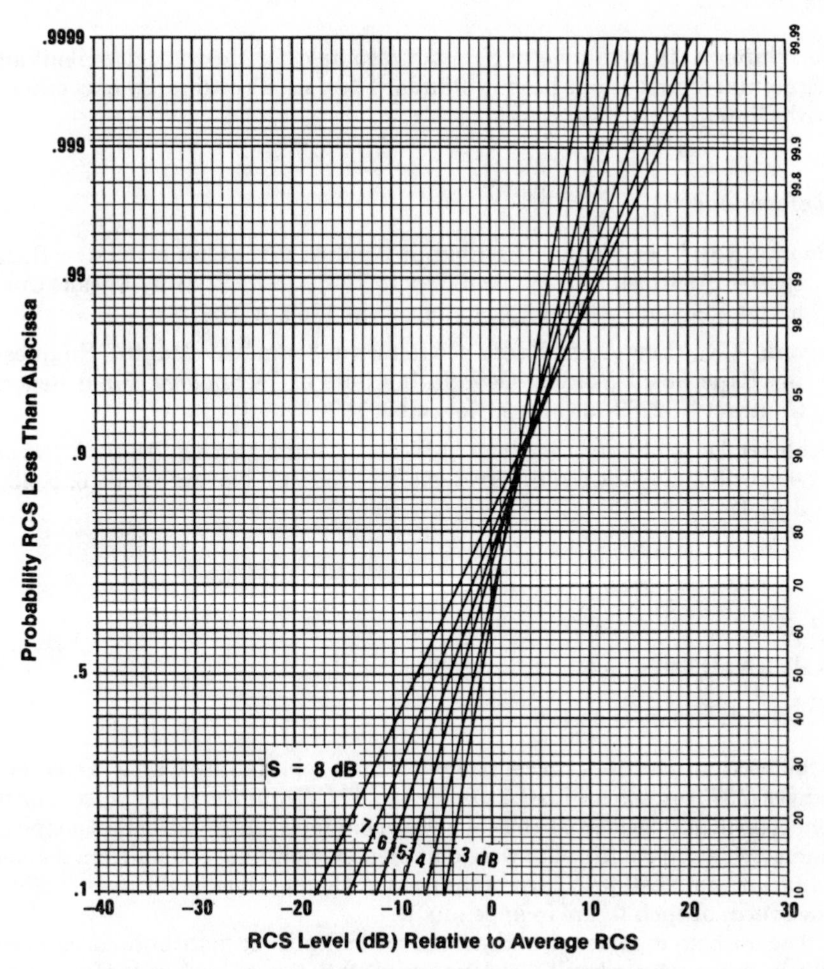

Figure A-2. Lognormal Cumulative Distributions.
Source: Rivers (1970).

likely with a lognormal than with a Weibull distribution. For example, the b = 0.6 curve exceeds its average by 14 dB with probability 10^{-4} (ordinate = 0.9999) but the S = 8 dB curve exceeds its average by 14 dB with probability 4×10^{-3} (ordinate equals 0.996). In other words, for this example the lognormal exceeds one given large RCS value 40 times as often as does the Weibull distribution exceed the same RCS value.

Figures A-1 and A-2 are useful for assisting with the visualization of an interesting and powerful relationship derived by Rivers (1970) for estimating the average RCS of clutter. This relationship follows: the average RCS of clutter is 3.5 ±0.5 dB less than the RCS at the 0.9 cumulative probability point. Rivers found that the relationship held not only for the wide range of sea clutter data available to him, but also held for theoretical Weibull and lognormal distributions having standard deviations ordinarily encountered with clutter.

References

Boothe, R.R, "The Weibull Distribution Applied to the Ground Clutter Backscatter Coefficient," Technical Report RE-69-15, U.S. Army Missile Command, Redstone Arsenal, Alabama, June 1969, AD 691109.

Rivers, W.K., "Low-Angle Radar Sea Return at 3-mm Wavelength," Engineering Experiment Station, Georgia Institute of Technology, Final Report, Contract N62269-70-C-0489, November 1970.

Schleher, D.C., "Radar Detection in Weibull Clutter," *IEEE Transactions on Aerospace and Electronic Systems*, pp. 736-743, November 1976; p. 435, July 1977.

APPENDIX B

Fluctuations in Tree Echo

This section describes measurement results reported by Currie, Dyer, and Hayes (1975) on trees with 9.5, 16, 35, and 95 GHz noncoherent pulse radars. Short-term amplitude distributions, frequency spectra (noncoherent doppler), and correlation coefficients were obtained by sampling echoes from a given range-azimuth cell with each of the radars. Currie, Dyer, and Hayes (1975) is the source of each figure in appendix B.

The authors reported that the measured amplitude distributions can usually be approximated well with lognormal statistics for each radar frequency, for HH and VV polarizations, and for deciduous trees or pines. Figures B-1 through B-4 give probabilities of occurrence for the various standard deviations obtained. The authors also compared the average of the standard deviations obtained for each frequency, and observed a slight increase of these averages with increased frequency. The largest "average" standard deviation occurred at 95 GHz where unusually large (for temporal statistics) standard deviations were sometimes measured. Little dependence of standard deviation on polarization was observed.

Hayes (1979) reported that tree echo at 95 GHz, although usually lognormal, is sometimes Weibull in shape. Then, the shape factor b (fig. A-1) was usually between 0.4 and 0.6. Also, according to Hayes (1979), shape factor b of unity had been observed, in less than one percent of the data, at very low wind speeds of less than 3 mph.

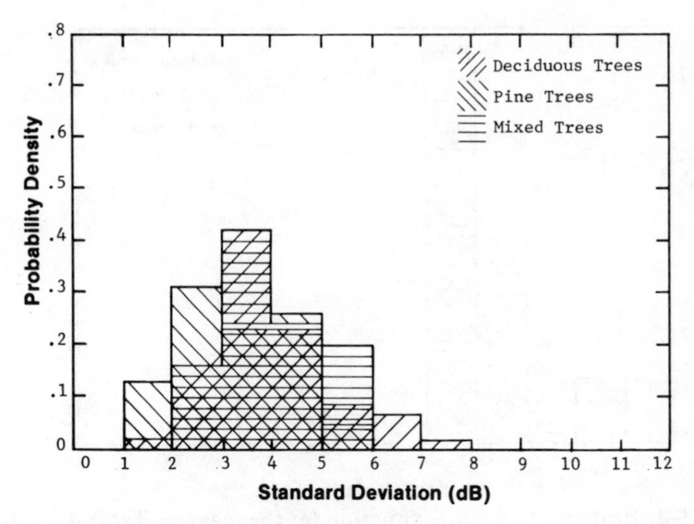

Figure B-1. Probability density function for the measured standard deviations of the amplitude distributions as a function of tree type; 9.5 GHz.

Source: Currie, Dyer, and Hayes (1975).

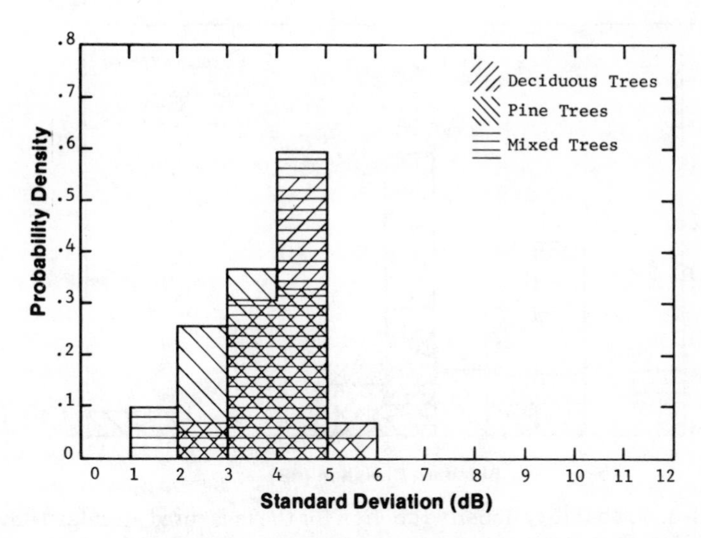

Figure B-2. Probability density function for the measured standard deviations of the amplitude distributions as a function of tree type; 16.5 GHz.

Source: Currie, Dyer, and Hayes (1975).

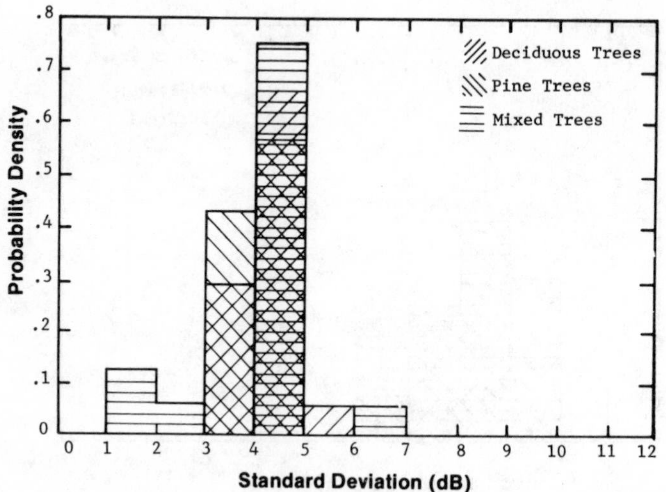

Figure B-3. Probability density function for the measured standard deviations of the amplitude distributions as a function of tree type; 35 GHz.

Source: Currie, Dyer, and Hayes (1975).

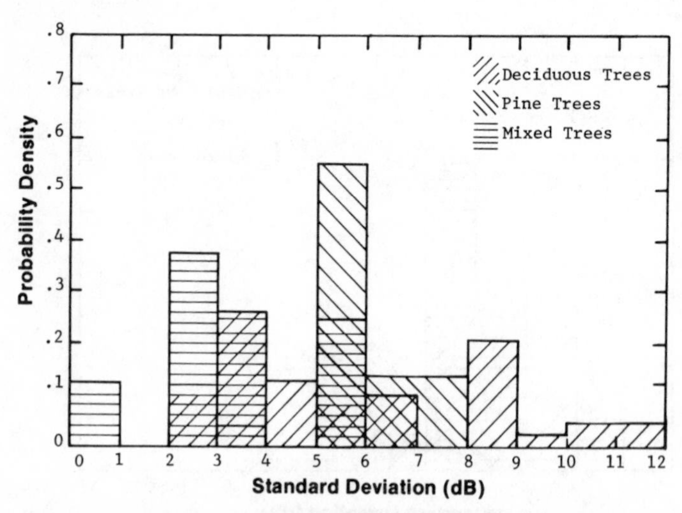

Figure B-4. Probability density function for the measured standard deviations of the amplitude distributions as a function of tree type; 95 GHz.

Source: Currie, Dyer, and Hayes (1975).

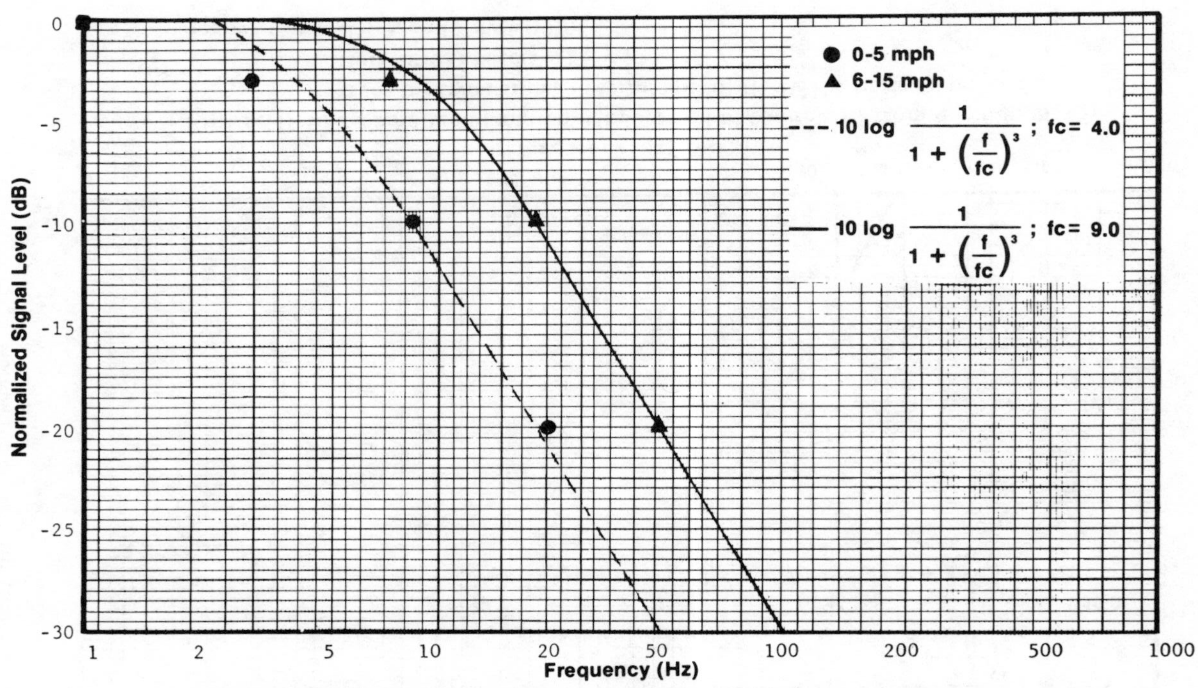

Figure B-5. Normalized frequency spectrum of the return from deciduous trees for two ranges of wind speed; 9.5 GHz, vertical polarization (linear receiver).

Source: Currie, Dyer, and Hayes (1975).

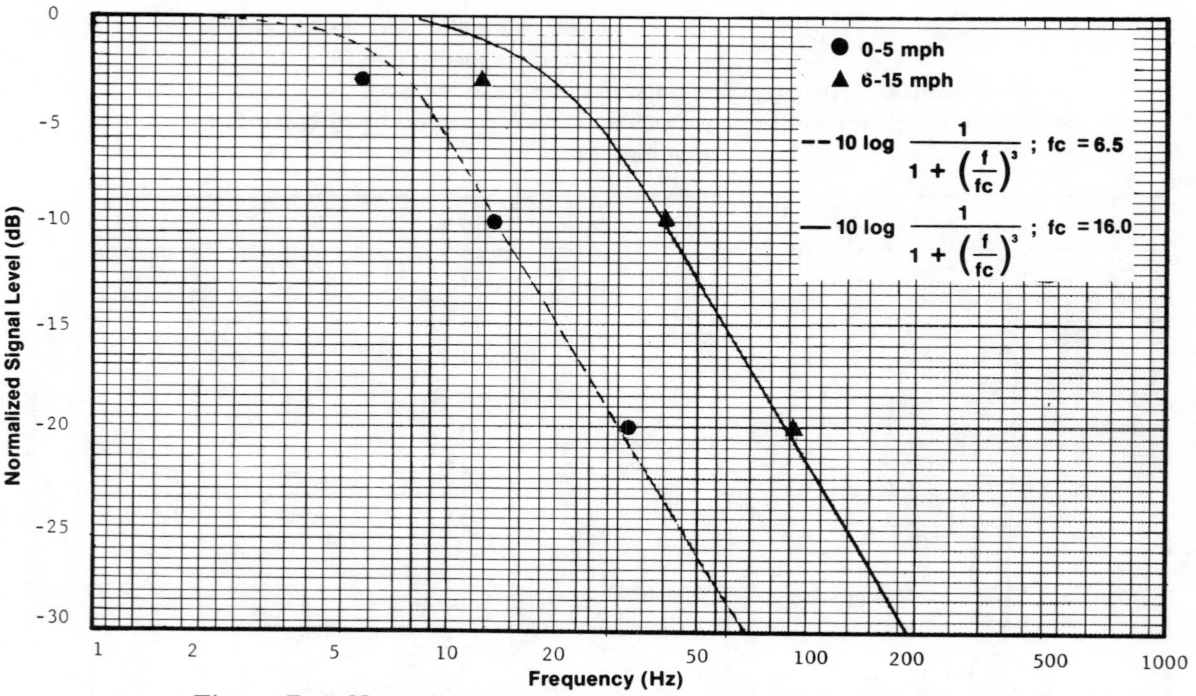

Figure B-6. Normalized frequency spectrum of the return from deciduous trees for two ranges of wind speed; 16.5 GHz, vertical polarization (linear receiver).

Source: Currie, Dyer, and Hayes (1975).

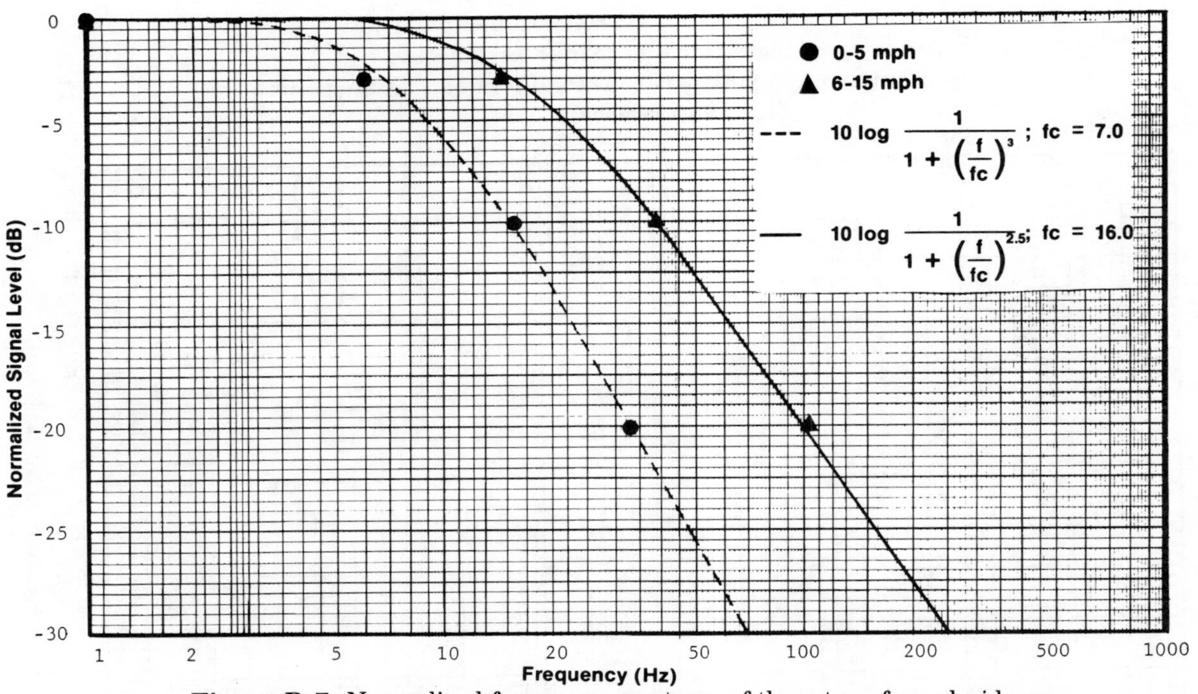

Figure B-7. Normalized frequency spectrum of the return from deciduous trees for two ranges of wind speed; 35 GHz, vertical polarization (linear receiver).

Source: Currie, Dyer, and Hayes (1975).

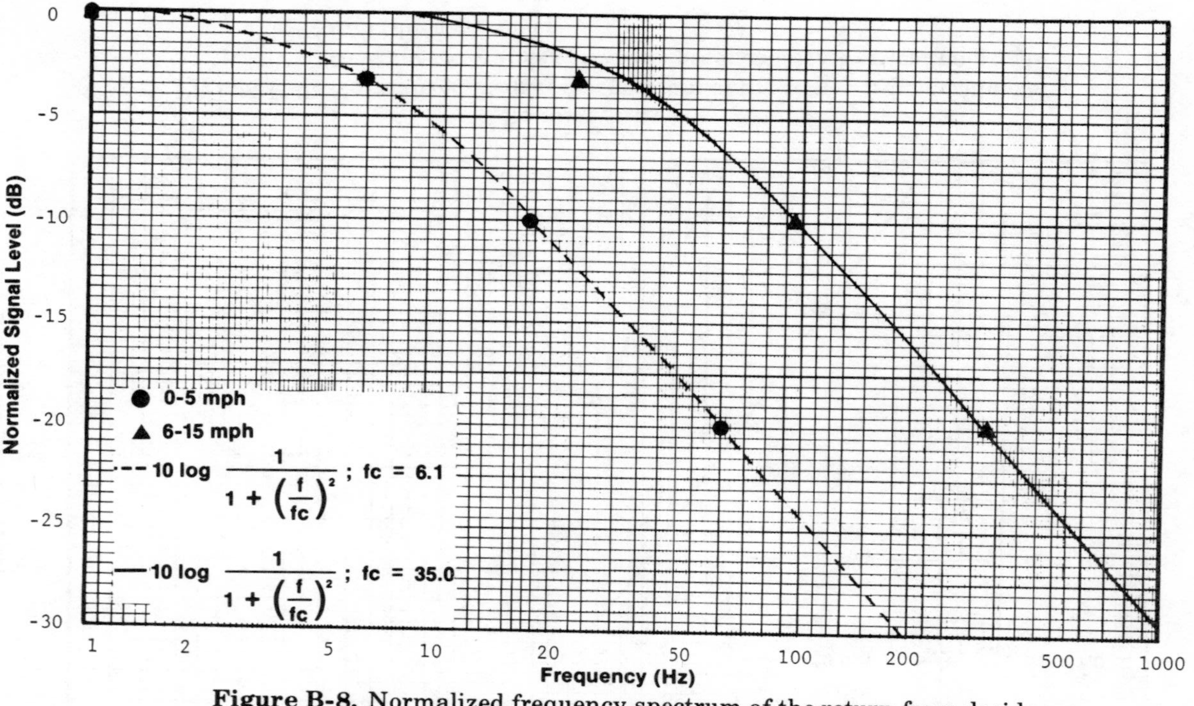

Figure B-8. Normalized frequency spectrum of the return from deciduous trees for two ranges of wind speed; 95 GHz, vertical polarization (linear receiver).

Source: Currie, Dyer, and Hayes (1975).

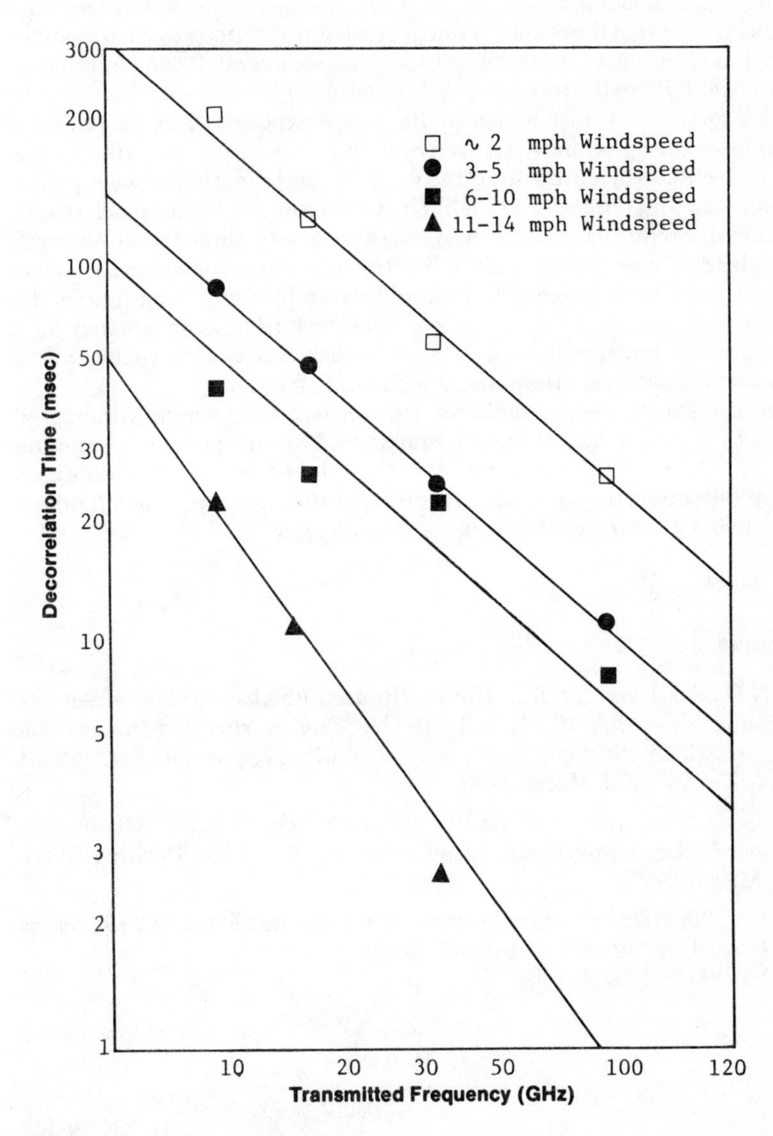

Figure B-9. Decorrelation time versus frequency and windspeed for return from deciduous trees.

Source: Currie, Dyer, and Hayes (1975).

Tree motion causes echo amplitude fluctuations, i.e., a noncoherent doppler spectrum. Thus, increases in wind speed cause increases in doppler spectra of tree echo. Figures B-5 through B-8 show the frequency spectra of tree echo that was measured at the output of linear receivers (Currie, *et al.* also reported spectra that were measured at the output of log receivers). It can be seen that the spectra of figures B-5 through B-8 do not drop off as fast with increases in frequency as does a Gaussian curve, the shape expected to be caused for a random positioning of many scatterers with about equal strength. As discussed in section 5.3, a drop-off of the doppler spectrum that is slower than Gaussian was first observed by Fishbein, Graveline, and Rittenbach (1967). Another feature characteristic of many scatterers with almost equal strength is a doppler frequency which varies directly with radar frequency. This feature is not valid for the tree echo as may be seen by comparing figures B-5 through B-8. For example, it can be seen that the half-power frequency for a given range of wind speeds, as indicated by the value of fc in each caption, increases at a slower rate than linear with radar frequency.

Figure B-9 shows the decorrelation time of tree echo versus wind speed observed by Currie *et al.* at the four frequencies. Through personal communications, N.C. Currie has informed me that the value of decorrelation time used for figure B-9 equals the time delay required for the auto-correlation function to decay to 0.3 of its maximum value (see sec. 5.3).

References

Currie, N.C., F.B. Dyer, and R.D. Hayes, "Radar Land Clutter Measurements at Frequencies of 9.5, 16, 35, and 95 GHz," Engineering Experiment Station, Georgia Institute of Technology, Technical Report No. 3, Contract DAAA-25-73-C-0256, March 1975.

Fishbein, W., S.W. Graveline, and O.E. Rittenbach, "Clutter Attenuation Analysis," U.S. Army Electronics Command, Technical Report ECOM-2808, March 1967.

Hayes, R.D., "95 GHz Pulsed Radar Return from Trees," *EASCON 79 Conference Record*, pp. 353-356, October 1979.

APPENDIX C

Low Angle Land Reflectivity

Figures C-1 through C-3 are graphs from Zehner and Tuley (1979) of X-band (3 cm) data. These figures were used by the authors with other low angle data for developing empirical models for land clutter. Since there are few available non-vegetated land data for grazing angles less than 5 degrees, Zehner and Tuley used all the Goodyear Aircraft Corporation (GAC, 1959) data runs (see sec. 6.8) from broken deserts and sparsely vegetated croplands for forming two composite figures (C-1 and C-2). Figure C-3 on trees, also taken from Zehner and Tuley (1979), is included for comparative purposes. In each figure the solid line depicts the least squares fit to the graphed data points under the assumption that

$$\sigma^{o}\ (dB) = 10 \log A + 10 \log B\theta,$$

where A and B are constants and θ is the incidence angle in degrees (with respect to horizontal). The constants so obtained by Zehner and Tuley are as follows:

(a) Figure C-1: $10 \log A = -30.7$ dB, $B = 0.83$
(b) Figure C-2: $10 \log A = -35.1$ dB, $B = 1.48$
(c) Figure C-3: $10 \log A = -35.1$ dB, $B = 0.70$

Only *HH* polarization data were available for figures C-1 and C-2. Figure C-1 contains data from two deserts and a dry lake. Each site for the figure C-1 data was described as being composed of flat, homogeneous patches of barren terrain with scattered rock and dry vegetation. Figure C-2 contains data from cotton seedlings, irrigated farmland, mature cotton, and meadows. Only three data points, of a total of 118 used, are from meadow land. According to Zehner and Tuley, actual RCS differences between leafy crops and grass or differences between tall and short vegetation may have existed but such possible differences could not be characterized with the data available.

The data of figure C-3 were originally reported by Currie, Dyer, and Hayes (1975) and are for X-band deciduous and coniferous tree echo during summer and fall. No identifiable differences were discernible between *HH* and *VV* polarization data and, as a consequence, figure C-3 is comprised of a mixture of such data. Zehner and Tuley state that the data in figure C-3 is for tree canopy and caution that a broadside illumination of exposed tree trunks will produce higher σ^{o} values.

Figure 6-18 (chap. 6) was also prepared from GAC (1959) data and includes two curves of σ^{o} versus θ for which the σ^{o} values, obtained for two specific airborne data runs, dip sharply to minimum values for θ between 1 degree and 4 degrees. These curves help to illustrate the variability and lack of uniformity that can exist in σ^{o} data at low angles. X-band data from the Swedish Defense Establishment, known as FOA, for cropland at small θ values for *HH* and *VV* polarizations are given in sections 5.2 and 6.8. For these data the differences between σ^{o} for *HH* and *VV* polarizations were reported to be

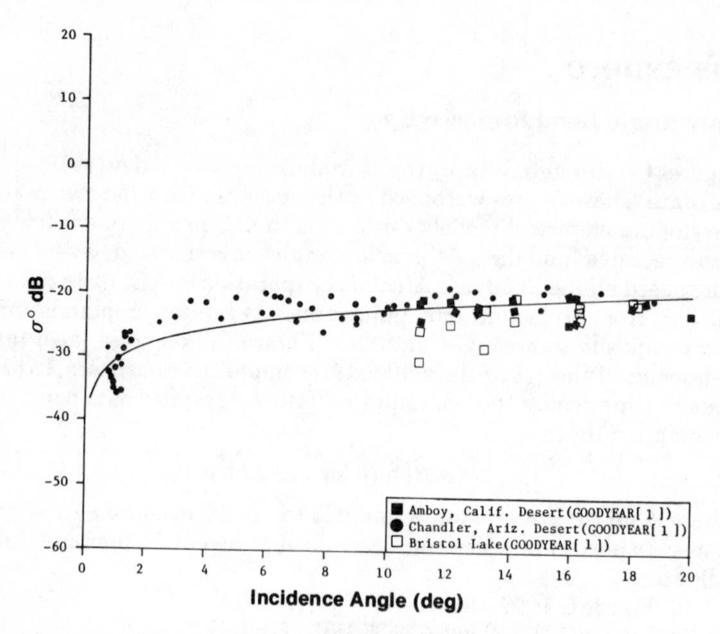

Figure C-1. Least squares fit to soil, sand, rock clutter data at X-band.
Source: Zehner and Tuley (1979)

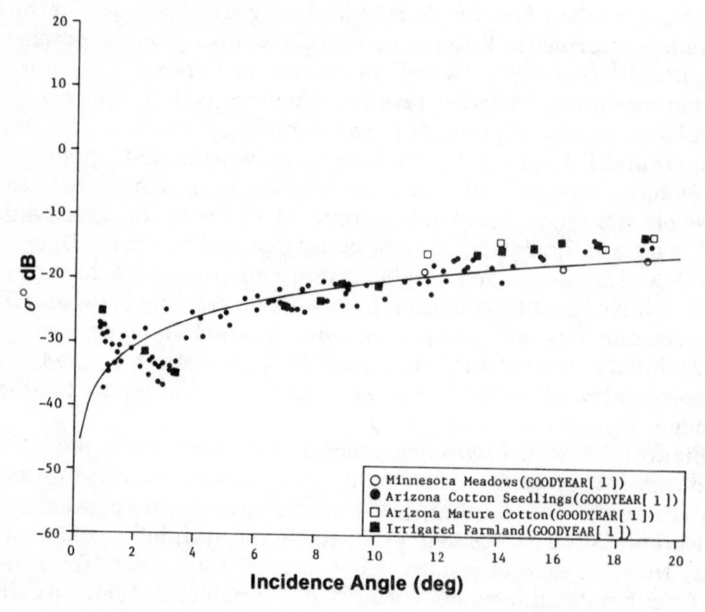

Figure C-2. Least squares fit to cropland, grass clutter data at X-band.
Source: Zehner and Tuley (1979).

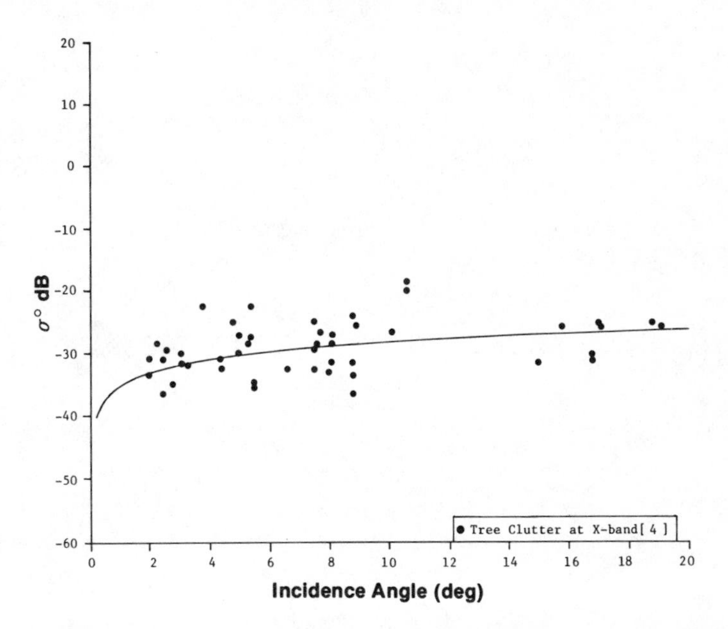

Figure C-3. Least squares fit to tree clutter data at X-band.
Source: Zehner and Tuley (1979).

negligible, and dips in σ° like those discussed for the GAC data were also observed (fig. 6-19). Thus, dips in σ° versus θ are sometimes observed for land that is relatively flat, and these dips have been observed for both *HH* and *VV* polarizations. The dips in σ° are presumed to be caused by small undulations in terrain slope.

In figures C-1 and C-2 the variability, or scatter, in σ° tends to increase for the smaller angles. This is consistent with the FOA study (sec. 5.2) which observed dramatic increases in the standard deviation of σ° values with reductions in θ for angles less than 5 degrees. This increased variability in σ°, i.e., increased standard deviation of the amplitude (spatial) distribution at low angles is considered basic to broken terrain.

References

Goodyear Aircraft Corporation, "Radar Terrain Study, Final Technical Report: Measurements of Terrain Backscattering Coefficients with an Airborne X-Band Radar," Contract NOas-59-6186-C, Report GERA-463, September 30, 1959.

Zehner, S.P. and M.T. Tuley, "Development and Validation of Multipath and Clutter Models for TAC Zinger in Low Altitude Scenarios," Engineering Experiment Station, Georgia Institute of Technology, Final Report, Contract F49620-78-C-0121, March 1979.

Author Index

Subject Index